Race and Ethnicity in Society

The Changing Landscape

FOURTH EDITION

ELIZABETH HIGGINBOTHAM
University of Delaware

MARGARET L. ANDERSEN
University of Delaware

CENGAGE
Learning®

Australia • Brazil • Mexico • Singapore • United Kingdom • United States

Race and Ethnicity in Society:
The Changing Landscape,
Fourth Edition
Elizabeth Higginbotham and
Margaret L. Andersen

Product Director: Marta
Lee-Perriard

Product Manager: Elizabeth
Beiting-Lipps

Content Developer: Lori
Bradshaw

Product Assistant: Chelsea
Meredith

Media Developer: John Chell

Marketing Manager: Kara
Kindstrom

Content Project Manager: Cheri
Palmer

Art Director: Michelle Kunkler

Manufacturing Planner: Judy
Inouye

Production, Composition, and
Illustration: MPS Limited

Text Researcher: Lumina
Datamatics

Copy Editor: Heather McElwain

Cover Designer: Michelle
DiMercurio

Cover Image: Suzy Higginbotham
Cameleon

Library of Congress Control Number: 2015944017

ISBN: 978-1-305-09389-8

Cengage Learning
20 Channel Center Street
Boston, MA 02210
USA

Cengage Learning is a leading provider of customized learning solutions with employees residing in nearly 40 different countries and sales in more than 125 countries around the world. Find your local representative at **www.cengage.com**.

Cengage Learning products are represented in Canada by Nelson Education, Ltd.

To learn more about Cengage Learning Solutions, visit **www.cengage.com**.

Purchase any of our products at your local college store or at our preferred online store **www.cengagebrain.com**.

Printed in the United States of America
Print Number: 01   Print Year: 2015

Table of Contents

Preface

The study of race and ethnicity is changing. High rates of immigration, a larger number of people identifying as multiracial, a civil rights framework encoded in law, overt racism less likely to be expressed in polite company, and the election of the nation's first African American president all make it seem that the United States has "moved beyond" race. Stark realities of racial inequality still persist, however, even though they are sometimes hidden from view. Racial inequality and its sister, ethnic inequality, are continuing realities of American life.

The United States is now a very multiracial society. New patterns of immigration since the late 1960s have brought new populations to the United States—populations that are changing the racial and ethnic composition of the nation. Greater acceptance of interracial unions is also producing more young people from mixed-race backgrounds—a phenomenon that is not new, but is now more accepted. Whereas the focus of study about race and ethnicity has long assumed a "Black/White" framework, different racial and ethnic groups now comprise the United States. This is especially obvious in the rather heated national discussions about immigrants and immigration policy.

This anthology is intended to introduce students to the study of race and ethnicity. The introductions and the diverse articles analyze the major topics and themes now framing the study of race in the United States. Because society is changing in its racial and ethnic makeup, current scholarship on race is also changing, focusing more on the social construction of race, complex racial and ethnic identities, but also continuing patterns of racial exclusion. This book departs from the traditional "smorgasbord approach" to studying race—an approach that documents the different histories of an array of groups, but where ethnicity is seen as only applying to Whites of European ancestry. Within this model, ethnicity has been perceived as mostly "white," race as "color." Such an approach no longer captures the complexities and dynamism of race and ethnicity in the United States.

Race and ethnicity in the United States are still mechanisms for sorting people, what the American Sociological Association has called "stratifying practices" (American Sociological Association 2003). Disparities in jobs, housing, health, education, public safety, and other facets of life can be largely attributed to race and ethnicity, although other social factors such as social class, gender, age, nationality, and immigration status also matter. Many people claim that the United States is now a color-blind society where race no longer matters because race appears to have lost some of its significance in organizing relationships and social institutions. Popular culture makes it appear that interracial relationships are common and that race no longer matters in shaping social relations. The dominant ideology is one of "color-blindness," as if to recognize race is to be a racist.

But, race does matter—and it matters a lot—in education, in workplaces, in communities, in health care, in courts and policing, in everyday interactions. Race also matters in our identities, complex as they are in such an evolving racial and ethnic environment. Race and ethnicity are present throughout day-to-day life; they are embedded in social institutions and in social relationships; and, they have real and recurring social and historical consequences. Neither race nor ethnicity go away just by denying that they are present.

We offer this anthology to those who want to understand the complexity of race and ethnicity in contemporary times. We have selected articles and written introductions that make the book's contents accessible to undergraduate readers. The book is intended primarily for courses on the sociology of race and ethnicity, although it can be used in other programs and courses, such as in ethnic and racial studies programs, education, political science, social justice, multicultural studies, and some humanities departments.

We have organized the fourth edition to reflect the different themes that underlie the study of race and ethnicity—consciously doing so around themes instead of around particular group experiences. The major themes of this book include:

- explaining that race is a social construction;
- showing the diversity of experiences that now constitute race and ethnicity in the United States;
- showing the connection between different racial identities and the social structures of race and ethnicity;
- understanding how racism works as a belief system rooted in societal institutions;
- providing a social structural analysis of racial inequality;
- providing a historical perspective on how the racial order has emerged and how it is maintained;
- examining how people have contested the dominant racial order; and,
- exploring current strategies for building a more just multiracial society.

NEW TO THE FOURTH EDITION

We have made substantial changes in the fourth edition, reflecting changes in society since the publication of the third edition and the newest developments in the scholarship on race and ethnicity. The fourth edition of *Race and Ethnicity in Society: The Changing Landscape* has also benefited tremendously from the comments of reviewers and those who used the previous editions of the book. As a result, we have reorganized the Table of Contents into seven major parts, adding a new section that grounds the book in the historical development of race and racism; changing how we present racial beliefs; and moving the section on race, class, and gender so it becomes it own major part in the Table of Contents. These specific changes are described further below (in the section on "Organization of the Book").

We have significantly shortened the book both to keep the cost down to students and because we recognize that most instructors can only teach two or three articles per week of the term. Still, we wanted the book to be long enough that instructors would have choices in the selections they use. We know that people will teach the course in different ways, but we have organized the book to start with the basic concepts moving on to more social structural and institutional analyses of race and ethnicity.

Of the forty readings included, there are **twenty-two new articles**. The new articles, like those we have used from the previous edition, have been selected for their importance and for their accessibility to undergraduate readers. The new articles are on such current topics as racism and popular culture, race and education, interracial romantic relationships, Latina immigrant labor, so-called "race-based" diseases, and the criminalization of immigrants, to name a few.

We have continued the feature called **Face the Facts**, which are graphic depictions of basic information relevant to the major section topics. This feature will help students read and interpret graphic presentations of data. This feature also includes critical thinking questions to help students analyze the facts they see depicted in the graphs. As an example, in the part on "How Did We Get Here and What is Changing?" we present two graphs—one on the percentage of immigrants in the U.S. population from 1890 to 2010, and a second chart on changes in where immigrants have come from since 1960. The **Think about It** feature that follows the two graphs asks students to describe the data presented and to think about the implications of the changes so noted. There is a "Face the Facts" feature at the conclusion of each of the part and section introductions.

ORGANIZATION OF THIS BOOK

This book is organized in seven major parts, one of which (Part VI, "The Continuing Significance of Race: Racial Inequality in Social Institutions") is subdivided into five different sections, highlighting particular institutions. Part I ("Race: What Does it Mean? The Social Construction of Race and Ethnicity") establishes the analytical frameworks that are now being used to think about race in society. The section examines the social construction of race and ethnicity. Together, the articles show the grounding of racial and ethnic categories in specific historical and social contexts and their fluidity over time.

Part II ("How Did We Get Here and What Is Changing?") explores both the historical patterns of inclusion and exclusion that have established racial and ethnic inequality. This part also explains some of the contemporary changes that are shaping contemporary racial and ethnic relations. Critical to this section is the concept of citizenship—that is, who belongs, both in the formal and informal sense—and how this is shaped by social policies and social beliefs.

Part III ("How Do We See Each Other? Beliefs, Representations, and Stereotypes") examines the most immediately experienced dimensions of race: beliefs and ideology. We think that students learn a lot about race by examining the manifestations of a racially stratified society in people's beliefs. This section also includes material on the current ideology of color-blind thinking and how it perpetuates racial inequality. Articles in this section also ask readers to think critically about the media and popular culture that they consume and the racial representations that are rampant but often taken for granted. We think that having students examine popular culture early in the book will make them more critical observers of the cultures in which they likely participate.

Part IV ("Why Can't We Just Get Along? Racial Identity and Interactions") examines racial identities—a topic that we think is especially interesting to students. In a racially diverse and changing society, many people now form racial identities that cross racial and ethnic boundaries. We include here a new article on "racial microaggressions" to help students see the sometimes inadvertent practices that also harm and insult people of color. This section also includes an article on White identity to remind students of the hidden privileges of dominant group status in a racially stratified society.

Part V ("Intersecting Inequalities: Race, Class, and Gender") is a new stand-alone part that incorporates articles from the earlier edition, but now includes a new piece on African American men in traditional women's work. We have noted from reviewers that some people want to use this book in courses on race, class, and gender so we feature this topic more in the fourth edition. Understanding the intersections of race, class, and gender is important throughout the book, but here we focus on the changed perspective that such an intersectional analysis brings to different topics. Although the major focus of this anthology is on race and ethnicity, we know that understanding inequality requires bringing gender, class, and sexuality into one's thinking about race.

Part VI ("The Continuing Significance of Race: Racial Inequality in Social Institutions") examines the major institutional structures in contemporary society and investigates patterns of racial inequality within these institutions. Persistent inequality in the labor market and in patterns of family life, housing and education, health care, and the workings of the criminal justice system are critical in shaping the life chances and well-being of different groups. The readings in this section, many of them new, will help students see how institutional racism operates.

Finally, Part VII ("Moving Forward: Building a Just Society") concludes the book by looking at projected changes in the landscape of race and ethnicity and efforts to support further progress toward racial justice. We selected articles about social change that show students how change can be both at the individual level and the level of society. We thereby ask students to think about the contexts in which they might find themselves working and living—beyond their college experience. Together, the articles in this section show the different ways that change can occur and the efforts needed to build a more just society.

PEDAGOGICAL FEATURES

The fourth edition includes **new introductions to each of the seven parts** of the book. The introductions frame each section and provide discussion of major concepts needed to interpret the articles. In cases, the introduction includes some historical framing of the issues covered in those parts.

We have included **brief introductions for each article**, intended to help students understand the major contribution of the article and its connection to the rest of the book. We have included **Discussion Questions** for each article in this book, with the goal of helping students grasp the major points of each argument. These questions can also be used for student paper assignments, research exercises, and class discussion.

Student Exercises are included at the end of each part. These will provide opportunities for student discussion both in and out of class and for possible projects. They have been revised to reflect the new additions and revisions to the book.

Face the Facts is a feature included at the conclusion of each part introduction and the different sections in Part VI. This feature includes graphs, drawn from census and other public data that help students visualize some of the contemporary evidence for racial inequalities. This feature will also help students interpret data shown in graphic form so that they can be better educated about the realities of race and ethnicity in America.

ACKNOWLEDGMENTS

We have benefited from the support and encouragement of many people who have either discussed the contents of the book with us or provided clerical and computer assistance or other forms of help that enabled us to complete/work on this book even in the midst of many other commitments. We thank Maxine Baca Zinn, Colette Gaiter, Valerie Hans, Richard Rosenfeld, Marilyn Whittington, and Ruth Zambrana for all the advice, help, and support they provide us. We also thank Suzy Higginbotham Cameleon for producing the art for our cover. We are glad to have her work featured here and are appreciative of her willingness to create something for us on very short notice. We also appreciate the support provided by the University of Delaware Morris Library and the Center for the History of Business, Technology, and Society at the Hagley Museum and Library. The editors at Cengage/Wadsworth have also been enthusiastic about this project. We welcome Elizabeth Beiting-Lipps, our new editor, to this project, and thank Mark Kerr, Marta Lee-Perriard, and Carolyn Henderson-Meier for their help during this editorial transition. A hearty thank you goes to Lori Bradshaw, Jill Traut, and Cheri Palmer who guided us through the production and permissions process; we could not have lived so calmly without their assistance. The suggestions of those who carefully reviewed the first drafts of this book were extremely valuable and have helped make this a stronger anthology, thus we thank them for taking the time to review.

Elizabeth Higginbotham
Margaret L. Andersen

About the Editors

Elizabeth Higginbotham (B.A., City College of the City University of New York; M.A., Ph.D., Brandeis University) is Professor Emerita of Sociology at the University of Delaware where she also held appointments in Black American Studies and Women and Gender Studies. She is the author of *Too Much to Ask: Black Women in the Era of Integration* (University of North Carolina Press, 2001) and co-editor of *Women and Work: Exploring Race, Ethnicity, and Class* (Sage Publications, 1997; with Mary Romero). She has also authored many articles in journals and anthologies on the work experiences of African American women, women in higher education, and curriculum transformation. While teaching at the University of Memphis, she received the Superior Performance in University Research Award for two consecutive years. Along with colleagues Bonnie Thornton Dill and Lynn Weber, she is a recipient of the American Sociological Association's Jessie Bernard Award and the Distinguished Contributions to Teaching Award for the work of the Center for Research on Women at the University of Memphis. She also received the Robin M. Williams Jr. Award from the Eastern Sociological Society, given annually to one distinguished sociologist. She served a term as vice president of the Eastern Sociological Society, and has held many elected and appointed leadership positions in the American Sociological Association.

Margaret L. Andersen (B.A., Georgia State University; M.A., Ph.D. University of Massachusetts, Amherst) is the Edward F. and Elizabeth Goodman Rosenberg Professor of Sociology at the University of Delaware, where she has also served in several senior administrative positions, including most recently as Vice Provost for Faculty Affairs and Diversity. She holds secondary appointments in Black American Studies and Women and Gender Studies. She is the author of several books, including (among others) *Thinking about Women,* recently published in its tenth edition; the best-selling anthology, *Race, Class and Gender* (co-edited with Patricia Hill Collins, now in its ninth edition); *Living Art: The Life of Paul R. Jones, African American Art Collector;* and *On Land and On Sea: A Century of Women in the Rosenfeld Collection.* She is a member of the National Advisory Board for Stanford University's Center for Comparative Studies in Race and Ethnicity, the past vice president of the American Sociological Association, and past president of the Eastern Sociological Society from which she received the ESS Merit Award. She has also received two teaching awards from the University of Delaware and the American Sociological Association's Jessie Bernard Award.

PART I

Race: What Does it Mean?
The Social Construction of Race and Ethnicity

ELIZABETH HIGGINBOTHAM AND MARGARET L. ANDERSEN

When studying racial inequality, one of the first questions to answer is, "What is race?" Most people assume that they can tell someone's race by just looking. This is because many people think of race in terms of skin color or other biological features that seemingly distinguish different groups. Is this a reasonable understanding of the concept?

Race has been used historically to differentiate groups, but why not use eye color or height or some characteristic other than color to categorize people into so-called races? The actual meaning of race lies not in people's physical characteristics, but in the historical treatment of different groups and the significance that society gives to what is believed to differentiate so-called racial groups. In other words, what is important about race is not biological difference, but the different ways groups of people have been treated in society.

This means that **race** is a social construction. This is one of the most important lessons you can learn by analyzing race in society. The idea of race has developed within the context of specific social institutions and historical practices in which groups defined as races have been exploited, controlled, and—in some cases—enslaved. Imagine this scenario: A social and economic system is created in which some people are forced into slavery based on their presumed inferiority.

The dominant group then creates a belief system that tries to justify this exploitation. The solution is to create "race" as an idea that can be used to "explain" the treatment and abuse of people perceived to be different.

Sociologists define a race as a group that is treated as distinct in society because of certain perceived characteristics that have been defined as signifying superiority or inferiority. Note that in this definition, perception, belief, and social treatment are the key elements defining race, not the actual physical characteristics of human groups. Furthermore, so-called races are created within a system of social dominance. Race is thus a social construction, not just an attribute of certain groups of people.

As Ann Morning explains, clear definitions of race are very hard to establish. This is because, first, the meaning of race shifts over time, as Morning shows in how the Census Bureau of the United States has repeatedly revised its classification system for counting people in racial groups. Second, race is difficult to define because it is a social, not a natural, construct. That is, the meaning of race emerges in the context of society and, as society changes, so does our understanding of race. Although you might think of race as a fixed attribute of a person, as this book will show, that is really not the case. How people perceive each other in terms of race, how they define their own identity, and how society treats people because of their perceived race are all matters of social definition. You might even want to read Morning's article again after reading all of the other articles in this book, because the book will show you—in many and varied ways—just how complex and shifting the social construction of race—and its malevolent partner, racism—are.

Still, the complex social reality of race does not stop people from thinking of it as biologically rooted. Scientists who have mapped human genetic makeup have concluded that there is no "race gene." Yes, there are certainly physical differences between people, some of which make people identifiable, but at the level of genetic composition, as Joseph L. Graves, Jr. ("The Race Myth") points out, there are far more similarities among people than there are differences, even when you take into account such things as eye color, skin color, and other physical characteristics. Scientists have concluded that genetic variation among human beings is indeed very small, and thus, that there is no biological basis for race.

The idea of race is, in fact, a relatively recent development in modern history. Abby Ferber ("Planting the Seed: The Invention of Race") takes us back to the establishment of race as a concept. Race emerged through the work of quasi-scientists in the eighteenth century—a time when White Europeans sought to explain and rationalize the exploitation of African, Asian, and indigenous people. She shows how the process of developing systems of racial classification is tangled up with the history of racism. To put it simply, racist thinking has emerged from the exploitation of people of color over the course of history. Ferber's article shows us how racism arose as science was emerging. Her essay will also make you think twice about using the term *Caucasian*, commonly used to refer to White people, once you learn the term's racist origins.

How, you might ask, does the concept of *race* differ from the concept of *ethnicity*? **Ethnic groups** are groups that share a common culture and have a shared identity. Ethnicity can thus stem from religion, national origin, or other shared characteristics. Important to this definition of an ethnic group is not just the shared culture, but also the sense of group belonging. Jewish people are an ethnic group, as are Italian Americans, Cajuns, and Irish American Catholics. Ethnicity can exist even within a so-called racial group (such as Jamaicans, Haitians, or Cape Verdeans among Black Americans).

Ethnic groups, however, can become racialized, as happened to Jewish people during the Nazi Holocaust in Europe. This was a complete social construction by Nazi leaders, yet one with devastating and murderous results. In the United States, particularly during the late nineteenth and early twentieth century when immigration rates were high, other groups, such as the Irish and then Italians, were also defined as "racially inferior" (Brodkin 1999). That we now do not consider Jews, Irish Americans, or Italian Americans to be "races" shows how the social construction of race can change over time. Some would say, however, that other groups—namely certain Latinos—are now in the process of becoming racialized. Why? You have to understand the concept of *racial formation* to answer this question.

Michael Omi and Howard Winant ("Racial Formation") define **racial formation** as the social and historical process by which racial categories are created. Historically in the United States, racial membership was determined by law—though, interestingly enough, the meaning of race varied from state to state. In Louisiana, for example, you were defined as Black if you had one Black ancestor out of thirty-two; in Virginia, it was one in sixteen; in Alabama, you were Black by law if you had any Black ancestry. By just crossing state borders, a person would legally change his or her race! Why was it so important to decide a person's race? There are many reasons, but laws against interracial marriage are a major part of the answer. If laws prevent people from marrying someone of a different race, then the government needs to classify people to prevent such relationships. This might seem rather arbitrary and pernicious to you now, but it is the social reality of race in U.S. history.

Racial formation means that social structures, not biology, define race. Because a group's perceived racial membership has been the basis for group oppression, this understanding of race shows how systems of authority and governance construct concepts of race. There are, of course, sometimes observable differences between individuals, but it is what these differences have come to mean in history and society that matters.

Together, the articles here show us that racism is not just a matter of individual attitudes and prejudices, although those surely are important, as we will see in the next part.

REFERENCE

Brodkin, Karen. 1999. *How Jews Became White Folks and What That Says about Race in America*. New Brunswick, NJ: Rutgers University Press.

FACE THE FACTS: THE U.S. CENSUS COUNTS RACE

Reproduction of Questions on Race and Hispanic Origin from Census 2010

→ NOTE: Please answer BOTH Question 8 about Hispanic origin and Question 9 about race. For this census, Hispanic origins are not races.

8. **Is Person 1 of Hispanic, Latino, or Spanish origin?**

   ☐ **No,** not of Hispanic, Latino, or Spanish origin

   ☐ Yes, Mexican, Mexican Am., Chicano

   ☐ Yes, Puerto Rican

   ☐ Yes, Cuban

   ☐ Yes, another Hispanic, Latino, or Spanish origin — *Print origin, for example, Argentinean, Colombian, Dominican, Nicaraguan, Salvadoran, Spaniard, and so on.* ↘

9. **What is Person 1's race?** *Mark* ☒ *one or more boxes.*

   ☐ White

   ☐ Black, African Am., or Negro

   ☐ American Indian or Alaska Native — *Print name of enrolled or principal tribe.* ↘

   ☐ Asian Indian   ☐ Japanese   ☐ Native Hawaiian
   ☐ Chinese   ☐ Korean   ☐ Guamanian or Chamorro
   ☐ Filipino   ☐ Vietnamese   ☐ Samoan
   ☐ Other Asian — *Print race, for example, Hmong, Laotian, Thai, Pakistani, Cambodian, and so on.* ↘   ☐ Other Pacific Islander — *Print race, for example, Fijian, Tongan, and so on.* ↘

   ☐ Some other race — *Print race.* ↘

SOURCE: U.S. Census Bureau. 2010. www.census.gov

Think about It: This is how race and ethnic identity are tabulated by the U.S. Census. What would you put? What does this form of questioning tell you about the government's role in the social construction of race and ethnicity?

1

Race

ANN MORNING

This first essay shows how the concept of race emerges in the context of social institutions and the classification systems that are developed within those institutions. Morning's argument underscores the point that race is a social construction and thus one that serves very particular purposes in society.

Race is part of everyday life in the United States. We're asked for our race when we fill out forms at school or work, when we visit a government agency or doctor's office. We read or hear about race in the daily news, and it comes up in informal discussions in our neighborhoods and social circles. Most of us can apply—to ourselves and the people around us—labels like white or Asian. Yet for all its familiarity, race is strangely difficult to define. When I've asked people to explain what race is, many have trouble answering.

Two uncertainties are widespread. First, there is confusion about the relationship between *race* and *ethnicity*. Are these different concepts? How often have you come across descriptions of someone's "ethnicity" that use terms like *white* and *black*? The public, the media, politicians, and scientists often use *race* and *ethnicity* interchangeably. Both terms have something to do with our ancestral origins, or "background," and we often find both linked to ideas about "culture." As historian David Hollinger points out, we often use the term *multicultural* to refer to racial diversity. In doing so, we presume that racial groups have different cultural beliefs or practices, even though the way we classify people by race has little to do with their behavior, norms, or values.

Defining race is also a challenge because we are unsure how it is related to biology. Are racial categories based on surface physical characteristics? Do they reflect unobserved patterns of genetic difference? If race is a kind of biological taxonomy, we are uncertain about exactly which traits anchor it.

Clear-cut definitions of race are surprisingly elusive. *The New Oxford American Dictionary*, for example, equates *race* with *ethnic group*, and links it to a wide range of possible traits: "physical characteristics," "culture," "history," and "language." The U.S. Census Bureau is another place to look for an authoritative definition. In contrast to the dictionary definition, the federal government rejects both culture and biology as relevant to race. This is apparent in its approach to racial enumeration and how it explains its definition. The U.S. Census makes the most visible use

SOURCE: Morning, Ann. 2005. "Race." *Contexts* 4 (Fall): 44–46.

of the official racial categories that the Office of Management and Budget (OMB) first promulgated in 1977 and revised in 1997. These standards require all federal agencies to use the following classifications in their data collection and analysis:

1. American Indian and Alaska Native

2. Asian

3. Black or African American

4. Native Hawaiian and Other Pacific Islander

5. White

The OMB deliberately refrains from naming Hispanics as a race, instead identifying them as an "ethnic group" distinguished by culture (specifically, "Spanish culture or origin, regardless of race"). The growing tendency among journalists, researchers, and the public is to treat Latinos as a *de facto* racial group distinct from whites, blacks, and others, but the government view is that cultural differences do not determine racial boundaries.

Biological differences are also declared irrelevant to the official standards. The Census Bureau maintains that its categories "do not conform to any biological, anthropological or genetic criteria." Instead, the bureau says that its classification system reflects "a social definition of race recognized in this country"—but it does not elaborate further on that "social definition." The Census Bureau and OMB see themselves as technical producers of race-based statistics largely for the purpose of enforcing civil rights laws, not as the arbiters of the meaning of race.

We may take the Census Bureau's reference to a "social definition" as a version of the social-scientific understanding of race as a "social construct." In other words, race is whatever we as a society say it is. The American Sociological Association took this view in its 2002 "Statement on the Importance of Collecting Data and Doing Social Scientific Research on Race," where it defined race as "a social invention that changes as political, economic, and historical contexts change." The association also noted that concepts of race usually involve valuations of "physical, intellectual, moral, or spiritual superiority or inferiority." Both are crucial observations about the type of idea that race is: It arises at particular moments and in particular places, and has long served to perpetuate deep social fissures. However, the constructivist position does not necessarily define the actual content of racial beliefs. Many kinds of classification schemes are socially constructed and serve as the basis for class systems. So what distinguishes racial categories from other taxonomies?

Sociologist Max Weber (1864–1920) offers a useful starting point for seeing the elements of ancestry, culture, and biology in terms of socially shaped belief. However, it is his definition of "ethnic groups"—not races—that provides the template. Like most scientists of his time, Weber felt that races stemmed from "common inherited and inheritable traits that actually derive from common descent." In his definition of *ethnicity*, however, he introduced the notion of "believed" rather than actual commonality, describing ethnic groups as "those human groups that entertain a subjective belief in their common descent because of similarities of physical type or of customs or both, or because of memories of colonization and migration; this belief must be important for the propagation of group formation; conversely, it does not matter whether or not an objective blood relationship exists." Substitute "races" for "ethnic groups," and strike "customs" but retain "physical type," and we have the basic ingredients for a comprehensive definition of race.

An emphasis on *belief* in common descent, as well as *perception* of similarity and difference, is crucial for a useful definition of race. Without them, we could not account for the traditional American "one-drop" system of racial classification, for example. According to this logic, a person with one black great-grandparent and seven white great-grandparents is a black person, because their "drop of black blood" means they have more in common with blacks than with whites. This shows how we base racial classifications on socially contingent perceptions of sameness and difference, not on some kind of "natural" calculus.

Finally, regardless of our personal views on the biological basis of race, we must recognize that physical characteristics—Weber's "inherited and inheritable traits"—are central to the concept. As historian George Fredrickson points out, the word *race*, deriving from the late 15th-century Spanish designation for Jewish and Muslim origins, came freighted with Christians' belief that such people embodied an innate, permanent, and negative essence. Although Spaniards had previously believed that infidels could become Christians through religious conversion, the suspicion that they could never truly make the transition took root after 1492, when many Jews and Muslims chose conversion over expulsion. This early notion of inherent and unchangeable difference gave rise to the understandings of race we share today. We believe, for example, that a black person can "act white" or that a white person can "act black," but that no behavior, shared ideas, or values can actually determine the race that one truly is.

With these elements in place, we can improvise on Weber to define races as *groupings of people believed to share common descent, based on perceived innate physical similarities*. This formulation addresses the relation of culture and biology to race. Culture is absent here as an explicit basis of racial membership, leaving a clear distinction between race and ethnicity. Though both refer to beliefs about shared origins, ethnicity is grounded in the discourse of cultural similarity, and race in that of biological commonality. In addition, this definition emphasizes the constructivist observation that although racial categories are ostensibly based on physical difference, they need not be so in reality. Even if we disagree about whether or not races are biological entities, we can agree that they are based on *claims* about biological commonality. As a result, this definition gives us a shared starting point for the most contentious debate about the nature of race today, namely, whether advances in genetic and biomedical research have proven the essentialist claim that races are identifiable, biologically distinct groups that exist independently of our perceptions and preconceptions.

Sociological literature often suggests that the general acceptance by scholars of the idea of race as a "social construct" gives it the status of "conventional wisdom." Yet resurgent claims about the biological nature of race are increasingly difficult to ignore. After something of a postwar hiatus in "race science" following international condemnation of the Nazis, the question of the biological basis of race has received renewed scientific attention in recent years. In March 2005, for example, the *New York Times* published a geneticist's essay asserting that "scientists should admit that there is such a thing as race." A variety of professional scientific journals and popular science magazines have taken up the question of whether "there is such a thing as race." In December 2003, the cover of *Scientific American* inquired "Does Race Exist?" *Nature Genetics* devoted most of a November 2004 supplement to the same question. *Science*, the *New England Journal of Medicine, Genome Biology*, and the *International Journal of Epidemiology*, among others, have also addressed the issue. In fact, the relationship between race and human biology is far from settled in scientific circles.

The argument that race is socially constructed rests on two simultaneous claims, although we sociologists have tended to focus on just one. The constructionist idea implies that race is a product of particular historical circumstances and also that it is *not* rooted in biological difference—it only claims to be. By working harder to demonstrate the former—for example, the historical variability of racial categories, their roots in particular social institutions, their divergence from one society to the next—we have turned away from investigating why racial boundaries do not correspond to physical differences. In our teaching and writing, we have not tried to explain why race may not be rooted in biology even though Americans are accustomed to being able to see race; they see, for example, who is Asian or who is white. Yet a comprehensive constructionist account must explain why we consider only some of the many kinds of differences between geographic groupings of human beings to be racial differences. We see racial differences between Norwegians and Koreans, for example, but we do not consider the differences between Norwegians and Portuguese to be racial.

Although sociologists may be reluctant to evaluate geneticists' and medical practitioners' pronouncements on human differences, feeling that this is not our turf, the arguments made today about race and biology lend themselves to sociological analysis. As the recent *New York Times* essay demonstrated, the current claims about distinct racial genetic profiles involve assumptions about group membership that social scientists are generally accustomed to questioning. We are used to investigating problems of sample construction and potential bias, exploring how assumptions about boundary lines affect our results. If we find, for example, that African Americans and European Americans have different probabilities of having a particular gene variant, does that prove the existence of black and white "races" to which they belong? If Latinos have yet another probability of having that gene variant, have we proved that they too constitute a racial group? How about Ashkenazi Jews? Does it matter if only one in a thousand genes displays such a pattern of variation? Does it matter how many people provided the DNA samples, or how they were located? Should we be suspicious that the red/white/yellow/black racial classification that scientists and census-takers use today is at heart the same framework that Linnaeus established in the 18th century without the aid of genome sequencing? In short, despite the complexity of the human genome and the tools that we now have to study it, the debate about the nature of race revolves around broader questions of logic and reasoning—which makes it that much more important to establish a comprehensive but flexible definition of race.

DISCUSSION QUESTIONS

1. How does Morning's discussion of race challenge the understanding of race as simply a fixed, biological category?
2. How do social beliefs shape our understanding of race?
3. Having read Morning's article, how would you now define race?

2

The Race Myth

JOSEPH L. GRAVES, JR.

The idea that race is a biological or genetic fact is wrong, as Graves here demon-strates. He shows that there is no such thing as a "race gene," and that there is vastly more genetic similarity among human beings than there are differences across so-called races. This article underscores, from current biological research, that race is not a thing so much as it is an idea that human beings have created.

Sometimes, the scientific investigation of one problem presents solutions to another. In 1986, scientists proposed a major undertaking: to sequence the entire human genome, to draw a map of the 100,000 or so genes that make humans distinct from apes or dolphins or squirrels. We knew that each of those genes could be found in a specific place on one of the twenty-three pairs of human chromosomes, but for the most part, we didn't know exactly where.

Remember those twisting ladders of DNA molecules from high-school biology? The DNA of each gene is made up of letters, which are pairs of molecule combinations. There are three billion letters in the human genome. At the Technology Center of Silicon Valley (now called the Tech Museum for Innovation), scientists built a model of what this looks like. Think of a spiral staircase and each step as a telephone directory. Now, wind a second staircase around the first to make a double spiral. Every phonebook is a gene, and the contents are the letters. If you have 100,000 phonebooks, and the information in them isn't listed alphabetically, and you aren't sure which phonebook goes where along the spirals, you've got a big project on your hands.

Producing all three billion letters of the human genome might be the great-est achievement in the history of biology, possibly of all science. The human genome map could tell us about the origin of human disease, the general function of our bodies, and possibly what it means to be human. A number of theoretical and technical problems had to be solved to sequence the genome, not the least of which was whose genome should be read. The scientists directly involved with the genome project immediately began to wrestle with the prob-lem of human genetic variation and the concept of race.

SOURCE: Graves, Jr., Joseph L. 2004. *The Race Myth: Why We Pretend Race Exists in America.* New York: Dutton.

FOR EVERY DISCOVERY, A CONTROVERSY

Celera Genomics was the only major private corporation in the quest to map the human genome. In February 2001, Celera's CEO, Craig Venter, touched off a minor firestorm when he commented that "race is not a scientific concept."[1] He knew that it wasn't possible to distinguish people who were ethnically African American, Chinese, Hispanic, or white at the genome level. Celera's sequencing of the human genome showed that the average pair of human beings who are not close relatives differ by 2.1 million genetic letters out of those 3 billion, yet only a few thousand of those differences account for the biological differences between individuals.

Venter argued that we all are essentially identical twins at the level of the genome. Celera used DNA extracted from five volunteers, three women and two men, who were ethnically African American, Chinese American, Hispanic, and European American. Their results showed that at the DNA level you could clearly tell the females from the males (due to the genetic differences in the X and Y chromosomes), but you could not identify the race of the individual from the DNA.

Venter's comment should not have been controversial. The Celera study only confirmed at the molecular level something population geneticists and physical anthropologists had recognized for well over fifty years: the nonexistence of biological races in the human species. Still, some prominent biologists felt compelled to attack Venter and defend the race concept in biology. Among them was James Crow (yes, that is his name), who in early 2002 defended the legitimacy of identifying races in humans. In the publication of the National Academy of Arts and Sciences, *Daedalus*, he commented: "Whenever an institution or society singles out individuals who are exceptional or outstanding in some way, racial differences will become more apparent. That fact may be uncomfortable, but there is no way around it."[2]

In support of this notion, rather than citing scientific evidence, he gave as an example a social phenomenon: the overrepresentation of African Americans in track and field, and their underrepresentation in physics and engineering, relative to Asian Americans. His exact comment was: "A stopwatch is colorblind." It's amazing when geneticists of James Crow's stature demonstrate this blind spot when it comes to human variation and the concept of race. It still happens because the social construction of race is deeply ingrained in the thinking of most American intellectuals, including biologists and medical practitioners.

… The fact is that no biological races exists in modern humans. In the next few pages, I'll explain how a race is defined and why there isn't enough variation in humans for our differences to qualify as races.

WHAT DOES IT TAKE TO BE A RACE?

To qualify officially as a biological race (or subspecies or variety), an animal or plant has to meet one of two requirements:

1. It can have its own distinct genetic lineage, meaning that it evolved in enough isolation that it never (or rarely) mated with individuals outside its borders, or

2. The genetic distance between one population and another has to be significantly greater than the genetic variability that exists within the populations themselves.

The first requirement is pretty straightforward, but the second takes some explaining. Think of it as a formula, and we're looking for two percentages to plug into it: one for distance and one for variability. To do that, we have to understand how geneticists measure variety, and we have to define genetic distance and genetic variability as painlessly as possible.

VARIETY IS THE SPICE OF GENETICS

We're all pretty familiar with DNA by now: It's the DNA molecules along our chromosomes that determine what makes us the same as other animals, what makes us different from them, and what makes us different from each other.

Along each chromosome, there are specific parts—somewhere between twenty-five thousand and forty thousand of them—that control our traits. At each of these parts there are two chemical messages that code for a trait, like brown eyes or the ability to roll your tongue. Some messages are dominant over others, so that if you get a message from Mom that says "brown" and a message from Dad that says "blue," your eyes are going to be brown. Many more messages have a mixing effect, producing physical features that fall somewhere in the middle of both parents, so that, say, your nose might be bigger than your mother's but smaller than your father's.

The range of possible combinations of these messages is pretty mindboggling: One egg cell or one sperm cell can have 8,388,608 possible combinations of chromosomes, so for a couple producing a child, the number of potential chromosome arrangements for that child would be 8,388,608! This means that there is a tremendous amount of diversity that can be produced by even one pair of parents, yet on average the physical traits of the offspring still resemble a mixture of both of the parents.

No one's saying that these differences amount to races, of course, otherwise, you'd be a different race from your parents. The large number of combinations of traits just means that, any way you look at it, human genetics is complex.

MEASURING VARIATION: WE'RE NOT ALL THAT DIFFERENT

There are ways to compare the complexity of one group of humans with another, no matter how you want to define *group*. A group could be all the people who were born and raised on Maui, or a number of people who have Down's syndrome. With all the possible variety, where do we start measuring variation?

Let's look again at those messages in our DNA. About 33 percent of the spots on the chromosomes where they live allow a lot of possibilities for a particular trait—people can have brown, blue, green, hazel, or golden eyes, for instance. In the remaining 67 percent of the spots, only one type of message is allowed, so that nearly every human being will have identical messages for a particular trait at that spot (wrong messages in these spots can result in genetic disease). So, about a third of our messages are responsible for all of the variety we see in people worldwide.

That seems like a lot, but it's not all. From studying these messages for traits, scientists now know that two individuals from anywhere in the world can potentially share 86 percent of the traits out of that 33 percent. Doing the math, that leaves only 4.62 percent of our genetic makeup responsible for all our individuality. Put another way: the traits of an Irish businessperson, an African-American lawyer, and the prime minister of India are quite likely to be 95.38 percent identical. Geography does play a part, so that if two people are from the same continent, you can reduce the variability by another 10 percent, and if they're from the same village, you can reduce it by yet another 4 percent. So, variety is measured by looking at the percentage of our DNA that makes us unique. What's genetic distance?

GETTING WITHIN SHOUTING DISTANCE

Genetic distance is a statistic calculated by examining different groups of people. It is a measurement of how frequently the genetic messages for traits occur in populations. People who share larger proportions of the same messages across their genome are genetically close, and people who do not share the same messages are genetically more distant. For instance, let's examine the message that causes sickle cell anemia, a disease that people think of as associated with race. This message occurs in large numbers in people who live in tropical areas because, it turns out, if you have one sickle cell message and one normal message, you have a better chance of surviving malaria, a typical disease of the tropics. So, the sickle cell anemia message is in high frequencies in populations in western Africa, the Middle East, the Persian Gulf, the Mediterranean, and India. In the case of this message, someone from Ghana is genetically closer to someone from Syria than to someone from Kenya, because Kenyans (who live at high altitudes where there isn't much malaria) don't have a high frequency of sickle cell messages. This clearly shows that sickle cell anemia cannot be associated with any particular race.

This points to an important fact: Geographical distance does not necessarily equal genetic distance. In fact, assuming that two people are genetically different because they look like they came from different parts of the world can be really dangerous for their health. Why? Because things like people's blood type or their ability to accept transplanted organs are dictated by how genetically close they are, not necessarily by where their ancestors came from geographically.

Because people equate race with external physical characteristics, they assume more often than not that a person is more likely to find an organ donor among the members of their own supposed racial group. This misconception in biomedical research or clinical practice that insists on sticking to these false racial categories causes many errors and lost lives.

LOCATION, LOCATION, LOCATION?

If geographic distance doesn't necessarily equal genetic distance, does that make geographic distance irrelevant? No. There is still a strong correlation between geographic distance and genetic distance. After all, it's geographic isolation that is the biggest factor in the development of new traits. It's just a mistake to assume that one type of distance equals the other.

So, whether we're looking at geographically close or distant populations of humans, the question that really matters is still this: If all the genetic information from two populations that we think of as races is examined, how similar are they? And are their differences big enough to qualify them as separate races?

SIZE MATTERS: THE STANDARD FOR MEASURING GENETIC VARIATION

To figure this out, we would have to look at a lot more than a couple of indicators like a sickle cell message or blood types. We'd have to evaluate the genetic information from enough people across a big enough number of separated populations for the statistics to be meaningful. Fortunately for you and me, it's been done for us already.

This analysis has used a number of approaches, examining the big three sources: protein variation, nuclear DNA, and mitochondrial DNA. All of these techniques agree that anatomically modern humans are a young species—too young to have developed any significant genetic distances between populations before mass mobility on a global scale started blurring what few differences there were. The result is that there is no unambiguous way to describe biological races within our species.

It's not for lack of a standard: Biologists have studied genetic variation in a wide variety of organisms other than humans for over fifty years, and have described many geographical races or subspecies. The races we have identified outside of humans usually show about 20 percent total genetic distance between their populations, as is the case in, say, the various species of fruit flies.

We do not see anywhere near that much genetic variation in modern humans. The genetic distances in humans are statistically about ten times lower (2 percent) than the 20 percent average in other organisms, even when comparing the most geographically separated populations within modern humans. There is greater genetic variability found within one tribe of western African chimpanzees than exists in the entire human species! In fact, there has never been any degree of natural selection in modern humans equivalent even to the levels used to create the differences in the breeds of domestic pigeons or dogs. In order to support the existence of biological races in modern humans, you would have to use a very different sort of reasoning than has been applied to all other species of animals— and that would be bad science.

ONE REQUIREMENT BITES THE DUST

Okay, now we have the first number to plug into our formula for race: Genetic distance between populations of humans is about 2 percent. We still need the other number, for genetic variation within those populations.

The same types of studies that showed us that human genetic distances average only 2 percent have also shown that there is about 8.6 times more genetic variation within the classically defined racial groups than between them. Why? Because there is about 8.6 times more genetic variation between any given individual on the planet and another individual than there is between the populations they belong to. In other words, the variability that makes one African-American person different from another is greater than the variability between African Americans and Swedes or Tibetans or Amazonian tribes.

Remember we said that modern humans share 86 percent of their genetic variations, so that less than 5 percent of our genome is responsible for our individuality. Now, it's time to look back at the formula for genetic distance versus genetic variability. For any group of humans to be a race, their genetic distance from another group, around 2 percent, would have to be greater than their unique genetic variability, around 5 percent. Two is not bigger that five.

It's a fact that if you add the 10 percent that accounts for shared variations between populations on the same continent, plus the 4 percent shared by local populations, you end up with about 1 percent genetic variability. Since this is lower than the 2 percent statistic for genetic distance, this would technically meet our definition of a race. There are a couple of problems with that conclusion. For one, we would be accepting percentages of variability in humans that are far below the percentages that are applied to all other organisms on the planet as evidence of race. For another, treating populations with 1 percent variability as distinct races would result in identifying over 1,000 races of humans. Following the argument to its logical conclusion, since families are even more genetically similar than geographic neighbors, we could say that every individual family is a race, or every cultural group that tries to keep itself separate (such as the Amish) is a race. In other words, the distinction becomes meaningless.

That's really the point. The majority of genetic variation in humans occurs between individuals, without regard to membership in socially constructed race. And none of the unique variations we see approach the minimum levels used to identify races in other species.

HOMAGE TO ALEX HALEY: OUR COMMON ROOTS

Our formula didn't work, so we know that populations of humans just don't meet the required distance-to-variability ratio for being separate races. Identifying a race by physical characteristics such as skin color or eye shape is as invalid as saying that all people who are tall or who have straight hair or who are pigeon-toed constitute separate races. Let's look now at the other requirement for race identification, genetic lineage.

There is no evidence that any group of humans now in existence—geographically or in socially defined races—has an evolutionary or genetic line of descent that is distinct from other groups.

Every person is descended from two parents, four grandparents, and eight great-grandparents. With each generation back into our ancestry, the number of lineal ancestors doubles (if all the ancestors are unrelated). If we go back twenty-four generations into our past, about fifteen hundred years ago, we each would have had 33,554,432 ancestors! At that time, the world population was around 206,000,000 people. That means that if we all really had separate, unrelated ancestors, every single one of us would be descended from about 16 percent of the world's population in the year 502 A.D. Obviously the math doesn't work, which strongly suggests that not all of our ancestors came from independent families and that many individuals must have contributed to many family lineages.

In a place like the United States, where the population has been intermarrying for a long time, and has been engaged in interracial sex and conception outside of marriage, it is fairly easy to show that the socially described races do not exist as separate lines of genetic descent. Still, we can and should ask if there ever was a time when the world's populations were truly independent.

In this sense, an independent lineage would be a population that was isolated from mixing genetically with other humans—one that never mated and bred outside its own group. In the animal kingdom, kangaroos are an example of this: The geographic isolation of Australia and its surrounding islands was so complete that this particular species of marsupial developed there and nowhere else. Since we know that race is an intermediate step toward becoming a species, what we're looking for genetically is the human equivalent of a marsupial on its way to becoming a kangaroo....

ALL ROADS LEAD TO ...

More reliable methods of studying human populations can be used, such as *neighbor joining* techniques, which allow relative genetic distances between groups to be shown. The data from these show that humans vary in small increments from group to group, rather than differing drastically depending on where they originated.

Humans have been on the move since we first left sub-Saharan Africa 80,000 years ago, combining our DNA at the same time as we were adapting genetically to our local conditions. Distinct genetic lineages cannot be traced within or between human populations because we've been mixing with each other since we first evolved. We're the only species of the *Homo* genus to have survived into modernity—an analogy would be if the only members of *Canis* to survive were domestic dogs (no more coyotes, jackals, or wolves). Genetically, we're not even separate breeds: We're all mutts.

AREN'T PHYSICAL DIFFERENCES PROOF THAT RACES EXIST?

We kind of look like separate breeds, though. Observable physical features are what people point to first to justify their belief in different races. After everything you've read so far, you know that's a dead horse but, just to make the point, let's beat it a little longer.

What determines physical traits in populations? Genes do this in combination with environmental factors. So, if genetic variation can't be divvied up evenly into races, neither can physical traits. Skin color, hair type, body stature, blood groups, or tendencies to get certain diseases do not alone or in combination define the racial groups that have been socially constructed in North America.

It's absolutely true that these physical traits vary among geographical populations. What most people don't realize is the way that they vary. For example, Sri Lankans of the Indian subcontinent, Nigerians, and aboriginal Australians share a dark skin tone, but differ in hair type, facial features, and genetic predisposition for disease. If you try to use characteristics such as height, body proportions, skull measurements, hair type, and skin color to create a tree showing how human populations are related, you get a tree that doesn't match the measured genetic relatedness and known evolutionary history of our species.

A tree like that would say, "all short, extremely dark-skinned people with thick curly hair are the same race," and would link Papuans from New Guinea and aboriginal Australians most closely to sub-Saharan Africans. We know, however, from genetic analyses, that Papuans and aboriginal Australians are the group most genetically distant from sub-Saharan Africans. And that makes total sense, because sub-Saharan Africans and aboriginal Australians are two of the most geographically separated human groups, so intermarriage has been minimal. Australia, after all, was cut off from human migration for most of its history.

We also know that within sub-Saharan African populations, everything from skin color to skull types to total genetic diversity is more variable than in any other of the world's populations. In other words, a person from the Congo and a person from Mali are more likely to be different genetically from each other than either is from a person from Belgium. Yet, if everyone from this region got up and moved to the United States, we'd call them all African Americans and see them as members of the same race.

Physical traits fail to define races because local populations produce traits that adapt to climate and other environmental factors wherever those factors occur. This means that, however genetically or geographically distant they are, tropical populations will have physical traits that match tropical conditions, like the sickle cell message. Kenyans and Peruvians will have greater lung capacities and red blood cell counts from living at high altitudes. These features are completely independent of the other genetic aspects of their physical makeup, and cannot be used to determine membership in a socially defined race.

TAKING IT BACK HOME

America has perfected the concept of socially defined races. On first inspection, it might have seemed that these races had biological legitimacy. The English colonists, the native peoples they called *Indians*, and the West Africans they brought as slaves all came from different places along the range of genetic diversity. It is entirely possible that the social construction of race in America would have proceeded differently if the full spectrum of the world's populations had immigrated to America along with western Europeans and Africans. Certainly, things would have been different if the prior history of the world had allowed these groups to come together under conditions fostering social equality.

But, as we know, cultural rather than biological traits were used to define our races. The rules of cultural evolution differ from those of biological evolution, but sadly scientists, doctors, philosophers, and law makers have for the most part, not yet acknowledged the difference. We must stop masking the real social issues with racist ideologies in order to build a truly just society.

NOTES

1. Fox, Maggie. "First Look at Human Genome Shows How Little There Is." *The Washington Post*, February 11, 2001.

2. Crow, James F. "Unequal by Nature: A Geneticist's Perspective on Human Differences." *Daedalus*, Winter 2002: 85.

DISCUSSION QUESTIONS

1. What does Graves mean when he says that the majority of genetic variation in humans occurs between individuals, not between races?

2. Over the years, many have spent a great deal of time and effort arguing that there is a biological basis to race, but most have been proven wrong, and now we know that there is no genetic basis for race. Why do you think there has been so much more attention given to presumed biological bases for race, compared to social bases for race?

3

Planting the Seed

The Invention of Race

ABBY L. FERBER

The history of the concept of race is deeply linked to racist thinking, particularly as it emerged in some of the quasi-scientific notions created in the sixteenth and seventeenth centuries. Ferber shows how racist thought is a relatively recent historical development, one that stems from the exploitation of human groups by others. In this history of racist thought, many have tried to use science (illegitimately) to try to justify such exploitation.

My students are always surprised to learn that race is a relatively recent invention. In their minds, race and racial antagonisms have taken on a universal character; they have always existed, and probably always will, in some form or another. Yet this fatalism belies the reality—that race is indeed a modern concept and, as such, does not have to be a life sentence.

Winthrop Jordan has suggested that ideas of racial inferiority, specifically that blacks were savage and primitive, played an essential role in rationalizing slavery.[1] There was no conception of race as a physical category until the eighteenth century.[2] There was, however, a strong association between blackness and evil, sin, and death, long grounded in European thought. The term "race" is believed to have originated in the Middle Ages in the romance languages, first used to refer to the breeding of animals. Race did not appear in the English language until the sixteenth century and was used as a technical term to define human groups in the seventeenth century. By the end of the eighteenth century, as emphasis upon the observation and classification of human differences grew, "race" became the most commonly employed concept for differentiating human groups according to Northern European standards. Audrey Smedley argues that because "race" has its roots in the breeding of animal stock, unlike other terms used to categorize humans, it came to imply an innate or inbred quality, believed to be permanent and unchanging.[3]

Until the nineteenth century, the Bible was consulted and depended upon for explanations of human variation, and two schools of thought emerged. The first asserted that there was a single creation of humanity, monogenesis, while the

SOURCE: From Ferber, Abby L. 1998. *White Man Falling: Race, Gender, and White Supremacy.* Lanham, MD: Rowman & Littlefield (pp. 27–43). Reprinted by permission of Rowman & Littlefield Publishers, Inc.

second asserted that various human groups were created separately, polygenesis. Polygenesis and ideas about racial inferiority, however, gained few believers, even in the late 1700s when the slave trade was under attack, because few were willing to support doctrines that conflicted with the Bible.[4]

While European Americans remained dedicated to a biblical view of race, the rise of scientific racism in the middle of the eighteenth century shaped debate about the nature and origins of races.[5] The Enlightenment emphasized the scientific practices of observing, collecting evidence, measuring bodies, and developing classificatory schemata. In the early stages of science, the most prevalent activity was the collection, examination, and arrangement of data into categories. Carolus Linnaeus, a prominent naturalist in the eighteenth century, developed the first authoritative racial division of humans in his *Natural System*, published in 1735.[6] Considered the founder of scientific taxonomy, he attempted to classify all living things, plant and animal, positioning humans within the matrix of the natural world. As Cornel West demonstrates, from the very beginning, racial classification has always involved hierarchy and the linkage of physical features with character and cultural traits.[7] For example, in the descriptions of his racial classifications, Linnaeus defines Europeans as "gentle, acute, inventive ... governed by customs," while Africans are "crafty, indolent, negligent ... governed by caprice."[8] Like most scientists of his time, however, Linnaeus considered all humans part of the same species, the product of a single creation.

Linnaeus was followed by Georges Louis Leclerc, Comte de Buffon, who is credited with introducing the term "race" into the scientific lexicon. Buffon also believed in monogenesis and in his 1749 publication *Natural History*, suggested that human variations were the result of differences in environment and climate. Whiteness, of course, was assumed to be the real color of humanity. Buffon suggested that blacks became dark-skinned because of the hot tropical sun and that if they moved to Europe, their skin would eventually lighten over time. Buffon cited interfertility as proof that human races were not separate species, establishing this as the criterion for distinguishing a species.

Buffon and Johann Friedrich Blumenbach are considered early founders of modern anthropology. Blumenbach advanced his own systematic racial classification in his 1775 study *On the Natural Varieties of Mankind*, designating five human races: Caucasian, Mongolian, Ethiopian, American, and Malay. While he still considered races to be the product of one creation, he ranked them on a scale according to their distance from the "civilized" Europeans.[9] He introduced the term "Caucasian," chosen because he believed that the Caucasus region in Russia produced the world's most beautiful women. This assertion typifies the widespread reliance upon aesthetic judgments in ranking races....

The science of racial classifications relied upon ideals of Greek beauty, as well as culture, as a standard by which to measure races. Race became central to the definition of Western culture, which became synonymous with "civilization."[10]...

The history of racial categorizations is intertwined with the history of racism. Science sought to justify *a priori* racist assumptions and consequently rationalized and greatly expanded the arsenal of racist ideology. Since the eighteenth century, racist beliefs have been built upon scientific racial categorizations and the linking of social and cultural traits to supposed genetic racial differences. While some social critics have suggested that contemporary racism has replaced biology with a concept of culture, the [1994] publication of *The Bell Curve*[11] attests to the staying power of these genetic notions of race. Today, as in the past, racism weaves together notions of biology and culture, and culture is assumed to be determined by some racial essence.

Science defined race as a concept believed to be hereditary and unalterable. The authority of science contributed to the quick and widespread acceptance of these ideas and prevented their interrogation. Equally important, the study of race and the production of racist theory also helped establish scientific authority and aided discipline building. While the history of the scientific concept of race argues that race is an inherent essence, it reveals, on the contrary, that race is a social construct. Young points out that "the different Victorian scientific accounts of race each in their turn quickly became deeply problematic; but what was much more consistent, more powerful and long-lived, was the cultural construction of race."[12]

Because race is not grounded in genetics or nature, the project of defining races always involves drawing and maintaining boundaries between those races. This was no easy task. It is important to pay attention to the construction of those borders: how was it decided, in actual policy, who was considered white and who was considered black? What about those who did not easily fit into either of those categories? What were the dangers of mixing? How could these dangers be avoided? These issues preoccupied policy makers, popular culture, and the public at large....

Throughout the second half of the nineteenth century, discussion of race and racial purity grew increasingly popular in both academic and mainstream circles as Americans developed distinctive beliefs and theories about race for the first time. As scientific beliefs about race were increasingly accepted by the general public, support for the one-drop rule became increasingly universal. Popular opinion grew to support the belief that no matter how white one appeared, if one had a single drop of black blood, no matter how distant, one was black....

Throughout the history of racial classification in the West, miscegenation and interracial sexuality have occupied a place of central importance. The science of racial differences has always displayed a preoccupation with the risks of interracial sexuality. Popular and legal discourses on race have been preoccupied with maintaining racial boundaries, frequently with great violence. This [essay] suggests that racial classification, the maintenance of racial boundaries, and racism are inexorably linked. The construction of biological races and the belief in maintaining the hierarchy and separation of races has led to widespread fears of integration and interracial sexuality....

The history of racial classification, and beliefs about race and interracial sexuality, can be characterized as inherently white supremacist. White supremacy has been the law and prevailing worldview throughout U.S. history, and the ideology of what is today labeled the white supremacist movement is firmly rooted in this tradition. Accounts that label the contemporary white supremacist movement as fringe and extremist often have the consequence of rendering this history invisible. Understanding this history, however, is essential to understanding and combating both contemporary white supremacist and mainstream racism.

NOTES

1. Jordan, Winthrop. 1969. *White over Black.* Chapel Hill: University of North Carolina Press.

2. Banton, Michael, and Jonathan Harwood. 1975. *The race concept.* New York: Praeger; Mencke, John G. 1979. *Mulattoes and race mixture: American attitudes and images, 1865–1918.* Ann Arbor, Mich.: University Microfilms Research Press.

3. Smedley, Audrey. 1993. *Race in North America: Origin and evolution of a worldview.* Boulder, Colo.: Westview Press.

4. Banton and Harwood 1975, 19.

5. Banton and Harwood 1975, 24.

6. West, Cornel. 1982. *Prophesy deliverance! An Afro-American revolutionary Christianity.* Philadelphia: Westminster Press.

7. West 1982.

8. West 1982, 56.

9. Smedley 1993, 166.

10. Young, Robert J. C. 1995. *Colonial desire: Hybridity in theory, culture and race.* New York: Routledge.

11. Hernstein, Richard J., and Charles Murray. 1994. *The bell curve: Intelligence and class structure in American life.* New York: Free Press.

12. Young, 1995, p. 94.

DISCUSSION QUESTIONS

1. What does Ferber mean when she writes that "the history of racial categorizations is intertwined with the history of racism"?

2. What role have science and religion played in the social construction of racism?

4

Racial Formation

MICHAEL OMI AND HOWARD WINANT

The concept of racial formation, first developed by Omi and Winant, has become central to the sociological study of race. It refers to the social and historical processes by which groups come to be defined in racial terms, and it specifically locates those processes in state-based institutions, such as the law.

In 1982–83, Susie Guillory Phipps unsuccessfully sued the Louisiana Bureau of Vital Records to change her racial classification from black to white. The descendant of an 18th-century white planter and a black slave, Phipps was designated "black" in her birth certificate in accordance with a 1970 state law which declared anyone with at least 1/32nd "Negro blood" to be black.

The Phipps case raised intriguing questions about the concept of race, its meaning in contemporary society, and its use (and abuse) in public policy. Assistant Attorney General Ron Davis defended the law by pointing out that some type of racial classification was necessary to comply with federal recordkeeping requirements and to facilitate programs for the prevention of genetic diseases. Phipps's attorney, Brian Begue, argued that the assignment of racial categories on birth certificates was unconstitutional and that the 1/32nd designation was inaccurate. He called on a retired Tulane University professor who cited research indicating that most Louisiana whites have at least 1/20th "Negro" ancestry.

In the end, Phipps lost. The court upheld the state's right to classify and quantify racial identity....

Phipps's problematic racial identity, and her effort to resolve it through state action, is in many ways a parable of America's unsolved racial dilemma. It illustrates the difficulties of defining race and assigning individuals or groups to racial categories. It shows how the racial legacies of the past—slavery and bigotry—continue to shape the present. It reveals both the deep involvement of the state in the organization and interpretation of race, and the inadequacy of state institutions to carry out these functions. It demonstrates how deeply Americans both as individuals and as a civilization are shaped, and indeed haunted, by race.

Having lived her whole life thinking that she was white, Phipps suddenly discovers that by legal definition she is not. In U.S. society, such an event is indeed catastrophic. But if she is not white, of what race is she? The state claims

SOURCE: From Omi, Michael, and Howard Winant. 1994. *Racial Formation in the United States. From the 1960s to the 1990s.* 2nd ed., pp. 53–61. New York: Routledge. Reprinted by permission of Routledge/Taylor & Francis Books, Inc.

that she is black, based on its rules of classification ... and another state agency, the court, upholds this judgment. But despite these classificatory standards which have imposed an either-or logic on racial identity, Phipps will not in fact "change color." Unlike what would have happened during slavery times if one's claim to whiteness was successfully challenged, we can assume that despite the outcome of her legal challenge, Phipps will remain in most of the social relationships she had occupied before the trial. Her socialization, her familial and friendship networks, her cultural orientation, will not change. She will simply have to wrestle with her newly acquired "hybridized" condition. She will have to confront the "Other" within.

The designation of racial categories and the determination of racial identity is no simple task. For centuries, this question has precipitated intense debates and conflicts, particularly in the U.S.—disputes over natural and legal rights, over the distribution of resources, and indeed, over who shall live and who shall die.

A crucial dimension of the Phipps case is that it illustrates the inadequacy of claims that race is a mere matter of variations in human physiognomy, that it is simply a matter of skin color. But if race cannot be understood in this manner, how can it be understood? We cannot fully hope to address this topic—no less than the meaning of race, its role in society, and the forces which shape it—in one [article], nor indeed in one book. Our goal in this [article], however, is far from modest: we wish to offer at least the outlines of a theory of race and racism.

WHAT IS RACE?

There is a continuous temptation to think of race as an essence, as something fixed, concrete, and objective. And there is also an opposite temptation: to imagine race as a mere illusion, a purely ideological construct which some ideal non-racist social order would eliminate. It is necessary to challenge both these positions, to disrupt and reframe the rigid and bipolar manner in which they are posed and debated, and to transcend the presumably irreconcilable relationship between them.

The effort must be made to understand race as an unstable and "decentered" complex of social meanings constantly being transformed by political struggle. With this in mind, let us propose a definition: race is a concept which signifies and symbolizes social conflicts and interests by referring to different types of human bodies. Although the concept of race invokes biologically based human characteristics (so-called "phenotypes"), selection of these particular human features for purposes of racial signification is always and necessarily a social and historical process. In contrast to the other major distinction of this type, that of gender, there is no biological basis for distinguishing among human groups along the lines of race.... Indeed, the categories employed to differentiate among human groups along racial lines reveal themselves, upon serious examination, to be at best imprecise, and at worst completely arbitrary.

If the concept of race is so nebulous, can we not dispense with it? Can we not "do without" race, at least in the "enlightened" present? This question has been posed often, and with greater frequency in recent years.... An affirmative answer would of course present obvious practical difficulties: it is rather difficult to jettison widely held beliefs, beliefs which moreover are central to everyone's identity and understanding of the social world. So the attempt to banish the concept as an archaism is at best counterintuitive. But a deeper difficulty, we believe, is inherent in the very formulation of this schema, in its way of posing race as a *problem*, a misconception left over from the past, and suitable now only for the dustbin of history.

A more effective starting point is the recognition that, despite its uncertainties and contradictions, the concept of race continues to play a fundamental role in structuring and representing the social world. The task for theory is to explain this situation. It is to avoid both the utopian framework which sees race as an illusion we can somehow "get beyond," and also the essentialist formulation which sees race as something objective and fixed, a biological datum. Thus we should think of race as an element of social structure rather than as an irregularity within it; we should see race as a dimension of human representation rather than an illusion. These perspectives inform the theoretical approach we call racial formation.

Racial Formation

We define *racial formation* as the sociohistorical process by which racial categories are created, inhabited, transformed, and destroyed. Our attempt to elaborate a theory of racial formation will proceed in two steps. First, we argue that racial formation is a process of historically situated projects in which human bodies and social structures are represented and organized. Next we link racial formation to the evolution of hegemony, the way in which society is organized and ruled. Such an approach, we believe, can facilitate understanding of a whole range of contemporary controversies and dilemmas involving race, including the nature of racism, the relationship of race to other forms of differences, inequalities, and oppression such as sexism and nationalism, and the dilemmas of racial identity today.

From a racial formation perspective, race is a matter of both social structure and cultural representation. Too often, the attempt is made to understand race simply or primarily in terms of only one of these two analytical dimensions.... For example, efforts to explain racial inequality as a purely social structural phenomenon are unable to account for the origins, patterning, and transformation of racial difference.

Conversely, many examinations of racial difference—understood as a matter of cultural attributes à la ethnicity theory, or as a society-wide signification system, à la some poststructuralist accounts—cannot comprehend such structural phenomena as racial stratification in the labor market or patterns of residential segregation.

An alternative approach is to think of racial formation processes as occurring through a linkage between structure and representation. *Racial projects do the ideological "work" of making these links. A racial project is simultaneously an interpretation, representation, or explanation of racial dynamics, and an effort to recognize and redistribute resources along particular racial lines.* Racial projects connect what race means in a particular discursive practice and the ways in which both social structures and everyday experiences are racially *organized*, based upon that meaning. Let us consider this proposition, first in terms of large-scale or macro-level social processes, and then in terms of other dimensions of the racial formation process.

Racial Formation as a Macro-Level Social Process

To interpret the meaning of race is to frame it social structurally. Consider, for example, this statement by Charles Murray on welfare reform:

> My proposal for dealing with the racial issue in social welfare is to repeal every bit of legislation and reverse every court decision that in any way requires, recommends, or awards differential treatment according to race, and thereby put us back onto the track that we left in 1965. We may argue about the appropriate limits of government intervention in trying to enforce the ideal, but at least it should be possible to identify the ideal: Race is not a morally admissible reason for treating one person differently from another. Period....

Here there is a partial but significant analysis of the meaning of race: it is not a morally valid basis upon which to treat people "differently from one another." We may notice someone's race, but we cannot act upon that awareness. We must act in a "color-blind" fashion. This analysis of the meaning of race is immediately linked to a specific conception of the role of race in the social structure: it can play no part in government action, save in "the enforcement of the ideal." No state policy can legitimately require, recommend, or award different status according to race. This example can be classified as a particular type of racial project in the present-day U.S.—a "neoconservative" one.

Conversely, *to recognize the racial dimension in social structure is to interpret the meaning of race.* Consider the following statement by the late Supreme Court Justice Thurgood Marshall on minority "set-aside" programs:

> A profound difference separates governmental actions that themselves are racist, and governmental actions that seek to remedy the effects of prior racism or to prevent neutral government activity from perpetuating the effects of such racism....

Here the focus is on the racial dimensions of social structure—in this case of state activity and policy. The argument is that state actions in the past and present have treated people in very different ways according to their race, and thus the government cannot retreat from its policy responsibilities in this area. It cannot suddenly declare itself "color-blind" without in fact perpetuating the same type of differential, racist treatment..... Thus, race continues to signify difference and structure inequality. Here, racialized social structure is immediately linked to an interpretation of the meaning of race. This example too can be classified as a particular type of racial project in the present-day U.S.—a "liberal" one.

To be sure, such political labels as "neoconservative" or "liberal" cannot fully capture the complexity of racial projects, for these are always multiply determined, politically contested, and deeply shaped by their historical context. Thus encapsulated within the neoconservative example cited here are certain egalitarian commitments which derive from a previous historical context in which they played a very different role, and which are rearticulated in neoconservative racial discourse precisely to oppose a more open-ended, more capacious conception of the meaning of equality. Similarly, in the liberal example, Justice Marshall recognizes that the contemporary state, which was formerly the architect of segregation and the chief enforcer of racial difference, has a tendency to reproduce those patterns of inequality in a new guise. Thus he admonishes it (in dissent, significantly) to fulfill its responsibilities to uphold a robust conception of equality. These particular instances, then, demonstrate how racial projects are always concretely framed, and thus are always contested and unstable. The social structures they uphold or attack, and the representations of race they articulate, are never invented out of the air, but exist in a definite historical context, having descended from previous conflicts. This contestation appears to be permanent in respect to race.

These two examples of contemporary racial projects are drawn from mainstream political debate; they may be characterized as center-right and center-left expressions of contemporary racial politics.... We can, however, expand the discussion of racial formation processes far beyond these familiar examples. In fact, we can identify racial projects in at least three other analytical dimensions: first, the political spectrum can be broadened to include radical projects, on both the left and right, as well as along other political axes. Second, analysis of racial projects can take place not only at the macro-level of racial policy-making, state activity, and collective action, but also at the micro-level of everyday experience. Third, the concept of racial projects can be applied across historical time, to identify racial formation dynamics in the past.

DISCUSSION QUESTIONS

1. What do Omi and Winant mean by *racial formation*? What role does the law play in such a process? How is this shown in the history of the United States?

2. What difference does it make to conceptualize race as a property of social structures versus as a property (or attribute) of individuals?

Student Exercises

1. Your instructor will divide your class (or some other grouping) into two groups, one of which is designated the Blues and the other the Greens. Over a period of a week, the Greens should serve the Blues in any way the Blues ask—such as carrying their books, running errands for them, delivering meals to places they designate, or any other job that the Blues design. (Because this is a course assignment, you should be reasonable in your demands.)

    As the week progresses, observe how the Blues act among themselves and in front of the Greens. Also observe how the Greens act among themselves and in front of the Blues. What attitudes do the two groups develop toward each other and toward themselves? How do they talk publicly about the other group? Do classmates begin to generalize about the assumed characteristics of the two different groups? What does the experiment reveal about the *social construction of race*?

2. Have members of your class describe the ethnic background of their family members. You can describe such things as when and how your family arrived in the United States, ethnic traditions that your families may observe, and whether ethnic pride is a part of your family experience. After hearing from classmates belonging to different ethnic groups, list what you learned about ethnicity from listening to these different experiences. Is ethnicity more significant for some groups than others? Is ethnicity more important to some generations within a family than others? Why? What does this teach you about the *social construction of ethnicity* in society?

PART II

How Did We Get Here and What Is Changing?

ELIZABETH HIGGINBOTHAM AND
MARGARET L. ANDERSEN

Understanding how race and ethnicity were critical factors in the building and expansion of the United States helps you grasp the nature of racial inequalities today. In Part I, we learned that race is a social construction, a contemporary way of thinking about the racial distinctions between human beings. Historically race was invented to establish racial hierarchies that imposed distinctions between groups. Whether such distinctions were considered biological or cultural differences varies over time. These perceived racial differences and the classification systems that were developed to support them presumably gave legitimacy to structures of racial inequality, at least in the eyes of dominant groups. The invention of race happened over long periods of time in an effort to justify the emerging power arrangements and economic exploitation. Racial thinking allowed certain practices to appear "natural" to those who benefited from them, including such histories as the conquering of Native American lands by Europeans that disrupted established ways of living and working among indigenous populations; the enslavement and transportation of people from one continent to another; and the importation and then exclusion of different laboring groups, such as Chinese immigrants. Historians can see today how people with political and economic power shaped economic, social, and political arrangements to their own benefit, while cloaking their objectives and interests in the smokescreen of so-called racial differences.

This section examines some of the historical patterns that have shaped the racial landscape today. The racial landscape continues to emerge and change. Racial and ethnic inequality is often held in place by the assignment or denial of the rights of citizenship. Citizenship enables people to participate fully in society. As Supreme Court Chief Justice Earl Warren wrote in 1958, dissenting on a denationalization case, "Citizenship is man's basic right, for it is nothing less than the right to have rights" (Ngai 2003: 10).

The founding fathers envisioned the United States as a nation for White settlement, even though many inhabitants were not White. Race included some and excluded others through the adoption of various laws and practices. One example is the Nationalization Act in 1790, which offered citizenship to "free white persons" who were of good moral character.

Race was central to thinking about the nature of the country and the development of its citizenry. "We the people" in the U.S. Constitution did not refer to an inclusive group; rather, the founding fathers were deliberate in extending rights only to White men with property. As a group, American Indians—the indigenous populations—were seen as outside the developing society. The Constitution also identified "other persons," that is, enslaved people, most of whom were of African descent. This initial national vision of people of color as unworthy of citizenship meant that they had to battle for inclusion in the nation. The terms of inclusion would keep shifting, but the history of marginalization continues to shape race and ethnic relations today.

The rules of citizenship—that is, who is granted the rights of citizenship and how they achieve it—continue to shape the relationships among groups in a society. You can think of citizenship in two ways. First, there is an actual system of rights (the right to vote, the right to sit on a jury, the right to own property, and so forth). Second, citizenship is also symbolic—meaning the right of belonging to a nation or a community (Glenn 2002). For example, at the ceremony that opened the Lincoln Memorial in Washington, DC in 1922, Black Americans had to witness the event from behind a rope. The pattern of segregation, like separate schools for people of color and the denial of access to libraries and museums, communicated that they were not full citizens in the nation. The Lincoln Memorial was the site for the 1964 March on Washington, where people of all colors protested for the passage of a Civil Rights Act that would end much discrimination and communicate the notion of equal citizenship.

In the United States, citizenship developed along historic lines of racial and ethnic hierarchies, giving some groups access to social and political power and others less or none. Disenfranchised people, whose labor was essential for the development of the country, were not granted citizenship rights and thus had little leverage in shaping the social institutions that influenced their lives. Instead, disenfranchised racial groups had to struggle for inclusion in the larger political body, a struggle that is ongoing. The legacy of exclusion has limited the access of people of colors to political, economic, and social resources, while simultaneously shaping the dominant group's thinking about the place of racially different groups in society.

In the early twentieth century, although most were literate, Japanese immigrants could not become citizens and vote. Consider the case of Takao Ozawa, originally from Japan. Ozawa arrived in California in 1894, graduated from high school, and attended the University of California, Berkeley. After that, he worked for an American company, living with his family in the U.S. territory of Honolulu. Because he was not legally classified as White, he could not become a U.S. citizen, though he very much wanted to and filed for citizenship in 1914. Ozawa argued that he was a "true American," a person of good character who neither drank, smoked, nor gambled. His family went to an American church; his children went to American schools; he spoke only English at home and raised his children as Americans. In 1922, the U.S. Supreme Court denied his eligibility for citizenship based on the claim that he was not "Caucasian" (Takaki 1989; *Ozawa v. United States*, 260 U.S. 178, 1922).

As a group, the Japanese immigrants found their employment options limited; many worked to serve their own community. Using their agricultural skills to develop farms outside of cities, Japanese farmers provided fruits and vegetables for the growing urban population. As their settlements grew, so did opposition from the native and immigrant White population. Californians first passed laws in 1913 limiting Japanese people to being able to lease land for only three years. Later the Alien Land Act of 1920 prohibited them from leasing or owning land, so Japanese people either changed occupations or had White people purchase land for them (Takaki 1989). During World War II, 120,000 Japanese aliens and Japanese American citizens on the mainland United States were ordered from their homes and placed in internment camps away from the Pacific coast.

The patterns of exclusion that have marked the experiences of people of color have not been absent in the experiences of White ethnic groups. Yet, White ethnic groups could become citizens, even if they faced periods of exclusion and denigration. Slowly they became incorporated into the political and economic establishment. Once they secured citizenship, they were politically active and often shaped the politics of the early twentieth century, especially in cities.

Unlike the experiences of eastern and southern European immigrants, American Indians, African Americans, Asian immigrants, and other people of color faced extended hardships because of their lack of rights. Outside of agriculture, people of color found their employment prospects limited, often meaning that all family members worked. White immigrant men, on the other hand, could secure industrial jobs, thus earning higher wages and gaining economic security for their families.

The rights of people of color to full citizenship have been won through hard work and political struggle. The many movements for racial justice, including the civil rights movement initiated by African Americans—but also those organized by Native Americans, Asian Americans, Mexican Americas, and other Latinos—have sought to redistribute rights, privileges, and rewards to those who have been denied these basic citizenship rights. National laws, along with state and municipal actions, have in many respects dramatically changed employment, education, and public accommodations to create new opportunities for people of color. The entrance since the late 1960s of many new immigrants, who bring their own conceptions of race and ethnicity, is also changing U.S. society. People of color have made many accomplishments, but others still lack full rights as citizens. This book helps decipher the realities of our post-civil rights nation, by providing some historical background.

This section on the history of shifting racial hierarchies begins with Jacqueline Jones ("A Dreadful Deceit"). Jones interrogates the creation myth of our national origins, detailing how "race" is a fiction invented to advance certain economic and political objectives. Over time, racial categories were used to justify privileges for some, while enforcing hard work, denial of rights, and poverty for others. By the twentieth century, these racial meanings appeared fixed. Scholarship such as Jones's illuminates the political and economic origins of "racial differences" and how they operate.

Evelyn Nakano Glenn ("Citizenship and Inequality") explores how people of color battled the limitations of the Constitution and opened up options for others. She finds that, even when laws are changed, racial practices continue to shape social policy, making full participation as citizens difficult. Her selection shows the contested nature of citizenship and its relationship to labor exploitation.

C. Matthew Snipp ("The First Americans: American Indians") explains how early national leaders considered American Indians to be candidates for extinction. Thus, Native Americans' path to incorporation into the nation involved removal from their original homelands and, later, forced assimilation. Their inability to speak for themselves as citizens of the United States made them vulnerable to laws that were supposed to help, but really harmed American Indian people. Only in the twentieth century could tribes gain some measure of control and real sovereignty to improve their status, but their rights are still contested in many states.

The final two articles in this part look at more recent developments, as the racial landscape has profoundly shifted for Asian Americans and African Americans. Ellen D. Wu ("Imperatives of Asian American Citizenship") looks at the first group. In the mid-twentieth century, the descendants of Chinese and Japanese immigrants—citizens by birthright—were still excluded from many social institutions. Second-generation immigrants faced residential segregation and many attended segregated schools, and they were subjected to laws in many states prohibiting them from marrying White people. As adults, most Asian Americans could only secure employment within their own communities. By the 1970s, Asian Americans had become the "model minority," achieving a level of economic stability and respect that was not forthcoming earlier. Wu examines how this transformation was not only about the abilities of Asian Americans, but also about the integration that was finally offered to some people of color.

Wendy Leo Moore and Joyce M. Bell ("Embodying the White Racial Frame: The (In)Significance of Barack Obama") bring us to the present, asking what the election of President Barack Obama means regarding citizenship and participation. Many people have declared that the nation is now "postracial," a place where people are able to succeed based on their own abilities, not the rigid hierarchies of the past. The election of the first African American president leads many to think that civil rights legislation has been successful in fully integrating people of color. Yet, we have witnessed increasing racial tensions and divisions since President Obama took office in 2009. Wendy Leo Moore and Joyce M. Bell discuss how, even with legislation prohibiting discrimination, racism remains embedded in most social institutions. Furthermore, they argue that we still operate under a "White racial frame" that obscures structural disadvantages for people of color (Feagin 2009). Understanding these shifting arrangements helps us think about the work necessary to build a society where we participate as equals, thereby protecting all of our citizens.

REFERENCES

Feagin, Joe R. 2009. *The White Racial Frame*. New York: Routledge.

Glenn, Evelyn Nakano. 2002. *Unequal Freedom: How Race and Gender Shaped American Citizenship and Labor*. Cambridge, MA: Harvard University Press.

Ngai, Mae M. 2003. *Impossible Subjects: Illegal Aliens and the Making of Modern America*. Princeton, NJ: Princeton University Press.

Takaki, Ronald. 1989. *Strangers from a Different Shore: A History of Asian Americans*. New York: Penguin.

FACE THE FACTS: CHANGES IN THE FOREIGN-BORN WITHIN THE U.S. POPULATION

Foreign-Born U.S. Population as a Percentage of Total Population, 1890 to 2010

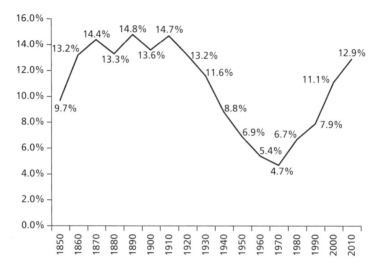

Foreign-Born U.S. Population by Region of Birth, 1960 to 2010

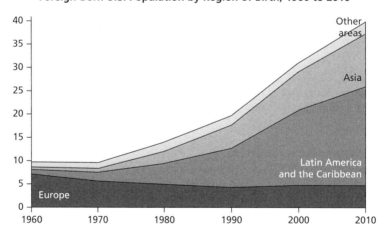

SOURCE: U.S. Census Bureau. 2014. *Historical Census Statistics on the Foreign-Born Population of the United States, 1850–2000*. Washington, DC: U.S. Department of Commerce. www.census.gov; U.S. Census Bureau. 2012. *The 2012 Statistical Abstract*. Washington, DC: U.S. Department of Commerce. www.census.gov.

Think about It:

1. Both of these graphs show the historic changes in the character of the U.S. population over time. Looking at both, what observations would you make about the makeup of the population over time?

2. What factors do you think have influenced the changes that you see?

5

A Dreadful Deceit

JACQUELINE JONES

Jones challenges the conception of race as "real" by looking at the origins of the concept in the U.S. colonies where the fiction of "race" was developed. She asks us to question this fiction that has been used to structure life options for different groups, advantaging some and disadvantaging others.

Like countless other cultures and countries throughout the world, the United States has its own creation myth—its own unique, dramatic story intended to explain where we came from and who we are today. In the case of the United States, this story holds that the nation was conceived in "racial" differences and that over the last four centuries these self-evident differences have suffused our national character and shaped our national destiny. The American creation story begins with a violent, self-inflicted wound and features subsequent incremental episodes of healing, culminating in a redemption of sorts. It is, ultimately, a triumphant narrative, one that testifies to the innate strength and moral rectitude of the American system, however imperfect its origins.

According to this myth, the first Europeans who laid eyes on Africans were struck foremost by their physical appearance—the color of their skin and the texture of their hair—and concluded that these beings constituted a lower order of humans, an inferior race destined for enslavement. During the American Revolution, patriots spoke eloquently of liberty and equality, and though their lofty rhetoric went unfulfilled, they inadvertently challenged basic forms of racial categorization. And so white Northerners, deriving inspiration from the Revolution, emancipated their own slaves and ushered in a society free of the moral stain of race-based bondage. The Civil War destroyed the system of slavery nationwide, but new theories of scientific racism gave rise to new forms of racial oppression in the North and South. Not until the Civil Rights Acts of 1964 and 1965 did the federal government dismantle state-sponsored race-based segregation and thus pave the way for better race relations. Though hardly an unmitigated triumph, the election of Barack Obama as president in 2008 signaled the dawn of a postracial society and offered a measure of the distance the country had traveled since slavery prevailed in British North America.

Yet America's creation myth is just that—a myth, one that itself rests entirely on a spurious concept: for "race" itself is a fiction, one that has no basis in

SOURCE: Jones, Jacqueline. 2013. *A Dreadful Deceit: The Myth of Race from the Colonial Era to Barack Obama's America.* New York: Basic Books.

biology or any long-standing, consistent usage in human culture. As employed in the popular rendition of America's national origins, the word and its various iterations mask complex historical processes that have little or nothing to do with the physical makeup of the people who controlled or suffered from those processes.

The ubiquity of the term *race* in modern discourse indicates that early-twenty-first-century Americans adhere to this creation myth with remarkable tenacity—in other words, that they believe that race is real and that race matters. In fact, however, like its worldwide counterparts, the American creation myth is the product of collective imagination, not historical fact, and it exists outside the realm of rational thought. Americans who would scoff at the notion that meaningful social or temperamental differences distinguish brown-eyed people from blue-eyed people nevertheless utter the word "race" with a casual thoughtlessness. Consequently, the word itself helps to sustain not only the creation myth but also all the human misery that the myth has wrought over the centuries. In effect, the word perpetuates—and legitimizes—the notion that some kind of inexorable primal prejudice has driven history and that, to some degree at least, the United States has always been held hostage to "racial" differences.

Certainly, the bitter legacies of historic injustices endure in concrete, blatant form. Today certain groups of people are impoverished, exploited in the workplace, or incarcerated in large numbers. This is the case not because of their "race," however, but rather because at a particular point in U.S. history certain other groups began to invoke the myth of race in a bid for political and economic power. This myth has served as a tool that one group can use to ratchet itself into a position of greater advantage in society, and a justification for the economic inequality and the imbalance in rights and privileges that result.

Perhaps the greatest perversity of the idea of race is how meaningless it truly is. Strikingly malleable in its contours, depending on the exigencies of the moment, race is a catchall term, its insidious reach metastasizing in response to any number of competitions—for political rights, scarce resources, control over cheap labor, group security. At times, within this constellation of "racial" ideas, physical appearance receded into nothingness—for example, when the law defined a person's race according to his or her "reputation" or when a mother's legal status as a slave decreed that her offspring would remain enslaved, regardless of their own skin color.

The indistinctness of this idea has given it a twisted trajectory. Throughout American history, members of the white laboring classes witnessed firsthand the struggles of their black coworkers and rivals and yet still maintained that "race" constituted a great divide between them. After the Civil War, on the campaign stump, white politicians charged that black people were incapable of learning, even while the descendants of slaves were rapidly gaining in literacy rates compared to poor whites. In the early twenty-first century, many Americans disavow the basic premise of racial prejudice—the idea that blacks and whites were somehow fundamentally different from each other—and yet scholars, journalists, and indeed Americans from all walks of life persist in categorizing and labeling groups according to those same, discredited principles....

The myth of race is, at its heart, about power relations, and in order to understand how it evolved, we must avoid vague theoretical and ahistorical formulations and instead ask, Who benefited from these narratives of racial difference, and how, where, and under what conditions? Race signifies neither a biological fact nor a primal prejudice, and it lacks the coherence of a robust political ideology; rather, it is a collection of fluid, contingent mythologies borne of (among other imperatives) fighting a war, assembling a labor force, advancing the designs of demagogues, organizing a labor union, and preserving voting and public schooling as privileges reserved for some, rather than as rights shared by all....

For the first century and a half or so of the British North American colonies, the fiction of race played little part in the origins and development of slavery; instead, that institution was the product of the unique vulnerability of Africans within a roiling Atlantic world of empire-building and profit-seeking. Not until the American Revolution did self-identified "white" elites perceive the need to concoct ideas of racial difference; these elites understood that the exclusion of a whole group of native-born men from the body politic demanded an explanation, a rationalization. Even then, many southern slaveholders, lording over forced-labor camps, believed they needed to justify their actions to no one; only over time did they begin to refer to their bound workforces in racial terms. Meanwhile, in the early nineteenth-century North, race emerged as a partisan political weapon, its rhetorical contours strikingly contradictory but its legal dimensions nevertheless explicit. Discriminatory laws and mob actions promoted and enforced the insidious notion that people could be assigned to a particular racial group and thereby considered "inferior" to whites and unworthy of basic human rights. Black immiseration was part and parcel of white privilege, all in the name of—the myth of—race.

By the late twentieth century, transformations in the American political economy had solidified the historic liabilities of black men and women, now in the form of segregated neighborhoods and a particular social division of labor within a so-called color-blind nation. The election of the nation's first black president in 2008 produced an outpouring of self-congratulation among Americans who heralded the dawn of a postracial society. In fact, the recession that began that year showed that, although explicit ideas of black inferiority had receded (though not entirely disappeared) from American public discourse, African Americans continued to suffer the disastrous consequences spawned by those ideas, as evidenced by disproportionately high rates of poverty, unemployment, home foreclosures, and incarceration....

By the time of the Revolution, the institution of bondage had made a small group of planters fabulously wealthy. Those in the South Carolina Lowcountry failed even to contemplate the hypocritical dimensions of their own revolutionary rhetoric of liberty and equality; they felt no compunction to rationalize human bondage, preferring to see the institution as one upon which their own self-interest depended. Slavery provided the ruling class with the labor that fueled a privileged way of life, of which richly furnished mansions, fashionable clothing, lavish dinners, and fancy carriages were key components. Race as a justification for black subordination developed only gradually as other slaveholders—Thomas Jefferson most notably—decided that revolutionary times demanded a theory of social difference that would posit black intellectual inferiority and rationalize black exclusion from the body politic of the new nation....

Meanwhile, in the post-Revolutionary North, whites of all classes and backgrounds aimed to preserve their political privileges by attaching the stigma of slavery to recently freed people of color. This contradictory, fictional narrative held that the former slaves and their descendants were by nature poor and dependent, but also at the same time predatory and dangerously capable of depriving whites of access to good jobs and even political rights. A rapidly transforming economy in the Northeast threatened the social and financial stability of white householders and provoked attacks on blacks, who served as scapegoats for the factory system's degradation of skilled labor....

By the time of the Civil War, myths of race had hardened to presume a large underclass of dark-skinned menial laborers, men and women "naturally" inferior to whites in intelligence and moral sensibility. Yet these prejudices could not account for the growing number of people of African descent who were well educated and accomplished—and in some cases as "white" as white supremacists themselves. As a result, the notion of race became more flexible and abstract.... By the latter part of the century, it was clear that, despite the massive loss of life during the Civil War era, the American partisan political system would fail to address, or redress, systematic forms of discrimination leveled at black people. In fact, both the Democratic and the Republican Parties catered only to their own constituents—white men who could vote—and acted on the presumption that no partisan good could come of challenging racial ideas.

From the early nineteenth century onward, control over public schooling in its various forms became a means by which whites could reserve for themselves specific privileges and economic opportunities. In the South, even rudimentary education such as literacy instruction seemed to threaten a political economy that relied on a large, subordinate agricultural labor force. In 1900 Mississippi politicians used the heavy hand of the state, and manipulated the fury of the lynch mob, to promote, once again, a baldly contradictory set of ideas—that first, black people were incapable of learning, and second, black people must be prevented from learning at all costs....

In the second half of the twentieth century, a small group of industrial workers sought to weld diverse laboring classes into a unified movement; yet they had to contend with racial ideologies that not only were well entrenched and granted "scientific" credibility, but also were perceived by whites regardless of class as necessary to their own well-being and the future well-being of their children. Transformations in the economy again produced wrenching dislocations for all workers, prompting whites to reprise their nineteenth-century practice of scapegoating blacks as the root cause of white unemployment.... By this time, generations of discriminatory laws and policies had produced whole communities marked by concentrated poverty—neighborhoods bereft of good schools, adequate health care, and decent jobs for men and women with little in the way of skills or formal education....

In a study devoted to the resonance of race, defining the terms *white* and *black* is essential. Like notions of race, these two remarkably dichotomous descriptors emerged from a fundamental imbalance in power among social groups. On slave ships transporting men, women, and children from their homelands to the New World, European captors became white and their African captives became black. Over time these two adjectives each took on multiple meanings, with white signifying "someone free and descended from free forebears," and black signifying "someone enslaved or descended from slaves." These contrasting associations were the result of political processes, when people in power foisted a "black" identity on people devoid of legal protection, and later, when members of the latter group embraced the term "black" as an act of solidarity among themselves. The terms "black" and "white" are used ... then, with the understanding that they distinguish between groups of people according to heritage, legal status, or collective self-identification, depending on the immediate context.

In the early twenty-first century, the words "race," "racism," and "race relations" are widely used as shorthand for specific historical legacies that have nothing to do with biological determinism and everything to do with power relations.... Racial mythologies are best understood as a pretext for political and economic opportunism both wide ranging and specific to a particular time and place. If this explication of the American creation myth leads to one overriding conclusion, it is the power of the word "race" to distort our understanding of the past and the present—and our hopes for a more just future—in equal measure.

DISCUSSION QUESTIONS

1. How does Jones's research challenge how you might have learned to think about race and racial categories in your early schooling?
2. If race is a fiction, how does it still operate to place people in society?

6

Citizenship and Inequality

EVELYN NAKANO GLENN

Glenn explores the concept of citizenship as it has developed in reference to racial and ethnic groups in American society. She points out that citizenship includes both formal rights and the informal sense of belonging—forms of citizenship that have historically been denied to different groups in different ways. Her article shows how central the construction of citizenship is to being seen as having full "personhood."

HISTORICAL DEBATES ABOUT CITIZENSHIP IN THE UNITED STATES

... The concept of citizenship is, of course, historically and culturally specific. The modern, western notion of citizenship emerged out of the political and intellectual revolutions of the seventeenth and eighteenth centuries, which overthrew the old feudal orders. The earlier concept of society organized as a hierarchy of status, expressed by differential legal and customary rights, was replaced by the idea of a political order established through social contract. Social contract implied free and equal status among those who were party to it. Equality of citizenship did not, of course, rule out economic and other forms of inequality. Moreover, and importantly, equality among citizens rested on the inequality of others living within the boundaries of the community who were defined as noncitizens. The relationship between equality of citizens and inequality of noncitizens had both rhetorical and material dimensions. Rhetorically, the "citizen" was defined and, therefore, gained meaning through its contrast with the oppositional concept of the "non-citizen" as one who lacked the essential qualities needed to exercise citizenship. Materially, the autonomy and freedom of the citizen were made possible by labor (often involuntary) of non-autonomous wives, slaves, children, servants, and employees....

A specifically sociological conception of citizenship as membership is offered by Turner (1993:2), who defines it as "a set of practices (judicial, political, economic, and cultural) which define a person as a 'competent' member of

SOURCE: From Glenn, Evelyn Nakano. 2000. "Citizenship and Inequality." *Social Problems* 47(1):1–20. Copyright © 2000 by The Society for the Study of Social Problems, Inc. All rights reserved. Reprinted by permission.

society...." Focusing on social practice takes us beyond a juridical or state conception of citizenship. It points to citizenship as a fluid and decentered complex that is continually transformed through political struggle.

Membership entails drawing distinctions and boundaries for who is included and who is not. Inclusion as a member, in turn, implies certain rights in, and reciprocal obligations toward, the community. Formal rights are not enough, however; they are only paper claims unless they can be enacted through actual practice. Three leading elements in the construction of citizenship, then, are membership, rights and duties, and conditions necessary for practice.

These three elements formed the major themes that have run through debates, contestation, and struggles over American citizenship since the beginning. First, membership: who is included or recognized as a full member of the imagined community (Anderson 1983), and on what basis? Second, what does membership mean in terms of content: that is, what reciprocal rights and duties do citizens have? Third, what are the conditions necessary for citizens to practice citizenship, to actually realize their rights and carry out their responsibilities as citizens?

MEMBERSHIP

Regarding membership, there are two major strains of American thought regarding the boundaries of the community. One tradition is that of civic citizenship, a definition based on shared political institutions and values in which membership is open to all those who reside in a territory. The second is an ethno-cultural definition based on common heritage and culture, in which membership is limited to those who share in the heritage through blood descent (Smith 1989; Kettner 1978).

Because of its professed belief in equality and natural rights (epitomized by the Declaration of Independence), the United States would seem to fit the civic model. However, since its beginnings, the U.S. has followed both civic and ethno-cultural models. The popular self-image of the nation, expressed as early as the 1780s, was of the United States as a refuge of freedom for those fleeing tyranny. This concept, later elaborated in historical narratives and sociological accounts of America as a nation of immigrants, blatantly omitted Native Americans, who were already here, Mexicans, who were incorporated through territorial expansion, and Blacks, who were forcibly transported. This exclusionary self-image was reflected at the formal level in the founding document of the American polity, the U.S. Constitution. The authors of the Constitution, in proclaiming a government by and for "we the people," clearly did not intend to include everyone residing within the boundaries of the U.S. as part of "the people." The Constitution explicitly named three groups: Indians, who were identified as members of separate nations, and "others," namely slaves; and finally, "the people": only the latter were full members of the U.S. community (Ringer 1983).

Interestingly, the Constitution was silent as to who was a citizen and what rights and privileges they enjoyed. It left to each state the authority to determine qualifications for citizenship and citizens' rights, e.g., suffrage requirements, qualifications for sitting on juries, etc. Individuals were, first, citizens of the states in which they resided and only secondarily, through their citizenship in the state, citizens of the United States. The concept of national citizenship was, therefore, quite weak.

However, the Constitution did direct Congress to establish a uniform law with respect to naturalization. Accordingly, Congress passed a Naturalization Act in 1790, which shaped citizenship policy for the next 170 years. It limited the right to become naturalized citizens to "free white persons." The act was amended in 1870 to add Blacks, but the term "free white persons" was retained. As Ian Haney Lopez (1996) has documented, immigrants deemed to be non-white (Hawaiians, Syrians, Asian Indians), but not Black or African, were barred from naturalization. The largest such category was immigrants from China, Japan, and other parts of Asia, who were deemed by the courts to be "aliens ineligible for citizenship." This exclusion remained in force until 1953....

It was Black Americans, both before and after the Civil War, who were the most consistent advocates of universal citizenship. Hence, it is fitting that the Civil Rights Act of 1865 and the Fourteenth Amendment, ratified in 1867 to ensure the rights of freed people, greatly expanded citizenship for everyone. Section 1 of the Fourteenth Amendment stated that "All persons born or naturalized in the United States and subject to the jurisdiction thereof, are citizens of the United States and of the State wherein they reside. No State shall make or enforce any law which shall abridge the privileges or immunities of citizens of the United States; nor shall any State deprive any person of life, liberty, or property, without due process of law; nor deny to any person within its jurisdiction, the equal protection of the law."

In these brief sentences, the Fourteenth Amendment established three important principles for the first time: the principle of national citizenship, the concept of the federal state as the protector and guarantor of national citizenship rights, and the principle of birthright citizenship. These principles expanded citizenship for others besides Blacks. To cite one personal example, my grandfather, who came to this country in 1894, was ineligible to become a naturalized citizen because he was not white, but his daughter, my mother, automatically became a citizen as soon as she was born. Birthright citizenship was tremendously important for second and third generation Japanese Americans and other Asian Americans who otherwise would have remained perpetual aliens, as is now the case with immigrant minorities in some European countries. Foner (1998) was on the mark when he said that the Black American struggle to expand the boundaries of freedom to include themselves, succeeded in changing the boundaries of freedom for everyone.

The battle was not won, once and for all, in 1867. Instead, the nation continued to vacillate between the principle of the federal government having a duty to protect citizens' rights and states' rights. By the end of the Reconstruction period in the 1870s, the slide back toward states' rights accelerated as all branches of the federal government withdrew from protecting Black rights and allowed southern states to impose white supremacist regimes (Foner 1988). In the landmark 1896 *Plessy v. Ferguson* decision, the Supreme Court legitimized segregation based on the principle of "separate but equal." This and other court decisions gutted the concept of national citizenship and carved out vast areas of life, employment, housing, transportation, and public accommodations as essentially private activity that was not protected by the Constitution (Woodward 1974). It was not until the second civil rights revolution of the 1950s and 1960s that the federal courts and Congress returned to the principles of national citizenship and a strong federal obligation to protect civil rights....

MEANINGS OF RIGHTS AND RESPONSIBILITIES

Just as with the question of membership, there has not been a single understanding of rights and duties. American ideas on rights and duties have been shaped by two different political languages (Smith 1989). One, termed liberalism by scholars, grew out of Locke and other enlightenment thinkers. In this strain of thought, embodied in the Declaration of Independence, citizens are individual rights bearers. Governments were established to secure individual rights so as to allow each person to pursue private as well as public happiness. The public good was not an ideal to be pursued by government, but was to be the outcome of individuals pursuing their own individual interest. The other language was that of *republicanism*, which saw the citizen as one who actively participated in public life. This line of thought reached back to ancient Greece and Rome where republicanism held that man reached his highest fulfillment by setting aside self-interest to pursue the common good. In contrast to liberalism, republicanism emphasizes practice and focuses on achieving institutions and practices that make collective self-government possible.

There has been continuing tension and alternation between these two strains of thought, particularly around the question of whether political participation is essential to, or peripheral to, citizenship. In the nineteenth century, when many groups were excluded from participation, the vote was a mark of standing; its lack was a stigma, a badge of inferiority, hence, the passion with which those denied the right fought for inclusion. Three great movements—for universal white manhood suffrage, for Black emancipation, and for women's suffrage—resulted in successive extensions of the vote to non-propertied white men, Black men, and women, between 1800 and 1920 (Roediger 1991; Foner 1990; DuBois 1978). According to Judith Shklar (1991), once the vote became broadly available, it ceased to be a mark of status. Judging by participation rates, voting is no longer an emblem of citizenship. One measure of the decline in the significance of the vote has been the precipitous drop in persons voting. In 1890, fully 80% of those who were eligible, voted. By 1924, four years after the extension of suffrage to women, participation had fallen to less than half of those registered, and it has been declining ever since....

In the U.S., participation is discouraged by not making registration easy. Requiring re-registration every time one moves is only one example of the obstacles placed upon a highly mobile population. While Americans today would object if the right to vote was taken away, the majority don't seem to feel it necessary to exercise the vote. The U.S. obviously has the capacity to make registration automatic or convenient, but has not made major efforts to do so. Needless to say, it is in the interest of big capital and its minions that most people don't participate politically.

CONDITIONS FOR PRACTICE

This leads to the third main theme, the actual practice of citizenship and the question of what material and social conditions are necessary for people to actually exercise their rights and participate in the polity.

The short answer to this question for most of American history has been that a citizen must be *independent*—that is, able to act autonomously.... Independence has remained essential, but its meaning has undergone drastic transformation since the founding of the nation.... In the nineteenth century, as industrialization proceeded apace and wage work became common even for skilled artisans and white collar men, the meaning of independence changed to make it more consistent with the actual situation of most white men. It came to mean, not ownership of productive property, but nothing more than ownership of one's own labor and the capacity to sell it for remuneration (Fraser and Gordon 1994). This formulation rendered almost all white men "independent," while rendering slaves and women, who did not have complete freedom to sell their labor, "dependent." It was on the basis of the independence of white working men that the movement for white Universal Manhood Suffrage was mobilized (Roediger 1991). By the 1830s, all states had repealed property requirements for suffrage for white men, while simultaneously barring women and Blacks (Litwack 1961).

By the late 19th century, capitalist industrialization had widened the economic gap between the top owners of productive resources and the rest, making more apparent the contradiction between economic inequality and political democracy. Rising levels of poverty, despite the expansion in overall wealth, raised the question of whether low-income, unemployment, and/or lack of access to health care and other services, diminished citizenship rights for a large portion of the populace. Growing economic inequality also raised the issue of whether some non-market mechanism was needed to mitigate the harshness of inequities created by the market.

One response during the years between World War I and World War II was rising sentiment for the idea of what T. H. Marshall called social citizenship. In Marshall's words (Marshall 1964:78), social citizenship involves "the right to a modicum of economic security and to share in the full social heritage and to live the life of a civilized being, according to the standards prevailing in the society."

Economic and social unrest after World War I spurred European states to institute programs to ensure some level of economic security and take collective responsibility for "dependents"—the aged, children, the disabled, and others unable to work.... The significance of the "redistributive" mechanisms of the Welfare State, according to Marshall, was that they enabled working-class people to exercise their civil and political rights.... Compared to Western Europe, the concept of social citizenship has been relatively weak in the U.S. Welfare state researchers have pointed out that, although 1930s New Deal programs such as Social Security and unemployment insurance greatly increased economic security, they continued a pattern of a two-tiered system of social citizenship from the 1890s (Nelson 1990; Skocpol 1992; Fraser and Gordon 1993; Fraser and Gordon 1994). The upper tier consisted of "entitlements" based on employment or military service, e.g., unemployment benefits, old-age insurance, and disability payments, which were relatively generous and did not require means-testing. The lower tier consisted of various forms of "welfare," such as Aid to Dependent Children (changed to Aid to Families with Dependent Children—AFDC—in 1962), which were relatively stingy and entailed means testing and surveillance by the state.

White men, as a class, have drawn disproportionately on first tier rights by virtue of their records of regular and well-paid employment. White women, more often, had to rely on welfare, which is considered charity, a response to dependence, rather than a just return for contributions. Latino and African American men were generally excluded from employment-based benefits because of their concentration in agriculture, day jobs, and other excluded occupations; Latina and African American women, in turn, have often been denied even second tier benefits (Oliver and Shapiro 1995; Mink 1994). Moreover, in contrast to the situation in most European countries, there has been little sense of collective responsibility for the care of dependents. Thus, raising children is not recognized as a contribution to the society and, therefore, as a citizenship responsibility that warrants entitlements such as parental allowances and retirement credit, which are common in Europe (Glenn 2000; Sainsbury 1996).

As with the other issues of citizenship, the 1980s and 1990s [saw] a neoconservative turn with a concerted effort to roll back even attenuated social citizenship rights. Government funding of social services has been vilified for draining money from hardworking citizens to support loafers and government bureaucrats who exert onerous control over people's lives. And, after years of attacks on Black and other single mothers "dependent" on welfare, Congress passed the Personal Responsibility and Work Opportunity Act in 1996, which dismantled AFDC and replaced it with grants to the states that limited eligibility for federal benefits for a lifetime maximum of five years. The aim was, as Representative Richard Armey put it, that the poor return to the natural safety nets —family, friends, churches, and charities. States instituted stringent work requirements and limited total lifetime benefits. By limiting total years that they can stay on welfare, proponents argued the new regulations would wean single mothers from unhealthy dependence on the state. In contrast to an overly indulgent government, the market is expected to exert a moral force, disciplining them and forcing them to be "independent" (Roberts 1997; Boris 1998)....

Thus, we must ask: if people have a responsibility to earn, then, don't they also have a corresponding right to earn?—i.e., to have a job and to earn enough to support themselves at a level that allows them, in Marshall's (1964:78) words, "to participate fully in the cultural life of the society and to live the life of a civilized being according to the standards prevailing in the society."

REFERENCES

Anderson, Benedict. 1983. *Imagined Communities: Reflections on the Origins and Spread of Nationalism*. London: Verso.

Boris, Eileen. 1998. "When work is slavery." *Social Justice* 25, 1:18–44.

DuBois, Ellen Carol. 1978. *Feminism and Suffrage*. Ithaca: Cornell University Press.

Foner, Eric. 1988. *Reconstruction: America's Unfinished Revolution*. New York: Harper and Row.

———— 1990. "From slavery to citizenship: Blacks and the right to vote." In *Voting and the Spirit of American Democracy*, ed. Donald W. Rogers. Urbana and Chicago: University of Illinois Press.

———— 1998. *The Story of American Freedom*. New York: Norton.

Fraser, Nancy, and Linda Gordon. 1993. "Contract versus charity: Why is there no social citizenship in the United States?" *Socialist Review* 212, 3:45–68.

———— 1994. "A genealogy of dependence: Tracing a keyword of the U.S. welfare state." *Signs* 19, 2:309–336.

Glenn, Evelyn. 2000. "Creating a caring society." *Contemporary Sociology* 29, 1:84–94.

Haney Lopez, Ian F. 1996. *White By Law: The Legal Construction of Race*. New York: New York University Press.

Kettner, James. 1978. *The Development of American Citizenship, 1608–1870*. Chapel Hill: University of North Carolina Press.

Litwack, Leon F. 1961. *North of Slavery: The Negro in the Free States, 1790–1860*. Chicago: University of Chicago Press.

Marshall, T. H. 1964. *Class, Citizenship, and Social Development*. Garden City, NY: Doubleday and Company.

Mink, Gwendolyn. 1994. *The Wages of Motherhood: Inequality in the Welfare State, 1917–1942*. Ithaca: Cornell University Press.

Nelson, Barbara. 1990. "The origins of the two-channel welfare state: Workman's compensation and mother's aid." In *Women, the State and Welfare*, ed. Linda Gordon, 123–151. Madison: University of Wisconsin Press.

Oliver, Melvin L., and Thomas M. Shapiro. 1995. *Black Wealth/White Wealth: A New Perspective on Racial Inequality*. New York: Routledge.

Ringer, Benjamin B. 1983. *We the People and Others: Duality and America's Treatment of Its Racial Minorities*. New York: Tavistock.

Roberts, Dorothy. 1997. *Killing the Black Body: Race, Reproduction, and the Meaning of Liberty*. New York: Pantheon.

Roediger, David. 1991. *The Wages of Whiteness: Race and the Making of the American Working Class*. London: Verso.

Sainsbury, Diane. 1996. *Gender, Equality and Welfare States*. Cambridge, UK: Cambridge University Press.

Shklar, Judith. 1991. *American Citizenship: The Quest for Inclusion*. Cambridge, MA: Harvard University Press.

Skocpol, Theda. 1992. *Protecting Soldiers and Mothers: The Origins of Social Policy in the United States*. Cambridge, MA: Harvard University Press.

Smith, Rogers M. 1989. "'One United People': Second-class female citizenship and the American quest for community." *Yale Journal of Law and the Humanities* 1:229–293.

Turner, Brian S. 1993. "Contemporary Problems in the Theory of Citizenship." In *Citizenship and Social Theory*, ed. Brian S. Turner, 1–19. London: Sage Publications.

Woodward, C. Vann, ed. 1974. *The Strange Career of Jim Crow*, 3rd rev. New York: Oxford University Press.

DISCUSSION QUESTIONS

1. What elements does Glenn identify as constructing citizenship?

2. During what era did social citizenship rights expand and why?

3. What do you see as your rights of citizenship and your obligations of citizenship?

7

The First Americans

American Indians

C. MATTHEW SNIPP

Snipp's analysis of different periods in the treatment of American Indians by the U.S. government shows how citizenship rights have been denied to Native Americans, even though the specific processes by which this has happened have changed over time.

By the end of the nineteenth century, many observers predicted that American Indians were destined for extinction. Within a few generations, disease, warfare, famine, and outright genocide had reduced their numbers from millions to less than 250,000 in 1890. Once a self-governing, self-sufficient people, American Indians were forced to give up their homes and their land, and to subordinate themselves to an alien culture. The forced resettlement to reservation lands or the Indian Territory (now Oklahoma) frequently meant a life of destitution, hunger, and complete dependency on the federal government for material needs.

Today, American Indians are more numerous than they have been for several centuries. While still one of the most destitute groups in American society, tribes have more autonomy and are now more self-sufficient than at any time since the last century. In cities, modern pan-Indian organizations have been successful in making the presence of American Indians known to the larger community, and have mobilized to meet the needs of their people (Cornell 1988; Nagel 1986; Weibel-Orlando 1991). In many rural areas, American Indians and especially tribal governments have become increasingly more important and increasingly more visible by virtue of their growing political and economic power. The balance of this [reading] is devoted to explaining their unique place in American society.

THE INCORPORATION OF AMERICAN INDIANS

The current political and economic status of American Indians is the result of the process by which they were incorporated into Euro-American society (Hall 1989). This amounts to a long history of efforts aimed at subordinating an

SOURCE: From Pedraza, Silvia, and Rubén G. Rumbaut, eds. 1996. *Origins and Destinies: Immigration, Race, and Ethnicity in America.* Belmont, CA: Wadsworth.

otherwise self-governing and self-sufficient people that eventually culminated in widespread economic dependency. The role of the U.S. government in this process can be seen in the five major historical periods of federal Indian relations: removal, assimilation, the Indian New Deal, termination and relocation, and self-determination.

Removal

In the early nineteenth century, the population of the United States expanded rapidly at the same time that the federal government increased its political and military capabilities. The character of Indian-American relations changed after the War of 1812. The federal government increasingly pressured tribes settled east of the Appalachian Mountains to move west to the territory acquired in the Louisiana Purchase. Numerous treaties were negotiated by which the tribes relinquished most of their land and eventually were forced to move west.

Initially the federal government used bargaining and negotiation to accomplish removal, but many tribes resisted (Prucha 1984). However, the election of Andrew Jackson by a frontier constituency signaled the beginning of more forceful measures to accomplish removal. In 1830 Congress passed the Indian Removal Act, which mandated the eventual removal of the eastern tribes to points west of the Mississippi River, in an area which was to become the Indian Territory and is now the state of Oklahoma. Dozens of tribes were forcibly removed from the eastern half of the United States to the Indian Territory and newly created reservations in the west, a long process ridden with conflict and bloodshed.

As the nation expanded beyond the Mississippi River, tribes of the plains, southwest, and west coast were forcibly settled and quarantined on isolated reservations. This was accompanied by the so-called Indian Wars—a bloody chapter in the history of Indian-White relations (Prucha 1984; Utley 1963). This period in American history is especially remarkable because the U.S. government was responsible for what is unquestionably one of the largest forced migrations in history.

The actual process of removal spanned more than a half-century and affected nearly every tribe east of the Mississippi River. Removal often meant extreme hardships for American Indians, and in some cases this hardship reached legendary proportions. For example, the Cherokee removal has become known as the "Trail of Tears." In 1838, nearly 17,000 Cherokees were ordered to leave their homes and assemble in military stockades (Thornton 1987, p. 117). The march to the Indian Territory began in October and continued through the winter months. As many as 8,000 Cherokees died from cold weather and diseases such as influenza (Thornton 1987, p. 118).

According to William Hagan (1979), removal also caused the Creeks to suffer dearly as their society underwent a profound disintegration. The contractors who forcibly removed them from their homes refused to do anything for "the large number who had nothing but a cotton garment to protect them from the sleet storms and no shoes between them and the frozen ground of the last stages of their hegira. About half of the Creek nation did not survive the migration and the difficult early years in the West" (Hagan 1979, pp. 77–81). In the West, a band of Nez Perce men, women, and children, under the leadership of Chief Joseph, resisted resettlement in 1877. Heavily outnumbered, they were pursued by cavalry troops from the Wallowa valley in eastern Oregon and finally captured in Montana near the Canadian border. Although the Nez Perce were eventually captured and moved to the Indian Territory, and later to Idaho, their resistance to resettlement has been described by one historian as "one of the great military movements in history" (Prucha 1984, p. 541).

Assimilation

Near the end of the nineteenth century, the goal of isolating American Indians on reservations and the Indian Territory was finally achieved. The Indian population also was near extinction. Their numbers had declined steadily throughout the nineteenth century, leading most observers to predict their disappearance (Hoxie 1984). Reformers urged the federal government to adopt measures that would humanely ease American Indians into extinction. The federal government responded by creating boarding schools and the allotment acts—both were intended to "civilize" and assimilate American Indians into American society by Christianizing them, educating them, introducing them to private property, and making them into farmers. American Indian boarding schools sought to accomplish this task by indoctrinating Indian children with the belief that tribal culture was an inferior relic of the past and that Euro-American culture was vastly superior and preferable. Indian children were forbidden to wear their native attire, to eat their native foods, to speak their native language, or to practice their traditional religion. Instead, they were issued Euro-American clothes, and expected to speak English and become Christians. Indian children who did not relinquish their culture were punished by school authorities. The curriculum of these schools taught vocational arts along with "civilization" courses.

The impact of allotment policies is still evident today. The 1887 General Allotment Act (the Dawes Severalty Act) and subsequent legislation mandated that tribal lands were to be allotted to individual American Indians, … and the surplus lands left over from allotment were to be sold on the open market. Indians who received allotted tribal lands also received citizenship, farm implements, and encouragement from Indian agents to adopt farming as a livelihood (Hoxie 1984, Prucha 1984).

For a variety of reasons, Indian lands were not completely liquidated by allotment, many Indians did not receive allotments, and relatively few changed their lifestyles to become farmers. Nonetheless, the allotment era was a disaster because a significant number of allottees eventually lost their land. Through tax foreclosures, real estate fraud, and their own need for cash, many American Indians lost what for most of them was their last remaining asset (Hoxie 1984).

Allotment took a heavy toll on Indian lands. It caused about 90 million acres of Indian land to be lost, approximately two-thirds of the land that had belonged to tribes in 1887 (O'Brien 1990). This created another problem that continues to vex many reservations: "checkerboarding." Reservations that were subjected to allotment are typically a crazy quilt composed of tribal lands, privately owned "fee" land, and trust land belonging to individual Indian families. Checkerboarding presents reservation officials with enormous administrative problems when trying to develop land use management plans, zoning ordinances, or economic development projects that require the construction of physical infrastructure such as roads or bridges.

The Indian New Deal

The Indian New Deal was short-lived but profoundly important. Implemented in the early 1930s along with the other New Deal programs of the Roosevelt administration, the Indian New Deal was important for at least three reasons. First, signaling the end of the disastrous allotment era as well as a new respect for American Indian tribal culture, the Indian New Deal repudiated allotment as a policy. Instead of continuing its futile efforts to detribalize American Indians, the federal government acknowledged that tribal culture was worthy of respect. Much of this change was due to John Collier, a long-time Indian rights advocate appointed by Franklin Roosevelt to serve as Commissioner of Indian Affairs (Prucha 1984).

Like other New Deal policies, the Indian New Deal also offered some relief from the Great Depression and brought essential infrastructure development to many reservations, such as projects to control soil erosion and to build hydroelectric dams, roads, and other public facilities. These projects created jobs in New Deal programs such as the Civilian Conservation Corps and the Works Progress Administration.

An especially important and enduring legacy of the Indian New Deal was the passage of the Indian Reorganization Act (IRA) of 1934. Until then, Indian self-government had been forbidden by law. This act allowed tribal governments, for the first time in decades, to reconstitute themselves for the purpose of overseeing their own affairs on the reservation. Critics charge that this law imposed an alien form of government, representative democracy, on traditional tribal authority. On some reservations, this has been an ongoing source of conflict (O'Brien 1990). Some reservations rejected the IRA for this reason, but now have tribal governments authorized under different legislation.

Termination and Relocation

After World War II, the federal government moved to terminate its longstanding relationship with Indian tribes by settling the tribes' outstanding legal claims, by terminating the special status of reservations, and by helping reservation Indians relocate to urban areas (Fixico 1986). The Indian Claims Commission was a special tribunal created in 1946 to hasten the settlement of legal claims that tribes had brought against the federal government. In fact, the Indian Claims Commission became bogged down with prolonged cases, and in 1978 the commission was dissolved by Congress. At that time, there were 133 claims still unresolved out of an original 617 that were first heard by the commission three decades earlier (Fixico 1986, p. 186). The unresolved claims that were still pending were transferred to the Federal Court of Claims.

Congress also moved to terminate the federal government's relationship with Indian tribes. House Concurrent Resolution (HCR) 108, passed in 1953, called for steps that eventually would abolish all reservations and abolish all special programs serving American Indians. It also established a priority list of reservations slated for immediate termination. However, this bill and subsequent attempts to abolish reservations were vigorously opposed by Indian advocacy groups such as the National Congress of American Indians. Only two reservations were actually terminated, the Klamath in Oregon and the Menominee in Wisconsin. The Menominee reservation regained its trust status in 1975 and the Klamath reservation was restored in 1986.

The Bureau of Indian Affairs (BIA) also encouraged reservation Indians to relocate and seek work in urban job markets. This was prompted partly by the desperate economic prospects on most reservations, and partly because of the federal government's desire to "get out of the Indian business." The BIA's relocation programs aided reservation Indians in moving to designated cities, such as Los Angeles and Chicago, where they also assisted them in finding housing and employment. Between 1952 and 1972, the BIA relocated more than 100,000 American Indians (Sorkin 1978). However, many Indians returned to their reservations (Fixico 1986). For some American Indians, the return to the reservation was only temporary; for example, during periods when seasonal employment such as construction work was hard to find.

Self-Determination

Many of the policies enacted during the termination and relocation era were steadfastly opposed by American Indian leaders and their supporters. As these programs became stalled, critics attacked them for being harmful, ineffective, or both. By the mid-1960s, these policies had very little serious support: Perhaps inspired by the gains of the Civil Rights movement, American Indian leaders and their supporters made "self-determination" the first priority on their political agendas. For these activists, self-determination meant that Indian people would have the autonomy to control their own affairs, free from the paternalism of the federal government.

The idea of self-determination was well received by members of Congress sympathetic to American Indians. It also was consistent with the "New Federalism" of the Nixon administration. Thus, the policies of termination and relocation were repudiated in a process that culminated in 1975 with the passage of the American Indian Self-Determination and Education Assistance Act, a profound shift in federal Indian policy. For the first time since this nation's founding, American Indians were authorized to oversee the affairs of their own communities, free of federal intervention. In practice, the Self-Determination Act established measures that would allow tribal governments to assume a larger role in reservation administration of programs for welfare assistance, housing, job training, education, natural resource conservation, and the maintenance of reservation roads and bridges (Snipp and Summers 1991). Some reservations also have their own police forces and game wardens, and can issue licenses and levy taxes. The Onondaga tribe in upstate New York have taken their sovereignty one step further by issuing passports that are internationally recognized. Yet there is a great deal of variability in terms of how much autonomy tribes have over reservation affairs. Some tribes, especially those on large and well-organized reservations have nearly complete control over their reservations, while smaller reservations with limited resources often depend heavily on BIA services....

CONCLUSION

Though small in number, American Indians have an enduring place in American society. Growing numbers of American Indians occupy reservation and other trust lands, and equally important has been the revitalization of tribal governments. Tribal governments now have a larger role in reservation affairs than ever in the past. Another significant development has been the urbanization of American Indians. Since 1950, the proportion of American Indians in cities has grown rapidly. These American Indians have in common with reservation Indians many of the same problems and disadvantages, but they also face other challenges unique to city life.

The challenges facing tribal governments are daunting. American Indians are among the poorest groups in the nation. Reservation Indians have substantial needs for improved housing, adequate health care, educational opportunities, and employment, as well as developing and maintaining reservation infrastructure. In the face of declining federal assistance, tribal governments are assuming an ever-larger burden. On a handful of reservations, tribal governments have assumed completely the tasks once performed by the BIA.

As tribes have taken greater responsibility for their communities, they also have struggled with the problems of raising revenues and providing economic opportunities for their people. Reservation land bases provide many reservations with resources for development. However, these resources are not always abundant, much less unlimited, and they have not always been well managed. It will be yet another challenge for tribes to explore ways of efficiently managing their existing resources. Legal challenges also face tribes seeking to exploit unconventional resources such as gambling revenues. Their success depends on many complicated legal and political contingencies.

Urban American Indians have few of the resources found on reservations, and they face other difficult problems. Preserving their culture and identity is an especially pressing concern. However, urban Indians have successfully adapted to city environments in ways that preserve valued customs and activities— powwows, for example, are an important event in all cities where there is a large Indian community. In addition, pan-Indianism has helped urban Indians set aside tribal differences and forge alliances for the betterment of urban Indian communities.

These alliances are essential, because unlike reservation Indians, urban American Indians do not have their own form of self-government. Tribal governments do not have jurisdiction over urban Indians. For this reason, urban Indians must depend on other strategies for ensuring that the needs of their community are met, especially for those new to city life. Coping with the transition to urban life poses a multitude of difficult challenges for many American Indians. Some succumb to these problems, especially the hardships of unemployment, economic deprivation, and related maladies such as substance abuse, crime, and violence. But most successfully overcome these difficulties, often with help from other members of the urban Indian community.

Perhaps the greatest strength of American Indians has been their ability to find creative ways for dealing with adversity, whether in cities or on reservations. In the past, this quality enabled them to survive centuries of oppression and persecution. Today this is reflected in the practice of cultural traditions that Indian people are proud to embrace. The resilience of American Indians is an abiding quality that will no doubt ensure that they will remain part of the ethnic mosaic of American society throughout the twenty-first century and beyond.

REFERENCES

Cornell, Stephen. (1988). *The Return of the Native: American Indian Political Resurgence.* New York: Oxford University Press.

Fixico, Donald L. (1986). *Termination and Relocation: Federal Indian Policy, 1945–1960.* Albuquerque, NM: University of New Mexico Press.

Hagan, William T. (1979). *American Indians.* Chicago, IL: University of Chicago Press.

Hall, Thomas D. (1989). *Social Change in the Southwest, 1350–1880.* Lawrence, KS: University Press of Kansas.

Hoxie, Frederick E. (1984). *A Final Promise: The Campaign to Assimilate the Indians, 1880–1920.* Lincoln, NE: University of Nebraska Press.

Nagel, Joanne. (1986). "American Indian Repertoires of Contention." Paper presented at the annual meeting of the American Sociological Association, San Francisco, CA.

O'Brien, Sharon. (1990). *American Indian Tribal Governments.* Norman, OK: University of Oklahoma Press.

Prucha, Francis Paul. (1984). *The Great Father.* Lincoln, NE: University of Nebraska Press.

Snipp, C. Matthew and Gene F. Summers. (1991). "American Indian Development Policies," pp. 166–180 in *Rural Policies for the 1990s*, edited by Cornelia Flora and James A. Christenson. Boulder, CO: Westview Press.

Sorkin, Alan L. (1978). *The Urban American Indian.* Lexington, MA: Lexington Books.

Thornton, Russell. (1987). *American Indian Holocaust and Survival: A Population History since, 1942.* Norman, OK: University of Oklahoma Press.

Utley, Robert M. (1963). *The Last Days of the Sioux Nation.* New Haven: Yale University Press.

Weibel-Orlando, Joan. (1991). *Indian Country, L.A.* Urbana, IL: University of Illinois Press.

DISCUSSION QUESTIONS

1. How do the four different periods of state policy that Snipp identifies reveal different ways that the state has managed American Indian affairs?

2. Explain what Snipp means when he concludes that American Indians have been resilient even in the face of massive state-based social control?

8

Imperatives of Asian American Citizenship

ELLEN D. WU

Wu helps us understand the origins and significance of the term model minority, and how it reflects a very different moment in the history of Asian Americans in this nation. She explores how an opening in the racial system has finally created a space for racial minorities to assimilate, but in limited ways.

This is *the* success story of *a* success story.

In December 1970, the *New York Times* ran a front-page article declaring Japanese and Chinese in the United States "an American success story." Both groups had witnessed "the almost total disappearance of discrimination ... and their assimilation into the mainstream of American life"—a situation that would have been "unthinkable twenty years ago." The *Times* opened with the biography of immigrant J. Chuan Chu as proof. When Chu arrived from China at the end of World War II, he had run into difficulty finding a place to live because of his "Oriental face." Two and half decades later, his race was no longer a handicap. A graduate of the University of Pennsylvania's school of engineering, Chu had risen through the ranks of Massachusetts-based Honeywell Information Systems, Inc. to a vice presidential position. "If you have the ability and can adapt to the American way of speaking, dressing, and doing things," said Chu, "then it doesn't matter any more if you are Chinese."[1]

Chu's experience was hardly unique. In interviews with dozens of Asian Americans, the *Times* heard little of discrimination in housing, education, and the realm of interpersonal interaction. Southerners even considered Asians to be white. By and large, Japanese and Chinese Americans no longer faced "artificial barriers" to high-status professions. Whereas most of the previous generation had had no choice but to toil in menial service work such as laundering and gardening, stars of Asian America's current cohort had achieved nationally distinguished careers: architects Minoru Yamasaki (of New York's World Trade Center fame) and I. M. Pei; multimillion-dollar investment management firm Manhattan Fund director Gerald Tsai; Nobel Prize–winning physicists Tsung Dao Lee and Chen Ning Yang; San Francisco State College president S. I. Hayakawa; and U.S.

SOURCE: Wu, Ellen D. 2013. The Color of Success: Asian Americans and the Origins of the Model Minority. Princeton: Princeton University Press.

senator Daniel Inouye. "The pig-tailed coolie has been replaced in the imagination of many Americans by the earnest, bespectacled young scholar," announced the *Times*. Hunter College junior Elaine Yuehy, the daughter of a laundryman, agreed. "My teachers have always helped me because they had such a good image of Chinese students. 'Good little Chinese kid,' they said, 'so bright and so well-behaved and hard-working,'" she recalled. Once despised by American society, "Orientals"—especially Japanese and Chinese, the two major Asian-ethnic populations at midcentury—had become its most exceptional and beloved people of color, its "model minority."[2]

Indeed, before the 1940s and 1950s, whites had deemed ethnic Japanese and Chinese unassimilable aliens unfit for membership in the nation. Americans had subjected so-called Orientals to the regime of Asiatic Exclusion, marking them as *definitively not-white*, and systematically shutting them out of civic participation through such measures as bars to naturalization, occupational discrimination, and residential segregation. Beginning in World War II, however, the United States' geopolitical ambitions triggered seismic changes in popular notions of nationhood and belonging, which in turn challenged the stronghold of white supremacy.[3] As a result, federal officials, behavioral scientists, social critics, and ordinary people worked in tandem to dismantle Exclusion. Yet such a decision posed a problem for America's racial order and citizenship boundaries. The social standing of Asian Americans was no longer certain, and the terms of their inclusion into the nation needed to be determined. A host of stakeholders resolved this dilemma by the mid-1960s with the invention of a new stereotype of Asian Americans as the model minority—a racial group distinct from the white majority, but lauded as well assimilated, upwardly mobile, politically nonthreatening, and *definitively not-black*.

This astounding transformation reflected the array of new freedoms accorded to Japanese and Chinese Americans by the state and society in the mid-twentieth century. Their emancipation entailed liberation from the lowly station of "aliens ineligible to citizenship," the legal turn of phrase with which lawmakers had codified Asian immigrants as external to American polity and society. Landmark state and federal litigation and legislation in the 1910s and 1920s both drew on as well as reinvigorated the social consensus that peoples of Asian ancestry were wholly incapable of assimilation, because they were racially and culturally too different from white Americans.[4] Under Exclusion, immigrants from Japan and China were subjected to a shock of discriminatory and dehumanizing limitations, from harsh restrictions on entry into the country to the denial of naturalized citizenship along with its attendant rights, including the franchise and property ownership.... Their American-born children, birthright citizens of the United States, fared little better. Often forced to attend segregated schools, their career options were narrowly bound to the same peripheral economic niches into which their parents were funneled: truck farming, gardening, domestic labor, restaurants, and laundries. The few who managed to earn professional degrees could only hope to find clients in their Little Tokyo and Chinatown ghettos. The vast majority, however, found it futile to aspire beyond their lot as "professional carrot washers," as one second-generation Japanese American put it, until the demise of Exclusion.[5] With the regime's abolition in the 1940s and 1950s, Asian Americans enjoyed access to previously forbidden areas of employment, neighborhoods, and associational activities. They also benefited from the federal government's relaxation of immigration restrictions and its revocation of their ineligibility to citizenship....

But Asian Americans discovered, too, that various authorities—both within and outside their ethnic communities—checked their autonomy to choose their own futures by pressuring them to behave as praiseworthy citizens. Some gladly complied, others inadvertently went along, and not a few refused to succumb to these demands. All found their lives conscripted into the manufacture of a certain narrative of national racial progress premised on the distinction between "good" and "bad" minorities.... What explains the drastic turnaround of the image of ethnic Japanese and Chinese, long regarded by many in the United States as the unalterably strange and despicable "pig-tailed coolie"? Put another way, how did the Asian American success story *itself* become a success story—literally front-page news—edging out other possibilities for understanding their place in the nation? And what did their crowning as model minorities ("the earnest, bespectacled young scholar") mean not only for themselves but also for all Americans?

Answering these queries begins with comprehending the model minority's debut as the unanticipated outcome of a series of intersecting political, social, and cultural imperatives—ethnic and mainstream, domestic and global—that impelled the radical restructuring of America's racial order in the mid-twentieth century. Excavating the origins and aftereffects of this formidable concept there- fore necessitates a consideration of the vicissitudes within Japanese and Chinese communities alongside the broader sweep of national and international historical change. In other words, connections between internal and external develop- ments are indispensable to uncovering the birth of this construct....

The evolution of the political philosophy known as liberalism was foremost among the dynamisms that set the stage for the coming of the model minority. Historians have pointed to liberalism's centrality as well as its "protean" character in U.S. history. The foundational tenets at the core of liberalism—freedom, rational self-interest, and a belief in human progress—have undergirded the nation's political life since the early days of the republic, but Americans have acted on them in ways that have changed decidedly over time. At the start of the twentieth century, the social and economic inequalities resulting from indus- trial capitalism motivated Progressive Era reformers to redefine liberalism from its nineteenth-century iteration (laissez-faire economics and limited government intervention) to one that valued an activist state attuned to the welfare of its citi- zens. The impulse to "tame capitalism" dominated liberal thinking through the Great Depression and early years of the New Deal. By the 1940s, however, lib- erals embraced new priorities cherishing the protection and promotion of the freedoms of individuals as well as groups....

Mobilization for World War II fostered the advent of *racial* liberalism: the growing belief in political and intellectual circles that the country's racial diver- sity could be most ably managed through the assimilation and integration of nonwhites. The ideology emphasized federal government intervention in orchestrating the social engineering necessary to achieve civil rights and equality of citizenship for minority groups.... Champions of racial liberalism—including many ethnic Japanese and Chinese themselves—pushed the notion that Asians might be something other than indelibly and menacingly alien, and that they deserved to be included in the national polity as bona fide citizens—a giant conceptual leap from the unanimity of previous decades.[6] Liberals of all races invested racial reform with grave urgency: the failure of the nation to live in accordance with its professed democratic ideals endangered the country's aspirations to world leadership. The United States' battles against fascism and then Communism meant that Asiatic Exclusion, like Jim Crow, was no longer tenable. Seeking global legitimacy, Americans moved to undo the legal framework and social practices that relegated Asians outside the bounds of the nation.... Certainly, Japanese and Chinese Americans had not lacked in attempts to attain substantive, full citizenship and respectable social standing in the late nineteenth and early twentieth centuries. But their efforts to claim unfettered inclusion only gained traction with the rise of racial liberalism and outbreak of global wars. Japanese and Chinese American fortunes, in short, were tied directly to the national identity politics of World War II and the Cold war....

International imperatives of the 1940s and 1950s anchored the nation's recasting of Asian Americans into *assimilating Others*—persons acknowledged as capable of acting like white Americans while remaining racially distinct from them. Unlike the progeny of turn-of-the-century southern and eastern European immigrants who melted into unambiguous whiteness in the crucibles of mass consumption, industrial unionism, New Deal ethnic pluralism, and military service, Japanese and Chinese did not disappear into whiteness after the end of Exclusion.... Instead, state authorities, academicians, cultural producers, and common folk renovated Asian America's perceived differences from liability to asset to benefit U.S. expansionism. In the throes of the worldwide decolonization movement, more precisely, Cold Warriors encountered the dilemma of differentiating their own imperium from the personae non gratae of the European empires. As nonwhites, the entrance of Asian Americans into the national fold provided a powerful means for the United States to proclaim itself a racial democracy and thereby credentialed to assume the leadership of the free world. The rearticulation of Asian Americans from ineradicable aliens to assimilating Others by outside interests bolstered the framing of U.S. hegemony abroad as benevolent—an enterprise that mirrored the move toward racial integration at home....

Above all, Japanese and Chinese Americans harbored a profound interest in characterizing anew their racial image and conditions of citizenship, and they often took the lead in this regard. By yoking U.S. officialdom's world-ordering logic to their own quests for political and social acceptance, they actively participated in the revamping of their racial difference. They made claims to inclusion based on the assumption of not only Americanness but also and particularly diasporic Japanese and Chinese identities. Recognizing that the Asian Pacific region loomed large on the U.S. foreign relations agenda, community representatives strategically typecast themselves, asserting that their own ancestries endowed them with innate cultural expertise that qualified them to serve as the United States' most natural ambassadors to the Far East. Therefore, they suggested, admitting people of Japanese and Chinese heritage to first-class citizenship made good diplomatic sense.

Equally decisively, Asian Americans' self-stereotyping convinced others not only because of its payoff for foreign relations but also because it corroborated the nation's cultural conservatism at midcentury. Ethnic Japanese and Chinese emissaries consistently touted their putatively "Oriental" attributes, such as the predisposition to harmony and accommodation, the reverence for family and education, and unflagging industriousness to enhance their demands for equality. These descriptions endorsed not only liberal assimilationist and integrationist imperatives but also the Cold War cultural emphasis on home life rooted in the strict division of gender roles. Self-representations of Japanese and Chinese American masculinity, femininity, and sexuality, purposefully conforming to the norms of the white middle class, were crucial to the reconstruction of aliens ineligible to citizenship into admirable—albeit colored—Americans....

Undeniably, this embrace of Cold War nationalism and traditional values was a politically charged calculation. Japanese and Chinese America were hardly monolithic entities in this period. Rather, they were rife with internal divisions, rival agendas, and disagreements about their collective futures—all of which helped dictate how they would make their way in American society after World War II. External pressures generated commonalities and a modicum of cohesiveness within the two communities, but they also provided the structures that enabled certain individuals and factions to achieve authority and influence. Demographic shifts also mattered. As U.S.-born, second-generation Japanese and Chinese came of age in the 1930s and 1940s, they began to vie with the immigrant elite for leadership positions and the privileges of representing themselves in the public sphere. The winning contenders were those whose politics hewed closest to the reigning dogmas of the day: liberal assimilationism, prowar patriotism, anti-Communism, and respectable heterosexuality. As ethnic spokespersons, the victors in these contests spun flattering portrayals of their peoples to dislodge deeply embedded "yellow peril" caricatures. At the same time, their tales of exemplary Asian American citizenship validated their own political choices and upheld intracommunity power arrangements in their favor.... [T]hese success story narrators beat out alternative voices including those of zoot-suiters, sexual deviants, draft resisters, those who renounced citizenship, leftists, Communists, and juvenile delinquents—the various entities who did not subscribe to postwar racial liberalism and political-cultural conservatism as the most suitable guidelines for encountering postwar American life....

In the mid-1960s, the assimilating Other underwent a subtle yet profound metamorphosis into the model minority: the Asiatic who was at once a model citizen and definitively not-black. The zenith of liberal racial reform—the 1964 Civil Rights Act and 1965 Voting Rights Act—also marked the beginnings of its collapse under the weight of both progressive and conservative critique.

The abolition of de jure apartheid had done little to alter the vast disparities between black and white incomes, housing, employment, and education. Participants in the African American freedom movement urgently pressed for lasting changes to, if not a complete overhaul of, the nation's—and the world's—existing structures of capitalist democracy. Liberals unnerved by blacks' wide-ranging, radical challenges to effect a meaningful redistribution of wealth and power held up Japanese and Chinese Americans as evidence of minority mobility to defend the validity of assimilation as well as integration. Conservatives who feared that black power would go even further than racial liberalism to destroy white supremacy also looked to Asians to salve what they viewed as the decline of "law and order"—wrought especially by black and brown peoples—in American society. Either way, Japanese and Chinese in the United States were catapulted to a new status as model minorities—living examples of advancement *in spite of* the persistent color line and *because of* their racial (often coded as cultural) differences.... In recirculating Asian American success stories, both liberals and conservatives grafted the now-familiar postwar tropes of Japanese and Chinese American conduct (patriotism, family values, accommodation, and so forth) onto the new imperative of taming the reach of the Civil Rights revolution....

Tracing the course of Japanese and Chinese American racialization in the mid-twentieth century provides a useful way to revise the standard narrative of democratic citizenship in the United States by linking inclusion to racism. The mythology of American democracy depicts liberal egalitarianism as a succession of triumphs over exclusions, and that the circle of those included in the polity as full members of society has continued to widen over time.[7] The ascendance of racial liberalism and its reforms, including the death of the Asiatic alien ineligible to citizenship, would seem to uphold this folklore. Yet the lifting of Exclusion did not result in a teleological progression toward the unmitigated inclusion of Asian Americans in the nation. Rather, the racial logic that politicians, scholars, and journalists deployed to invent the model minority generated new modes of exclusion. Their reliance on culture to explain postwar Asian American socio-economic mobility re-marked ethnic Japanese and Chinese as not-white, indelibly foreign others, compromising their improvements in social standing. This same reasoning also undergirded contentions that African Americans' cultural deficiencies was the cause of their poverty—assertions that delegitimized blacks' demands for structural changes in the political economy and stigmatized their utilization of welfare state entitlements. The history of the model minority therefore destabilizes the conceptual boundaries between exclusion and inclusion, allowing for a more complete understanding of how the United States and other liberal democracies devise, uphold, and justify social differences and inequalities, even as they expand their boundaries of inclusion and ostensibly progress toward the achievement of universal citizenship for all members.... Approaching the model minority as a simultaneously inclusive *and* exclusive reckoning supplies clues to how racism "reproduce[s] itself even after the historical conditions that initially gave it life have disappeared."[8]...

NOTES

1. "Orientals Find Bias Is Down Sharply in U.S.," *New York Times*, December 13, 1970, 1, 70.

2. Ibid.

3. On the concept of belonging, see Barbara Young Welke, *Law and the Borders of Belonging in the Long Nineteenth Century United States* (New York, 2010).

4. Yuji Ichioka, *The Issei: The World of First Generation Japanese Immigrations, 1885–1924* (New York, 1988), 210–43; Bill Ong Hing, *Making and Remaking Asian America through Immigration Policy* (Stanford, CA, 1993), 17–36; Mae M. Ngai, *Impossible Subjects: Illegal Aliens and the Making of Modern America* (Princeton, NJ, 2004), 37–50.

5. Taishi Matsumoto, "The Protest of a Professional Carrot Washer," *Kashu Mainichi*, August 4, 1937, cited in John Modell, *The Economics and Politics of Racial Accommodation: The Japanese of Los Angeles, 1900–1942* (Urbana, IL, 1977), 138.

6. Beyond the legal status signifying one's formal membership in a nation-state, I consider the category citizenship to encompass what T. H. Marshall describes as "full members[hip in a] society," Judith N. Shklar terms "social standing" in a community, Linda Bosniak conceives of as "identity" and "solidarity," and Leti Volpp depicts as "inclusion." T. H. Marshall, *Citizenship and Social Class and Other Essays* (Cambridge, UK, 1950), 8; Judith N. Shklar, *American Citizenship: The Quest For Inclusion* (Cambridge, MA, 1991), 2; Linda Bosniak, "Citizenship Denationalized," *Indiana Journal of Global Legal Studies* 7 (2000): 447–509; Linda Bosniak, *The Citizen and the Alien: Dilemmas of Contemporary Membership* (Princeton, NJ, 2006); Leti Volpp, "The Citizen and the Terrorist," *UCLA Law Review* 49 (2001–2): 1575–91.

7. Evelyn Nakano Glenn, *Unequal Freedom: How Race and Gender Shaped American Citizenship and Labor* (Cambridge, MA, 2002), 24.

8. Thomas C. Holt, *The Problem of Race in the 21st Century* (Cambridge, MA, 2000), 19–20.

DISCUSSION QUESTIONS

1. After reading Wu's article, what do you think about the term *model minority* and its uses?

2. What are the international factors that influenced the shifting image of Asian Americans during World War II and the Cold War?

3. Why does Wu think this new status for Asian Americans has elements of both inclusion and exclusion?

9

Embodying the White Racial Frame
The (In)Significance of Barack Obama

WENDY LEO MOORE AND JOYCE M. BELL

*Moore and Bell identify recent shifts in racial hierarchies. Does Obama's victory
mean we have overcome the past? Moore and Bell identify the paths that made
his election possible, but also how White privileges continue to racial inequality.*

INTRODUCTION

The victory of Barack Obama in the 2008 presidential election marked a significant historic moment in the United States. In a nation founded upon, and
owing much of its economic growth and prosperity to the institution of racialized slavery, the election of the first African American president was a powerful
symbolic challenge to a long and violent history of racial oppression and white
supremacy. Immediately after Obama's election, within both popular and scholarly discourse, claims that his election signified a "post-racial" America emerged.
Even among commentators who didn't go so far as to suggest that racial inequality or racism were now a thing of the past, Barack Obama's election was hailed
as evidence of the decreasing relevance of racism in the fabric of our society. In
support of this assertion, many commentators pointed to the fact that Obama
received widespread support from whites, without which his victory would not
have been possible. Yet the systemic racial inequality that characterized the
United States social structure on November 4, 2008, did not disappear, or even
lessen because of the election results. How then, as social scientists and public
policy analysts, do we reconcile a symbolic racial victory in American politics
and the resulting rhetoric surrounding it, which declares this victory as indicative
of racial progress, with the persistent and deep structural racial inequalities that
characterize the United States? Herein we suggest that, in fact, these phenomena
are not incompatible, but are instead deeply, and dangerously connected.

As Joe Feagin (2000) has documented, the United States, in all its major
institutions, is characterized by racism and racial inequality. As a result of a
history and legacy of legally-constructed and enforced racial oppression, deep

SOURCE: Moore, Wendy Leo, and Joyce M. Bell. 2010. "Embodying the White Racial Frame:
The (In)significance of Barack Obama." *Journal of Race and Policy* 6:123–138.

structural inequalities permeate U.S. society today (Bell, 2000; Harris, 1993; Haney López, 2006; Moore, 2008). The United States is the most residentially racially segregated country in the world, and this geographical segregation (which is the result of historical and contemporary legal, political and economic racist practices) corresponds to severe structural economic inequality (Conley, 2009; Jargowsky, 1997; Massey and Denton, 1993; Bell, 2000; Oliver and Shapiro, 2006). The history of racism, combined with contemporary structural inequalities, affects nearly every facet of American society.... Within such a clearly structurally unequal society, where does the suggestion of post-racialism" or a declining significance of racial inequality come from?

At least part of the answer to this question lies, we suggest, within a changing terrain of racial politics and ideology that emerged as dominant in the post-civil rights era. In reaction to the dramatic challenges of the Civil Rights Movement to the explicitly white racist structure and ideology of the United States, a new ideological framing of race emerged. The new post-civil rights ideological and discursive frame seemingly incorporates civil rights conceptions of color-blindness, equality and democracy, yet covertly protects white privilege, power, and wealth by rhetorically divorcing these concepts from the structural realities of racial inequality (Bell, 1987, 1992). Rhetorical manipulations of civil rights language makes it possible for individuals to assert, for example, an opposition to racial segregation in schools, and a simultaneous opposition to having their own children bussed in a desegregation plan (see Wellman, 1993). This is accomplished through color-blind and abstract liberalist discursive tactics which recognize race as a superficial set of cultural differences, which should be accepted and celebrated, but simultaneously denies that race has any real effect on the life chances of individuals or that race shapes social life in any significant way (see Bonilla-Silva, 2001, 2003; Carr, 1997; Crenshaw, 1997; DiTamaso et al., 2003; Doane, 2003; Gallagher, 2003). The notion of post-raciality as an outcome of the election of an African American president fits neatly within the cultural and political context of a society characterized by a major disjuncture in the dominant ideological narrative of racial inequality and the structural actualities of this inequality.

Thus the political atmosphere that enabled the election of Barack Obama is one that is connected to a broader politics of race—which is more nuanced and less obvious than that of previous historical eras. With color-blindness and abstract liberalism at the center of the dominant post-civil rights discourse, the overt racial hostilities of the Jim Crow era were transformed into a seemingly kinder, gentler form of racism, and one which is often less easy to articulate (see, for example Bonilla-Silva, 2001, 2003; Bell and Hartmann, 2007; Feagin, 2009; Moore, 2008; Wellman, 1993). Within this ideological framing, there is room for people of color to occupy positions of power as long as they espouse the central tenets of the color-blind, abstract liberalist position. In fact, this racial "diversity" secures credibility for the broader ideological frame precisely because it provides the appearance that color-blindness and abstract liberalism lead to the dismantling of white supremacy. Yet—and this is why "diversity" here must remain in quotations—the strategy by which individuals like Barack Obama gain mass white support takes place through the incorporation and presentation of the dominant white framed ideology (see Harvey Wingfield and Feagin, 2009); an ideology that actually functions to secure the reproduction of white privilege, power and wealth. Moreover, the symbolic power of the espousal of the tenets of a white-framed ideology by an African American man is dangerous to real racial progress in that it provides an illustrative tool to those who would assert that the United States is now a post-racial society.

The election of Barack Obama did nothing to dismantle the deeply embedded structural mechanisms that function to maintain racial inequality, yet immediately following his election the media declared his election an indication of racial progress. We suggest that this move is the natural extension of a color-blind, abstract liberalist white racial frame (see Feagin, 2009); one which has been damaging to policies designed to remedy racial inequality, and which functions to protect and preserve structures of inequality....

THE CHANGING TERRAIN OF
THE POLITICS OF RACE

Since the mid- to late 1970s, the political landscape has gradually morphed into one deeply connected to the politics of identity and the connection between identity and social structure. The affirmations of identity expressed by oppressed racial groups, as well as women of all races, and the challenges to structural inequalities that was fundamental to the Civil Rights Movement resulted in shifts in political discourse, and in particular in the politics of race (Marable, 1991; Omi and Winant, 1994). Michael Omi and Howard Winant (1994) note that the challenges to racial inequality during the Civil Rights Movement result in a white backlash, one aspect of which was the development of the "new right." The new right, unable to draw upon the old political narrative of natural (or biological) racial inferiority, needed to create new mechanisms by which to protect the racial status quo from further challenge from communities of color (Omi and Winant, 1994, 123). One of the central mechanisms employed by the new right was a new language of race, one which drew upon racially coded language. Thus we see conservative politicians decrying the rise in crime, often voiced over the face of a black man, and calling for a return to law and order; we see challenges to social welfare policies presented with the image of the African American woman portrayed as a "welfare queen" (Hill Collins, 2000), and we find a new rhetorical fervor concerning the dangers of a culture of poverty which deliberately signifies the poverty endemic in African American communities— in other words, a shift towards what has been called a "culture of poverty racism" (Bonilla Silva, 2001).

The tactics of the new right were remarkably successful, ushering in 12 consecutive years of Republican administration during the Reagan and Bush eras from 1981 to 1993 (and a total of 20 out of 28 years when including the George W. Bush terms) and opening space for a successful Republican takeover of

Congress in the 1994 mid-term elections. The broad support for this conservative Republican leadership came from a neo-conservative constituency which effectively co-opted the language of the Civil Rights Movement so that calls for "equality of opportunity" and "color-blindness," which had previously served to illuminate the disjuncture between the rhetorical ideals of democracy and the fact of structural racial inequality, were re-worked in such a way to stall further challenges to white privilege, power, and wealth (Crenshaw, 2007; Lipsitz, 2006). As Omi and Winant (1994, 131) note, "racial discrimination and racial equality—in the neo-conservative model—were problems to be confronted *only* [authors' emphasis] at the individual level, once legal systems of discrimination,... *de jure* segregation had been eliminated." Through political maneuvers like this, neo-conservatives were able to turn color-blindness from a concept meant to challenge both individual and structural racial oppression into a rhetorical tool of abstract liberalism by asserting concepts like "reverse discrimination" aimed at stalling affirmative action.

The political tactics of neo-conservatives ushered in an era of what Eduardo Bonilla-Silva (2003) has called "color-blind racism." In other words, the post-civil rights discourse has come to be dominated by an ideological framework that minimizes the relevance of race and racism, and discursively divorces structural racial inequality from historical and present day racism (Carr, 1997; Crenshaw, 1998). As Ian Haney Lopez argues, "the perversity of colorblindness [is that it] redoubles the hegemony of race by targeting efforts to combat racism while leaving race and its effects unchallenged and embedded in society, seemingly natural rather than the product of social choice" (2006, 125).

Yet while the post-civil rights politics of race, with its color-blind racist underpinnings, may have had its origins in new right political strategy, its effect was much broader. The discourse of abstract liberalism and color-blind racism spread well beyond the political constituencies of neo-conservatives (Omi and Winant, 1994). In fact, this discourse has come to operate as a central aspect of what Harvey Wingfield and Feagin term the "soft white racial frame" (2009, 19). The white racial frame can be defined as "an organized set of racialized ideas, emotions, and inclinations, as well as recurring or habitual discriminatory actions, that are consciously or unconsciously expressed in, and constitutive of, the routine operation and racist institutions of U.S. society" (Feagin, 2006, 23). With regard to understanding race in the United States, there has always existed a master frame, a powerfully dominant framework that centers on whiteness and a white perspective, thereby normalizing and justifying both white superiority and black inferiority. The master frame is cognitive and ideological, but also emotional and visual, and is often implicit so that it does the work of normalizing white privilege and white perspectives on the unconscious level as well as the conscious level. Furthermore, this master frame has been resilient, remaining dominant through centuries of challenge from communities of color, sometimes morphing and shifting, but remaining the pervasive organizing racial logic (Bell, 1987, 1992)....

In the political sphere, the white racial frame found its mirror in the Democratic party in what Omi and Winant (1994) have called neo-liberalism. They note that "neo-liberalism ... does not claim to be color-blind; indeed it argues that any effort to reduce overall inequality ... will disproportionately benefit those concentrated at the bottom of the socio-economic ladder, where racial discrimination has its most damaging effects" (1994, 148). However, they go on to note that "the neoliberal project avoids (as far as possible) framing issues or identities racially" (1994, 148). Here, Omi and Winant (1994) suggest that it was only by avoiding a direct claim of color-blindness, but identifying race only in the most abstract ways, and only in connection with a broader liberal social and economic agenda that will directly benefit whites, that Bill Clinton was able to secure victory after 12 years of Republican administration. This illustrates the pervasive, if subtle, influence of abstract liberalism and the white racial frame in U.S. politics. In the current political landscape, discussions of racial inequality that are not couched within the dominant frame are certainly outside the boundaries of polite conversation and would likely have negative impacts for politicians and political campaigns that engage them.

BLACK BODIES AND THE WHITE RACIAL FRAME

Even before the election of Bill Clinton in 1992, Republicans and their neo-conservative constituency recognized a need to distance themselves from allegations of racism. One of the central ways in which neo-conservatives did this was to engage in the exploitation of Black bodies in the proliferation of an abstract liberalist and color-blind racist ideology. Conservative strategists recognized the power in placing extremely conservative token African Americans into positions of power (Bell, 1994). This tactic provided legitimacy to claims that conservative ideology and discourse were non-racial *political* positions thereby turning the tables on liberal calls for affirmative action in employment, education and politics. This political project was illustrated, for example, when George Bush nominated Clarence Thomas, an African American man, to replace Thurgood Marshall, the first African American Supreme Court Justice to the United States Supreme Court calling into stark question the importance of descriptive versus substantive representation for African Americans.... Clarence Thomas had been appointed as Head of the Equal Employment Opportunity Commission (EEOC) in 1982 by then President Ronald Reagan, specifically for the purpose of changing and limiting the reach of the Commission with regard to racial discrimination in employment. As head of the EEOC, Thomas dramatically shifted the work of the Commission by taking an extremely individualistic view of racial discrimination, and limiting the action of the EEOC to only those cases where individual bad actors could be identified (Greene, 1989, 5). Class action suits and suits that were based upon alleged "pattern and practice" discrimination were no longer investigated by the EEOC (Greene, 1989, 5, 54). Having shown himself to be committed to an extremely limited view of discrimination, one which ignored entirely the present day structural consequences of the long history of racial oppression in the United States, Thomas became a natural choice for ultra-conservative President George Bush. It was clear in this move to replace an African American man who had articulated an expansive Constitutional construction in his race jurisprudence which recognized the structural realities of white supremacy, with an African American man who was politically and ideologically the polar opposite, particularly with regard to issues of race, that President Bush was engaged in both denying the realities of race, and using those realities as a political tool to protect white racial privilege, power, and wealth.

Conservatives garnered power by using African Americans who engaged in the discourse of the white racial frame to proliferate their message. As Derrick Bell (1994) has noted, the placement of extremely conservative African Americans in positions of power provides an easy retort for allegations of racism: a Black person says these things, so it cannot be racist. Thus, conservative African Americans acquired an authority *never* granted to progressive or critical African Americans (Bell, 1992, 1994). And when the Republicans came back into executive power, with the election of George W. Bush after Bill Clinton's tenure, their strategists seemed even more aware of the power of such exploitation of the bodies of people of color. Hence *Time* magazine made George W. Bush its 2004 "Person of the Year," in part by show-casing his appointment of the most "diverse" cabinet in presidential history, proclaiming his "The Benetton-Ad Presidency" … (referring to television and print advertisements of the time that showed a wide variety of racial and ethnic types).

To be clear, this political racial project was one that simultaneously ignored *and* participated in the political construction of race. In other words, as Chris Iijima (1997) has noted, race is not only a social and historical construct, but is also fundamentally a *politically* constructed identity. The social location of racial groups, the subordination and oppression of African Americans and people of color and the privileging and empowerment of whites, is one that was created and enforced by the state through force of law and politics (see also Harris, 1993 and Powell, 1997). United States "politics" has always been racialized and the political machinery of the U.S. government has been utilized in the construction of racial groups and the simultaneous construction of racial domination and subordination (Bell, 2000; Omi and Winant, 1994; Urofsky, 2001). Race is not, and has never been, an apolitical construct nor can political constructions in the United States be non-racial (Marable, 1991; Iijima, 1997). Understanding the political nature of race, it becomes clear that from a structural vantage point the effort among conservative whites to find and promote African Americans who espoused conservative political discourse was clearly an exploitation of Black bodies in the service of a broader political project that operated to maintain racial inequality in the midst of a changed social situation. There is no doubt that black conservative politicians believe in their political positions and are certainly responsible for their own politics and ideas. However, here we want to call attention to the crucial symbolic function that black espousers of white racial rhetoric play in the larger political and policy atmosphere.

The exploitation of Black bodies in the service of white supremacist positions is not a new phenomenon in United States racial history (see, for example

Hill Collins, 2000; Fanon, 1967). But the post-civil rights neo-conservative political project employed by neo-conservatives utilized the political gains achieved by people of color through the Civil Rights Movement to create a new form of black exploitation—one in which black political officials who espouse and symbolize the dominant white racial frame are put on display in such a way that validates this position on race relations. And though this tactic has been most extreme and obvious when practiced by political conservatives, it is not an uncommon tactic among white liberals (see Bell, 1994). We suggest that Barack Obama serves this same function within white liberal racial discourse by serving as a symbolic or token black body who espouses soft white racial framing.

Barack Obama, White Liberals, and Post-Civil Rights Politics of Race

As noted above, the white racial frame is a cognitive, emotional and visual master frame that structures understandings of race and racial equality broadly in the United States. We are arguing that the white racial frame has been employed not only in neo-conservative politics, but is also central to white liberalism. As Omi and Winant note, the election of Bill Clinton was largely a result of a movement of what they term "neoliberals" to put forward a message of abstract liberalism (1994, 152–158). In this political framing, Democratic strategists downplayed continuing significance of racism and structural racial inequality, promoted the idea of universalist and abstractly constructed liberal ideologies of equality, and emphasized the need for individual or "personal" responsibility. The power of these aspects of the white racial frame, and of color-blind racist ideologies, is that they give white liberals the ability to disassociate themselves from explicit racial hostility and animosity while simultaneously benefiting from structural white supremacy and white privilege. These rhetorical and ideological moves appease white liberal guilt over racial oppression while at the same functioning to reproduce structural racial inequality.... [T]hese discursive frames distort and ultimately stifle progressive racial policies aimed at dismantling structural racial inequality. First, however, it is important to situate the election or Barack Obama to the presidency within this ideological and political project.

As Harvey Wingfield and Feagin (2009) have noted in their analysis of Barack Obama's political campaign, Obama and his campaign staff framed their campaign, and Obama's ideological position, in much the same vein as Bill Clinton had. Obama's campaign manager David Plouffe continually explicitly noted that he did not want race to be a defining aspect of the campaign, and he counseled Obama to avoid making public statements explicitly about race (Harvey Winfield and Feagin, 2009, 40). Despite the clear and obvious fact that Barack Obama would be the first African American president in a country in which African Americans had been violently and systematically oppressed, the campaign downplayed race at every turn. Obama himself downplayed the issue of race by emphasizing his mixed racial heritage, and focusing on appeals to so-called universalistic ideologies of equality of opportunity.

What the election of an African American man who espouses the tenets of the white racial frame does for the frame is key; in the same way that extremely conservative African Americans are supported and utilized in key positions of power by white neo-conservatives in order to claim that their politics are non-racial, having an African American man who centers his political ideology in an abstract liberalist white racial frame enables and lends authority to the white liberal claim of "post-racialism." Furthermore, when the language and the ideological tenets of the white racial frame come from the mouth of an African American, it relieves white people of the perception that they are implicated in contemporary racial inequality or are accountable for past racial harms (Bonilla-Silva, 2003). It is not surprising, then, that Obama received significant support from white voters. As Eduardo Bonilla-Silva has argued about Obama, "the white left and right ... are willing to rally around any minority willing to deny collective and systemic racialized problems" (Bonilla Silva, 2009). It is our claim that Obama's presidency serves as a politically racialized symbol of authority for the white racial frame....

BARACK OBAMA, ABSTRACT LIBERALISM AND DIVERSITY

Responding to his own victory Barack Obama proclaimed:

> If there is anyone out there who doubts that America is a place where anything is possible, who still wonders if the dream of our founders is alive in our rime, who still questions the power of our democracy, tonight is your answer. Young and old, rich and poor, Democrat and Republican, black, white, Hispanic, Asian, Native American, gay, straight, disabled and not disabled.... We have been and always will be the United States of America.

In this emotionally powerful historical moment, President-elect Obama, by virtue of his words as well as his own racial identity, affirmed the color-blind notion that equality of opportunity is real in the United States, and that there are universalistic normative American values that transcend race (as well as age, class, political affiliation, sexuality, and ability). Here Obama utilizes the diversity element of the white racial frame as well as tacitly utilizing his own racial status in such a way to decry the universalist and abstract liberalist idea that upward mobility, even to the U.S. presidency, is an equally attainable possibility for all U.S. citizens (see Bonilla-Silva and Ray, 2009). Unfortunately, this statement blithely ignores the realities of the racial social structure, and in doing so Obama feeds directly into an already existing racialized legal frame.

    ... [A] central way in which color-blindness gets articulated in both ideology and action in the institutional contexts of American life is through the notion of diversity. Diversity as a racial project fits nicely into the dominant white racial frame because it simultaneously recognizes racial difference and avoids discussions of power, privilege and inequality. Part of the way that diversity works is to encourage the celebration of symbolic elements of difference that are in-line with a feel-good conception of race, which Joyce Bell and Doug Hartmann (2007) have called "happy talk." Harmony, then, is central to the diversity project in such a way that any discussion of discord among the "different" doesn't fit in the confines of the discourse.

In some ways Barack Obama is a central figurehead in the diversity project. Beyond his strategic framing of his "multi-cultural" heritage as post-racial, Obama's racial message has been very much in-line with diversity's "happy talk" tenor (see Eduardo Bonilla-Silva and Victor Ray, 2009). In his famous single speech on the subject of race during the 2008 campaign, *A More Perfect Union*, Obama certainly focuses on racial inequality. But the speech is constructed in such a way to suggest that despite all of the racial disparities we see, that he is convinced that we are fundamentally a society that is not divided by race. For example, in that speech he talks about his decision to run for president, saying,

> ... I believe deeply that we cannot solve the challenges of our time
> unless we solve them together, unless we perfect our union by under-
> standing that we may have different stories, but we hold common
> hopes; that we may not look the same and we may not have come from
> the same place, but we all want to move in the same direction—towards
> a better future for of children and our grandchildren.

This "vague promise of racial reconciliation," as Adolph Reed (2008) termed it, holds fast to a notion that racial harmony is possible, desirable, indeed on its way, without ever offering new or validating existing solutions for addressing past or present racial injustices. This rhetoric is particularly troubling because of that way that his racial identity shapes how people receive these messages. A black man, whose set of racial ideas is packaged in such a way to minimize our discomfort about racial inequality, thus works to validate the white racial frame. Regardless of the reasons behind his employment of the frame, the outcome is to reinforce a conception of race that, at its core, holds that there may be racial problems, but that we each, individually need to address our role in reaching our "common hopes." This is a conception that most often precludes structural remedies.

CONCLUSION

We suggest that the notion that the election of the U.S.'s first black president has transformed our society into one that is "post-racial" is much more insidious than being merely a troublesome liberal discursive move. Rather, we suggest that the combination of the symbolic significance of Obama's election, along with the discursive assertion of post-racialism has the potential to further entrench the severe and serious structural racial inequalities in U.S. society that continue to limit the life chances of people of color especially African Americans....

We fear that we are in a dangerous moment in U.S. history with regard to racial policy. The fact that we have a black man as president who is politically situated squarely within the white racial frame creates the ultimate justification for failing to create policy that promotes racial equality on a structural level. The very fact of his black body occupying the most powerful office in the country, indeed, the world, could be reason enough to claim that we have arrived— the ultimate proof of equality of opportunity. Combine this with Obama's espousal of a set of ideas that encourages that line of thinking, and his election, a truly exceptional event, becomes the grounds for claiming the end of racism— an idea that the millions of people of color in the United States who suffer because of racism cannot afford....

REFERENCES

Bell, Derrick. 1987. *And We Are Not Saved: The Elusive Quest for Racial Justice.* New York: Basic Books.

———. 1992. *Faces at the Bottom of the Well: The Permanence of Racism.* New York: Basic Books.

———. 1994. *Confronting Authority: Reflections of an Ardent Protester.* Boston: Beacon Press.

———. 2000. *Race, Racism and American Law.* Fourth Edition. New York: Aspen Press.

Bell, Joyce and Douglas Hartmann. 2007. "Diversity in Everyday Discourse: The Cultural Ambiguities and Consequences of 'Happy Talk'." *American Sociological Review* 72: 895–914.

Bonilla-Silva, Eduardo and Victor Ray. 2009. "When Whites Love a Black Leader: Race Matters in Obamamerica." *Journal of African American Studies* 13: 176–183.

———. 2001. *White Supremacy and Racism in the Post-Civil Rights Era.* Boulder, CO: Lynne Rienner Publishers, Inc.

———. 2003. *Racism without Racists: Colorblind Racism and the Persistence of Racial Inequality in the United States.* Lanham, MD: Rowman & Littlefield.

Carr, Leslie. 1997. *Colorblind Racism.* Thousand Oaks, CA: Sage.

Conley, Dalton. 2009. *Being Black, Living in the Red: Race, Wealth and Social Policy in America.* Berkeley: University of California Press.

Crenshaw, Kimberle W. 1997. "Colorblind Dreams and Racial Nightmares: Reconfiguring Racism in the Post-Civil Rights Era." In *Birth of a Nation'hood,* Toni Morrison and Claudia Brodsky Lacour, eds. New York: Pantheon Books.

Crenshaw, Kimberle. 1998. "Race, Reform, and Retrenchment: Transformation and Legitimation in Antidiscrimination Law." *Harvard Law Review* 101: 1331–1383.

Crenshaw, Kimberlé W. 2007. "Framing Affirmative Action." *Michigan Law Review* 105: 123–133. Retrieved from http://www.michiganlawreview.org/firstimpressions/vol105/crenshaw.pdf.

DiTamaso, Nancy, Rochelle Parks-Yancy, and Corinne Post. 2003. "White Views of Civil Rights: Color Blindness and Equal Opportunity." In *White Out: The*

Continuing Significance of Racism, Ashley Doane and Eduardo Bonilla-Silva, eds. New York: Routledge.

Doane, Ashley W. and Eduardo Bonilla-Silva. 2003. *White Out: The Continuing Significance of Racism.* New York: Routledge.

Fanon, Frantz. 1967. *Black Skin White Mask.* New York: Grove Press.

Feagin, Joe. 2000. *Racist America.* New York: Routledge.

———. 2006. *Systemic Racism.* New York: Routledge.

———. 2009. *The White Racial Frame.* New York: Routledge.

Gallagher, Charles A. 2003. "Color-Blind Privilege: The Social and Political Functions of Erasing the Color Line in Post Race America." *Race, Gender, and Class* 10(4): 22–37.

Greene, Kathanne. 1989. *Affirmative Action and Principles of Justice.* New York: Greenwood Press.

Haney López, Ian. 2006. *White by Law.* New York: New York University Press.

Harris, Cheryl. 1993. "Whiteness as Property." *Harvard Law Review*, 106: 1709–1789.

Harvey Wingfield, Adia and Joe R. Feagin. 2009. *Yes We Can? White Racist Framing and the 2008 Presidential Campaign.* New York: Routledge.

Hill Collins, Patricia. 2000. *Black Feminist Thought.* New York: Routledge.

Iijima, Chris. 1997. "The Era of We-construction: Reclaiming the Politics of Asian Pacific Identity and Reflections on the Critique of the Black/White Paradigm." *Columbia Human Rights Law Journal* 15: 47–90.

Jargowsky, Paul A. 1997. *Poverty and Place: Ghettos, Barrios, and the American City.* New York: Russell Sage Foundation.

Lipsitz, George. 2006. *The Possessive Investment in Whiteness.* Philadelphia: Temple University Press.

Marable, Manning. 1991. *Race, Reform, and Rebellion.* Jackson: University Press of Mississippi.

Massey, Douglas S. and Nancy A. Denton. 1993. *American Apartheid: Segregation and the Making of the Underclass.* Cambridge: Harvard University Press.

Moore, Wendy Leo. 2008. *Reproducing Racism: White Space, Elite Law Schools, and Racial Inequality.* Lanham, MD: Rowman & Littlefield.

Oliver, Melvin and Thomas Shapiro. 2006. *Black Wealth/White Wealth: A New Perspective on Racial Inequality.* New York: Routledge.

Omi, Michael and Howard Winant. 1994. *Racial Formation in the United States: From the 1960s to the 1990s.* New York: Routledge.

Powell, John. 1997. "The "Racing" of American Society: Race Functioning as a Verb Before Signifying as a Noun." *The Journal of Law and Inequality*, Vol. 15: 99–147.

Reed, Adolph. 2008. Obama, no. *The Progressive.* May. Retrieved from http://www.progressive.org/mag_reed0508

Urofsky, Melvin I. 1988. *A March of Liberty: A Constitutional History of the United States Since, 1865, volume II.* New York: McGraw Hill.

Wellman, David T. 1993. *Portraits of White Racism.* Second Edition. Cambridge: Cambridge University Press.

DISCUSSION QUESTIONS

1. What significant changes in the framing of race took place during the administrations of Ronald Reagan and George Bush (1981–1992)?

2. How did Bill Clinton's administration shift the discussion of racial remedies?

3. Barack Obama's victory was inspirational to many, but why do Moore and Bell caution us about the future of policies to address racial injustice?

Student Exercises

1. The American Anthropological Association has a website that provides details about race in the United States, www.understandingrace.org/history. Using that site, select a time period that is of interest to you, review the materials, and write up a summary that illuminates the nature of racial hierarchies at that era. What are the implications for various groups? How does this history help you see how race was embedded into the social structure? What are the necessary actions for social change?

2. Use the website on Immigration Research and Information, http://immigrationresearch-info.org/. How is this information different from how the topic of immigration is covered in the media? As you look at the community where you live, who are the immigrants, where do they come from, and are they on a path to equal citizenship with others?

PART III

How Do We See Each Other?
Beliefs, Representations, and Stereotypes

ELIZABETH HIGGINBOTHAM AND
MARGARET L. ANDERSEN

The beliefs that people hold are a powerful part of the persistence of racial inequality. How we think about race shapes how we see other people, how we see ourselves, and how we interpret the various images of race that bombard us through the media (whether or not we consciously recognize them). Social change in racial inequality is also made possible by changing how we think about race, especially if our thoughts about race otherwise anchor us in beliefs and attitudes that stifle, even if unwittingly, new possibilities for a more just society.

In this section, we present readings that explore different beliefs about race, especially in the contemporary context of alleged "color-blind" thinking. Are people color-blind? Is that even possible in a society that remains stratified and segregated by race and ethnicity? What does this mean for different groups in society? When you are attuned to the realities of race and racism, you will see images of race and ethnicity manifested everywhere—in the ever-present media, in popular culture, and even perhaps, in the logos of the sports teams you root for. Even food products—a package of butter, a jar of salsa, pancake mix, for example—display images of American Indians, Latinos/as, and African Americans that are highly stereotyped, although you may have never even taken conscious notice of such things. Once you begin to see and dissect such imagery, you realize how much these representations shape people's beliefs and, as a consequence, the persistence of racial and ethnic injustice. This section explores just a few of the ways that beliefs and representations of race are present in contemporary culture, thus laying the groundwork for thinking about how we can change.

First, learning a few basic concepts for understanding racial beliefs is important. When most people think about race beliefs, the term *prejudice* most likely comes to mind. Prejudice is an attitude that tends to denigrate individuals and groups who are perceived to be somehow different and undesirable. The social scientific definition of prejudice dates back to the 1950s and the work of psychologist Gordon Allport. Allport defined **prejudice** as "a hostile attitude directed toward a person or group simply because the person is presumed to be a member of that group and is perceived as having the negative characteristics associated with the group" (Allport 1954: 7).

Prejudice can be directed at many groups. One can be prejudiced against women or gays or athletes or foreigners—anyone who is perceived as a member of an "out-group"—that is, a group different from one's own. Although prejudice can be positive (as in thinking all women are nurturing), it is generally a negative attitude involving hostile or derogatory feelings as well as false generalizations about people in the so-called out-group.

Prejudice can also be expressed by any group, whether dominant or subordinate; thus, racial minorities may be prejudiced against other racial minorities or against members of the dominant group, just as more powerful people may be prejudiced against less powerful people. In other words, prejudice is a prejudgment and is the basis for much racial intolerance.

Prejudice rests on social **stereotypes**, that is, oversimplified beliefs about members of a particular social group. Stereotypes categorize people based on false generalizations along a narrow range of presumed characteristics, such as the belief that all Jewish people are greedy or that all blondes are dumb. Although stereotypes are perpetuated in many ways in society (e.g., in families, where parents teach children about other groups), the most influential are in popular culture—music, magazines, films, and television, among others. For example, because men of color are portrayed in the media as criminals, this is the most common way that they are stereotyped. Asian American women may be stereotyped as sexy and beguiling—an image repeatedly shown in magazines, videos, and other popular media. Stereotypes strongly shape how people come to define each other and, as such, are defined as *controlling images* (Collins 1990).

As an attitude, prejudice is distinct from discrimination, which is behavior. **Discrimination** is the negative and unequal treatment of members of a social group based on their perceived membership in a particular group. Although the term *discrimination* sometimes has a positive connotation (as in "she has a discriminating attitude"), such behavior generally is negative.

Prejudice and discrimination are typically thought of as related—prejudice causes discrimination—but things are not that simple. Many years ago, sociologist Robert Merton (1949) developed a four-square typology showing different ways that prejudice and discrimination are and are not related. Look at the following:

		PREJUDICE:	Positive (+)	Negative (–)
DISCRIMINATION:		Positive (+)	Case 1: Bigot (+ +)	Case 2: Nonprejudiced discriminator (– +)
		Negative (–)	Case 3: Prejudiced nondiscriminator (+ –)	Case 4: All-weather liberal (– –)

In Case 1, someone may be both prejudiced and discriminate—the classic bigot. Both prejudice and discrimination are overt, intentional, and hostile. In Case 4, someone may be free of prejudice and not discriminate (the person Merton called the "all-weather liberal"). In Cases 2 and 3, we see that prejudice and discrimination may not have a causal relationship. In Case 2, one may not be prejudiced, but still discriminate, such as a homeowner who holds no racial prejudice but will only buy a home in an all-White neighborhood "to protect their property value." Such people may say they hold no prejudice, but they might look out for their own interests and discriminate nonetheless, even without malice. In Case 3, someone may be prejudiced, but not discriminate—for example, when the law prohibits discrimination. A landlord may, for example, rent to a Black tenant despite holding prejudice. The point is that both prejudice and discrimination occur in a larger context—that of society as a whole. You can see from this typology that discrimination is caused as much by societal arrangements as from people's individual attitudes.

We think of prejudice as rooted in individual attitudes, but prejudice is not free-floating; it is linked to group positions in society (Blumer 1958; Bobo and Hutchings 1996). If prejudice is individual attitudes, what is racism? Racism certainly appears in individual attitudes (think of the racial bigot, as shown in Merton's typology), but racism is not just about thoughts. Racism is woven into the fabric of our society. **Racism**, different from prejudice, is a principle of social domination in which a group that is seen as inferior or different because of presumed biological or cultural characteristics is oppressed, controlled, and exploited—socially, economically, culturally, politically, or psychologically—by a dominant group (Wilson 1973).

Note the key elements of this definition:

- First, racism is a principle of domination, that is, embedded in this definition is the thought that racism involves one or more groups' subordinate position within a system of racial inequality. This means that the effects of the history of racism are cumulative and do not disappear easily through legally abolishing segregation or placing a few people of color in positions of influence.

- Second, the definition of racism emphasizes the word *presumed*. As we learned in Part I, race is not "real" in the biological or cultural sense, but it develops meaning through society and history. How people are perceived within a system of hierarchy and power is the key to understanding racism. Who gets the power to define different groups, and what are the means by which they do it? Law? Media? Schools? Those who shape how people are represented have enormous power to shape people's consciousness about race.

- Third, racism involves domination on a number of fronts: social, economic, political, cultural, and psychological. **Institutional racism** is the complex and cumulative pattern of racial advantage and disadvantage built into the structure of a society. Institutional racism, reflected in the prejudice and discrimination seen in a society, comprises more than an attitude or behavior. It is a system of power and privilege that gives the advantage to some groups over others. Thus you might say that prejudice is lodged in people's minds, but racism is lodged in society.

As several of the authors in Part III show, many people who benefit from institutional racism are often blind to the systemic advantage that it gives them. Thus, just being a White person will open some opportunities that might not be as readily available to others—independent of that White person's own attitudes and behavior. The invisibility of racial privilege to dominant groups is referred to as color-blind racism, the belief that race should be ignored and that race-conscious practices and policies should end only foster more racism. When dominant groups think that racism is no longer an issue, despite its ongoing reality, they are not likely to engage in practices or support policies that challenge racism (Brown et al. 2003; Bonilla-Silva 2013). To be color-blind in a society in which race still structures people's relationships, identities, and opportunities is to be blind to the continuing realities of race.

Charles A. Gallagher examines color-blind privilege in his essay "Color-Blind Privilege: The Social and Political Functions of Erasing the Color Line in Post-Race America." He points out that we live under the appearance of a multiracial society where race has actually become a commodity, something that White people can buy and display, while at the same time not challenging the privilege that underlies racial stratification. Products are mass-marketed using multiracial images and they sell across color lines, but such images legitimate color-blindness.

Gallagher's research shows that although many White people believe themselves to be color-blind, they are quick to defend the status quo.

Gallagher's work also shows how pervasive symbols of racial thinking are within everyday life. Stephanie A. Fryberg and Alisha Watts further examine such symbols in their research on the impact of American Indian mascots ("We're Honoring You, Dude: Myths, Mascots, and American Indians"). Contrary to popular beliefs that Indian mascots are "just for fun," Fryberg and Watts show the damaging effects such imagery has both for Native Americans, children especially, but also in fostering myths about Native American people.

Racial and ethnic prejudices cause harm in multiple ways—and against multiple groups. In the aftermath of the terrorist attacks on the United States on September 11, 2001, such hostilities have been especially directed at people of Arab descent and of the Muslim faith. Susan M. Akram and Kevin R. Johnson ("Race, Civil Rights, and Immigration Law after September 11, 2001: The Targeting of Arabs and Muslims") discuss the negative racial and ethnic profiling that has targeted Arabs and Muslims and often deprived them of the usual protections of citizenship.

The negative stereotypes and beliefs that are pervasive about different groups are often promulgated through taken-for-granted portrayals of racial-ethnic groups in popular culture. Danielle Dirks and Jennifer C. Mueller examine stereotypes in popular culture ("Racism and Popular Culture"), arguing that there is not a fixed meaning to race, but that popular culture is an apparatus through which cultural beliefs are produced and consumed.

How such racial beliefs influence people is cleverly examined in a research study by Andrew M. Penner and Aliya Saperstein ("How Social Status Shapes Race"). Penner and Saperstein show that stereotypes about specific groups actually change how people perceive another's race. People are more likely classified as being Black if they have been incarcerated, unemployed, or poor—a research finding that certainly challenges the idea that race is some fixed attribute that is unchanging and solely connected to some physical characteristics.

Altogether, the articles in this section underscore the importance of seeing race as an emergent and socially constructed idea. Race and racism, along with ethnicity, are manufactured through social, not natural, phenomena. Nonetheless, the effects on our beliefs, our images, and our understandings of each other are profound.

REFERENCES

Allport, Gordon. 1954. *The Nature of Prejudice*. Reading, MA: Addison-Wesley.

Blumer, Herbert. 1958. "Race Prejudice as a Sense of Group Position." *Pacific Sociological Review* 1 (Spring): 3–7.

Bobo, Lawrence, and Vincent L. Hutchings. 1996. "Perceptions of Racial Group Competition: Extending Blumer's Theory of Group Position to a Multiracial Social Context." *American Sociological Review* 25 (December): 951–972.

Bonilla-Silva, Eduardo. 2013. *Racism without Racists: Color-blind Racism and the Persistence of Racial Inequality in the United States*, 4th ed. Lanham, MD: Rowman & Littlefield.

Brown, Michael, Martin Carnoy, Elliott Currie, Troy Duster, David Oppenheimer, Marjorie M. Schultz, and David Wellman. 2003. *Whitewashing Race: The Myth of a Color-Blind Society*. New York: Oxford University Press.

Collins, Patricia Hill. 1990. *Black Feminist Thought: Knowledge, Consciousness, and the Politics of Empowerment*. Boston: Unwin Hyman.

Merton, Robert. 1949. "Discrimination and the American Creed." Pp. 99–126 in *Discrimination and the National Welfare*, edited by Robert W. MacIver. New York: Harper and Brothers.

Wilson, William Julius. 1973. *Power, Racism, and Privilege: Race Relations in Theoretical and Sociohistorical Perspectives*. New York: Macmillan.

FACE THE FACTS: AMERICANS' VIEWS OF SOCIETAL TREATMENT OF RACIAL-ETHNIC GROUPS, 2013

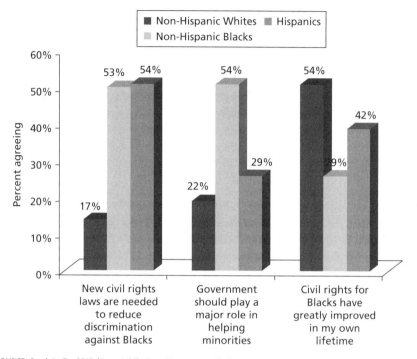

SOURCE: Saad, Lydia. 2013 (August 26). "Post Zimmerman Blacks are More Concerned about Civil Rights." *The Gallup Poll*. Princeton, NJ: Gallup Organization.

Think about It: These data are based on a random sample of adults in all fifty states and the District of Columbia. The survey was done by telephone interview. What differences do you see in the responses on non–Hispanic Whites, non–Hispanic Blacks, and Hispanics? How would you explain what you see?

10

Color-Blind Privilege

The Social and Political Functions of Erasing the Color Line in Post-Race America

CHARLES A. GALLAGHER

Popular culture is full of images that portray a multiracial society, but such representations encourage a new form of thinking—color-blind racism. Color-blind racism refers to the dominant belief that race no longer matters in shaping people's experiences—a belief that is contradicted by the reality of race in America.

The young white male sporting a FUBU (African-American owned apparel company "For Us By Us") shirt and his white friend with the tightly set, perfectly braided cornrows blended seamlessly into the festivities at an all-white bar mitzvah celebration. A black model dressed in yachting attire peddles a New England, yuppie boating look in Nautica advertisements. It is quite unremarkable to observe white, Asian or African-Americans with dyed purple, blond or red hair. White, black and Asian students decorate their bodies with tattoos of Chinese characters and symbols. In cities and suburbs young adults across the color line wear hip-hop clothing and listen to white rapper Eminem and black rapper 50 Cent. It went almost unnoticed when a north Georgia branch of the NAACP installed a white biology professor as its president. Subversive musical talents like Jimi Hendrix, Bob Marley and The Who are now used to sell Apple computers, designer shoes and SUVs. Du-Rag kits, complete with bandana headscarf and elastic headband, are on sale for $2.95 at hip-hop clothing stores and family–centered theme parks like Six Flags. Salsa has replaced ketchup as the best selling condiment in the United States. Companies as diverse as Polo, McDonalds' Tommy Hilfiger, Walt Disney World, Master Card, Skechers sneakers, IBM, Giorgio Armani and Neosporin antibiotic ointment have each crafted advertisements that show an integrated, multiracial cast of characters interacting and consuming their products in a post-race, color-blind world.

Americans are constantly bombarded by depictions of race relations in the media which suggest that discriminatory racial barriers have been dismantled.

SOURCE: From Gallagher, Charles A. 2003. "Color-Blind Privilege: The Social and Political Functions of Erasing the Color Line in Post-Race America." *Race, Gender & Class* 10: 22–37. Reprinted by permission.

Social and cultural indicators suggest that America is on the verge, or has already become, a truly color-blind nation. National polling data indicate that a majority of whites now believe discrimination against racial minorities no longer exists. A majority of whites believe that blacks have "as good a chance as whites" in procuring housing and employment or achieving middle class status while a 1995 survey of white adults found that a majority of whites (58%) believed that African Americans were "better off" finding jobs than whites (Gallup, 1997; Shipler, 1998). Much of white America now sees a level playing field, while a majority of black Americans sees a field which is still quite uneven.… The color-blind or race-neutral perspective holds that in an environment where institutional racism and discrimination have been replaced by equal opportunity, one's qualifications, not one's color or ethnicity, should be the mechanism by which upward mobility is achieved. Color as a cultural style may be expressed and consumed through music, dress, or vernacular but race as a system which confers privileges and shapes life chances is viewed as an atavistic and inaccurate accounting of U.S. race relations.

Not surprisingly, this view of society blind to color is not equally shared. Whites and blacks differ significantly, however, on their support for affirmative action, the perceived fairness of the criminal justice system, the ability to acquire the "American Dream," and the extent to which whites have benefited from past discrimination (Moore, 1995; Moore & Saad, 1995; Kaiser, 1995). This article examines the social and political functions colorblindness serves for whites in the United States. Drawing on interviews and focus groups with whites from around the country, I argue that colorblind depictions of U.S. race relations serve to maintain white privilege by negating racial inequality. Embracing a colorblind perspective reinforces whites' belief that being white or black or brown has no bearing on an individual's or a group's relative place in the socioeconomic hierarchy.

DATA AND METHOD

I use data from seventeen focus groups and thirty individual interviews with whites from around the country. Thirteen of the seventeen focus groups were conducted in a college or university setting, five in a liberal arts college in the Rocky Mountains and the remaining eight at a large urban university in the Northeast. Respondents in these focus groups were selected randomly from the student population. Each focus group averaged six respondents … equally divided between males and females. An overwhelming majority of these respondents were between the ages of eighteen and twenty-two years of age. The remaining four focus groups took place in two rural counties in Georgia and were obtained through contacts from educational and social service providers in each county. One county was almost entirely white (99.54%) and in the other county whites constituted a racial minority. These four focus groups allowed me to tap rural attitudes about race relations in environments where whites had little or consistent contact with racial minorities.…

COLORBLINDNESS AS NORMATIVE IDEOLOGY

The perception among a majority of white Americans that the socio-economic playing field is now level, along with whites' belief that they have purged themselves of overt racist attitudes and behaviors, has made colorblindness the dominant lens through which whites understand contemporary race relations. Colorblindness allows whites to believe that segregation and discrimination are no longer an issue because it is now illegal for individuals to be denied access to housing, public accommodations or jobs because of their race. Indeed, lawsuits alleging institutional racism against companies like Texaco, Denny's, Coke, and Cracker Barrel validate what many whites know at a visceral level is true: firms which deviate from the color-blind norms embedded in classic liberalism will be punished. As a political ideology, the commodification and mass marketing of products that signify color but are intended for consumption across the color line further legitimate colorblindness. Almost every household in the United States has a television that, according to the U.S. Census, is on for seven hours every day (Nielsen 1997). Individuals from any racial background can wear hip-hop clothing, listen to rap music (both purchased at Wal-Mart) and root for their favorite, majority black, professional sports team. Within the context of racial symbols that are bought and sold in the market, colorblindness means that one's race has no bearing on who can purchase a Jaguar, live in an exclusive neighborhood, attend private schools or own a Rolex.

The passive interaction whites have with people of color through the media creates the impression that little, if any, socio-economic difference exists between the races....

Highly visible and successful racial minorities like [former] Secretary of State Colin Powell and ... [former Secretary of State] Condoleezza Rice are further proof to white America that the state's efforts to enforce and promote racial equality have been accomplished.

The new color-blind ideology does not, however, ignore race; it acknowledges race while disregarding racial hierarchy by taking racially coded styles and products and reducing these symbols to commodities or experiences that whites and racial minorities can purchase and share. It is through such acts of shared consumption that race becomes nothing more than an innocuous cultural signifier. Large corporations have made American culture more homogenous through the ubiquitousness of fast food, television, and shopping malls but this trend has also created the illusion that we are all the same through consumption. Most adults eat at national fast food chains like McDonalds', shop at mall anchor stores like Sears and J.C. Penney's, and watch major league sports, situation comedies or television drama. Defining race only as cultural symbols that are for sale allows whites to experience and view race as nothing more than a benign cultural marker that has been stripped of all forms of institutional, discriminatory or coercive power. The post-race, color-blind perspective allows whites to imagine that depictions of racial minorities working in high status jobs and consuming the same products, or at least appearing in commercials for products whites desire or consume, is the same as living in a society where color is no longer used to allocate resources or shape group outcomes. By constructing a picture of society where racial harmony is the norm, the color-blind perspective functions to make white privilege invisible while removing from public discussion the need to maintain any social programs that are race-based.

How, then, is colorblindness linked to privilege? Starting with the deeply held belief that America is now a meritocracy, whites are able to imagine that the socio-economic success they enjoy relative to racial minorities is a function of individual hard work, determination, thrift and investments in education. The color-blind perspective removes from personal thought and public discussion any taint or suggestion of white supremacy or white guilt while legitimating the existing social, political and economic arrangements which privilege whites. This perspective insinuates that class and culture, and not institutional racism, are responsible for social inequality. Colorblindness allows whites to define themselves as politically and racially tolerant as they proclaim their adherence to a belief system that does not see or judge individuals by the "color of their skin." This perspective ignores, as Ruth Frankenberg puts it, how whiteness is a "location of structural advantage societies structured in racial dominance" (2001 p. 76).... Colorblindness hides white privilege behind a mask of assumed meritocracy while rendering invisible the institutional arrangements that perpetuate racial inequality. The veneer of equality implied in colorblindness allows whites to present their place in the racialized social structure as one that was earned.

OPPORTUNITY HAS NO COLOR

Given this norm of colorblindness it was not surprising that respondents in this study believed that using race to promote group interests was a form of (reverse) racism....

Believing and acting as if America is now color-blind allows whites to imagine a society where institutional racism no longer exists and racial barriers to upward mobility have been removed. The use of group identity to challenge the existing racial order by making demands for the amelioration of racial inequities is viewed as racist because such claims violate the belief that we are a nation that recognizes the rights of individuals, not rights demanded by groups....

The logic inherent in the colorblind approach is circular; since race no longer shapes life chances in a color-blind world there is no need to take race into account when discussing differences in outcomes between racial groups. This approach erases America's racial hierarchy by implying that social, economic and political power and mobility is equally shared among all racial groups. Ignoring the extent or ways in which race shapes life chances validates whites' social location in the existing racial hierarchy while legitimating the political and economic arrangements that perpetuate and reproduce racial inequality and privilege.

REFERENCES

Frankenberg, R. (2001). "The mirage of an unmarked 'whiteness'". In B. B. Rasmussen, E. Klineberg, I. J. Nexica & M. Wray (eds.) *The making and unmaking of whiteness*. Durham: Duke University Press.

Gallup Organization. (1997). "Black/white relations in the U.S." June 10, pp. 1–5.

Kaiser Foundation. (1995). *The four Americas: Government and social policy through the eyes of America's multi-racial and multi-ethnic society*. Menlo Park, CA: Kaiser Family Foundation.

Moore, D. (1995). "Americans' most important sources of information: Local news." *The Gallup Poll Monthly*, September, pp. 2–5.

Moore, D. & Saad, L. (1995). "No immediate signs that Simpson trial intensified racial animosity." *The Gallup Poll Monthly*, October, pp. 2–5.

Nielsen, A. C. (1997). *Information please almanac* (Boston: Houghton Mifflin).

Shipler, D. (1998). *A country of strangers: Blacks and whites in America*. New York: Vintage Books.

DISCUSSION QUESTIONS

1. How does Gallagher see color-blind racism as resulting from White people's privilege? How does privilege influence what White people can understand about racism?

2. In what ways does color-blind racism support the traditional American ideal that any individual can succeed if they only try hard enough?

11

We're Honoring You, Dude

Myths, Mascots, and American Indians

STEPHANIE A. FRYBERG AND ALISHA WATTS

In a series of experiments, Stephanie Fryberg and Alisha Watts challenge various myths about the popularity of American Indian sports mascots. Their research shows the impact of such negative and misleading stereotypes on Native American people. They also show how the images themselves distort people's understanding of the reality and variation in Native cultures.

Images of American Indians are prevalent in American culture—on television, in films, at sporting events, and even in schools. For example, you may encounter an American Indian mascot image when you attend a Washington Redskins football game, watch a University of North Dakota Fighting Sioux sporting event on television, or simply pass a T-shirt-clad sports fan on the street. Despite a long-standing national debate, many American high schools (e.g., Osceola High School Chieftains in Wisconsin, Napa High School Indians in California), universities (e.g., University of North Dakota Fighting Sioux, Florida State University Seminoles), and professional sports teams (e.g., Cleveland Indians, Atlanta Braves, Washington Redskins) continue to use American Indian team names and mascots.

The mascot imagery ranges from the Cleveland Indians Chief Wahoo mascot, which is a red-faced, big-nosed, grinning, cartoonlike character adorned with a headband and a feather, to the former University of Illinois Chief Illiniwek mascot, who is a European American male attired in traditional chief regalia (i.e., buckskin dress and leggings, moccasins, and a chief's headdress) who performs quasi-traditional dances. Across the country, many other sports teams use American Indian names and symbols. For example, the logo of the Atlanta Braves major league baseball team is a red tomahawk. To music identified as the Tomahawk War Chant, the Braves fans perform "The Chop," which is a repetitive bend of the arm at the elbow and is intended to evoke a swinging tomahawk.

Proponents of American Indian mascots argue that these images do not offend American Indians and that they stimulate interest in American Indian

SOURCE: Fryberg, Stephanie A., and Alisha Watts. 2010. "We're Honoring You, Dude: Myths, Mascots, and American Indians." Pp. 458–480 in *Doing Race: 21 Essays for the Twenty-First Century*, edited by Hazel Rose Markus and Paula M. L. Moya. New York: W.W. Norton.

culture and history. For example, the Honor the Chief Society, a group that was set up to defend the use of the Illinois mascot, claims that mascots elicit positive feelings for the team and for American Indians in general (2002). By contrast, opponents of using American Indian mascots argue that these images harm and offend, rather than honor, American Indians. They suggest that American Indian mascots are dehumanizing and restrict the full membership of American Indians in American society. Specifically, mascot opponents question the claim that American Indians feel honored when a European American man dresses up in traditional Indian regalia and dances around a football field or basketball court, or when the opposing teams chant "Scalp the Indians" or "Kill the Indians." These opposing views fuel an ongoing passionate national debate about the use of American Indian mascots.

BACKGROUND OF THE CONTROVERSY

While a wide range of both Native and non-Native organizations have taken strong positions against the use of American Indian mascots, a few organizations advocate keeping them. As noted above, the University of Illinois Honor the Chief Society contends that American Indian mascots honor and benefit American Indians, stating that "when people take the time to learn about the First Nation People for whom the State of Illinois was named, and reflect upon how Chief Illiniwek is portrayed by her flagship University, they are better able to recognize the difference between an athletic mascot and a time-honored symbol of tradition and respect" (2007). Additionally, Washington Redskins vice president Karl Swanson asserts that his team name "symbolizes courage, dignity and leadership" and that since the team does not use the name and image with negative intent, then it cannot be offensive (Price 2002).

Several predominantly American Indian organizations disagree with this perspective. The Society of Indian Psychologists, the National Congress of American Indians, the National Indian Education Association, and the Native American Journalists Association officially oppose the use of American Indian mascots. Moreover, a wide range of predominantly non-Native organizations, including the United States Commission on Civil Rights, the National Collegiate Athletic Association, the American Psychological Association, the American Anthropological Association, the American Sociological Association, the National Coalition for Racism in Sports and Media, and the North American Society for Sociology of Sport also officially oppose and recommend bans on the use of these mascots. For example, in 2001, the United States Commission on Civil Rights argued that mascots and related names, performances, and images are "disrespectful and offensive to American Indians and others who are offended by such stereotyping." Similarly, in 2005, the American Psychological Association declared that the use of American Indian mascots "undermines the ability of American Indian Nations to portray accurate and respectful images of their culture, spirituality, and traditions" and "is a detrimental manner of illustrating the cultural identity of American Indian people through negative displays and/or interpretations of spiritual and traditional practices."

Despite the increasing number of national organizations that officially oppose the use of American Indian mascots, the debate continues. Many people fail to grasp the significance of the American Indian mascot controversy. On the surface, American Indian mascots are simply mascots; importantly, however, they are not the same as animal mascots, such as the Chicago Bears, the Philadelphia Eagles, the Oregon State University Beavers, the University of California at Santa Cruz Banana Slugs, the University of Connecticut Huskies, or the University of Arizona Wildcats, or even other ethnic group mascots, such as the University of Notre Dame Fighting Irish or the Minnesota Vikings. American Indian mascots represent a group of people who historically have been, and in many instances continue to be, mistreated in American society. Historically, Irish Americans were also mistreated in American society, but contemporary Irish Americans fare quite well as a group and do not typically report much discrimination. Mascot proponents often overlook the unique dual status of American Indian mascots as both team representations and group representations.

The dual status of American Indian mascots is brought into high relief by asking, for example, whether African Americans would feel honored if a European American man painted his face black and ran around a field imitating a "tribesman" from West Africa or whether Christians would feel honored by someone running around wearing the liturgical vestments used by clergy during services. While many people might immediately feel outrage at the thought of such acts, these same people often feel no comparable outrage over the use of American Indian mascots. Why is this? One possible reason is that despite an abundance of research demonstrating the negative effects of stereotypical group representations on group members, many people continue to believe that mascots are just mascots. This myth takes a number of forms including "It's just a mascot," "If you don't believe the stereotypes, they can't hurt you," [and] "But we're honoring American Indians."...

MYTH 1: "IT'S JUST A MASCOT!"

American Indian mascots may seem relatively unimportant compared to other pressing national issues. In fact, those who support the use of American Indian mascots often argue that they are "just mascots" and that the entire mascot controversy is blown out of proportion. Those against the use of American Indian mascots argue that they are more than "just mascots"—that they are, in fact, powerful social representations that affect how American Indians are viewed and how American Indians view themselves.

The power of American Indian mascots is that they are widely circulated public or *social representations* of American Indians in American society. The psychologist Serge Moscovici defines social representations as widely distributed images, beliefs, and assumptions that help individuals know how to think and behave in their social worlds (1988). Social representations are everywhere— in institutions (e.g., schools, churches), in social structures and practices (e.g., the law, pedagogy, families, teams), and in everyday artifacts (e.g., television programs, films, posters, textbooks). They constitute the meaning systems that individuals use to orient themselves in social environments and provide a shared language for individuals to communicate with one another (Moscovici 1973/ 1988, 1984).

Social representations also provide answers to the questions "Who am I?" and "Who are we?" (i.e., Who is my group?). To answer these questions, people look to the available social representations of their group. They use these representations to help define themselves and to understand how they are defined by others. This is the social nature of being a person....

... American Indians are not widely represented in the public eye. Once reason this is true is that American Indians are a numerically small group. According to the 2000 census, American Indians constitute less than 1 percent (.9) of the U.S. population. In addition, American Indians are fairly invisible in and segregated from mainstream society. Approximately 40 percent of American Indians reside on Indian reservations (Family Education Network 2002), many of which are in fairly remote parts of the country, and roughly one-fourth of American Indians live in poverty, more than twice the national average (U.S. Census Bureau 2000). All of these factors contribute to decreased opportunities for American Indians and to their decreased visibility.

The American media also reflects this decreased visibility. In a two-week composite of prime time television programming in 2002, only six characters (.4 percent) were identified as American Indian (Mastro and Behm-Morawitz 2005), and in a composite week of television commercials, only nine characters (.4 percent) were identified as American Indian (Mastro and Stern 2003). Similar results were found with a content analysis of newspapers and films, revealing that approximately .2 percent of newspaper articles and popular films featured American Indians (Fryberg 2003).

One consequence of the decreased visibility of American Indians is that most Americans have no direct or personal experience with contemporary American Indians. As a result, most mainstream views of American Indians are formed and fostered by indirectly acquired information. For example, one of the authors of this essay (Stephanie Fryberg) has conducted several studies showing that individuals unfamiliar with American Indians characterize American Indians as they are portrayed in the media—as spiritual, in tune with nature, warriorlike, and as people with social problems. The majority of these representations are tied to sports teams....

The decreased visibility of American Indians, when combined with limited social representations, creates a situation in which the few prevalent representations of American Indians emerge as particularly strong communicators of how American Indians should appear and behave....

MYTH 2: "STICKS AND STONES MAY BREAK MY BONES, BUT [STEREOTYPES] WILL NEVER HURT ME."

When social representations characterize a group in a rigid or static manner, they are referred to as stereotypes. *Stereotypes* are powerful, hard-to-break mental links between a social group and a limited set of behaviors or traits. When people repeatedly see American Indians portrayed as mascots, they form automatic associations, or stereotypes, between American Indians and the common characteristics of American Indian mascots (e.g., aggressive, noble, violent, stoic, savagelike, spiritual). One popular myth about stereotypes is that if group members do not believe a stereotype or choose to ignore the stereotype, then it will not influence them. The implication is that if American Indians do not believe that American Indian mascot stereotypes are harmful, they will not be affected by exposure to the mascot stereotype. Some mascot advocates even suggest that American Indians who "allow" American Indian mascots to impact their well-being are mentally "weak" or at least overly sensitive. At issue, then, is whether negative stereotypes detrimentally affect the individuals who are the target of the stereotype even when individuals resist or contest the stereotype.

Social psychologists Claude Steele and Joshua Aronson (1995) have shown conclusively that negative stereotypes adversely impact the performance of individual group members. Their studies demonstrate that in a performance situation (e.g., taking a test), the concern that one's performance might confirm a stereotype about one's group, what is referred to as *stereotype threat*, can impair performance (see also Steele 1997; Steele et al. 2002)....

Since the initial stereotype threat studies described above, over 100 journal articles have confirmed that negative stereotypes have negative consequences for members of stereotyped groups, including Latinos, women in math, students from low socioeconomic status backgrounds, and European Americans in interracial interactions. Stereotype threat works by disrupting concentration....

Stereotype threat research also demonstrates that negative stereotypes detrimentally affect the stereotyped individuals, *even when the individuals do not consciously believe the stereotypes*....

MYTH 3: "BUT I'M HONORING AMERICAN INDIANS!"

If American Indian mascots are negative stereotypes, then believing that they could have a negative effect on American Indians is simple. In fact, ... stereotype threat theory argues that negative stereotypical representations harm members of the stereotyped groups. But what if American Indian mascots are perceived as positive stereotypical representations of American Indians? Proponents of American Indian mascots often argue that American Indian mascots positively portray, and thus honor, American Indians. Are these mascots positive, harmless representations in the way the mascot proponents suggest, or is it possible they are positive, yet harmful to American Indians?...

In a series of research studies, Fryberg and her colleagues (2008) asked whether seemingly positive representations of American Indians negatively influence how American Indians see themselves. To explore this possibility, they went to schools with relatively large American Indian populations.... The American Indian students in the study saw one of three social representations: an American Indian mascot image (Chief Wahoo, the Cleveland Indians Major League Baseball team mascot), a list of stereotypes about American Indians (e.g., three statistics about high rates of school dropout, depression, and suicide in the American Indian community), or a non-mascot social representation (a picture of the Pocahontas character from Disney's *Pocahontas* film).

To assess whether the three social representations bring to mind positive or negative associations, students saw one of the three American Indian representations and then wrote down the first five thoughts that came to mind. Analyses of these thoughts showed that American Indian high school students noted largely positive associations with Chief Wahoo and Pocahontas (i.e., approximately 80 percent of their listed thoughts were positive) and largely negative associations with the negative stereotypes (i.e., 91 percent of the thoughts they listed were negative).

Next, Fryberg and her colleagues explored whether the three American Indian social representations influenced American Indian students' everyday well-being. The researchers found that American Indian students who saw an image of Chief Wahoo or Pocahontas, the two American Indian social representations that brought to mind positive associations in the first study, reported lower self-esteem and lower community worth than students in the control group. Students who read the negative stereotypes also reported lower self-esteem and community worth than students in the control group, but students who saw Chief Wahoo and Pocahontas reported even lower self-esteem than students who saw the negative stereotypes....

Following exposure to the representations, students completed a *possible selves* measure. (In this line of research, "possible selves" are defined as the future goals or aspirations people hope to achieve or fear not achieving.) Specifically, participants were asked, "*Think a minute about next year and what you will be like this time next year. What do you expect you will be like? Write down at least four ways of describing yourself that will probably be true of you next year. You can write down ways you are now or ways you expect to become.*" After the study was complete, research assistants coded the number of achievement-related possible selves reported by students (e.g., getting good grades, graduating from college, getting a good job). They found that American Indian students exposed to any one of the three American Indian mascot representations (i.e., Chief Wahoo, Chief Illinewek, or the Haskell Indian) reported fewer future goals or aspirations related to achievement compared to students not exposed to a representation (control group) and to students exposed to the American Indian College Fund advertisement.

The research described here debunks the myth that positive representations must elicit positive or at least neutral outcomes....

SUMMARY OF FINDINGS: IMPLICATIONS FOR THE AMERICAN INDIAN MASCOT CONTROVERSY

The current controversy over the use of American Indian mascots is about more than "just mascots." From the point of view of the target—American Indians—Indian mascots do not honor their target. Instead they perpetuate stereotypes that negatively impact American Indians regardless of whether the representations are seen as positive or negative, and regardless of whether American Indians agree or disagree with being used as mascots. American Indian mascots and other related stereotypical representations of American Indians reinforce ideas of American Indians as "tee-pee-dwelling," spiritual beings who "commune with nature" or as aggressive, uncivilized, noble "savages" who existed in the eighteenth and nineteenth centuries (United States Commission on Civil Rights 2001). In this way, American Indian mascots prevent non-Natives from gaining an accurate understanding of both the historical and the contemporary experiences of American Indians. Mascots thus limit the way people view American Indians and restrict the variety of ways in which American Indians can view themselves....

WORKS CITED

American Psychological Association. 2005. APA resolution recommending the immediate retirement of American Indian mascots, symbols, images, and personalities by schools, colleges, universities, athletic teams, and organizations. www.apa.org/releases (accessed June 11, 2006).

Family Education Network. 2002. American Indians: Census facts. www.factmonster.com/spot/aihmcensus1.html (accessed October 10, 2005).

Fryberg, Stephanie A. 2003. Really? You don't look like an American Indian: Social representations and social group identities. PhD diss., Stanford University. *Dissertation Abstracts International 64*(1549), 3B.

Fryberg, Stephanie A., Hazel Rose Markus, Daphna Oyserman, and Joseph M. Stone. 2008. Of warrior chiefs and Indian princesses: The psychological consequences of American Indian mascots. *Basic and Applied Social Psychology 30*(3), 208–218.

Honor the Chief Society. 2002. Evidence demands a verdict. www.honorthechief.org/news_demand-verdict.html (accessed December 20, 2006).

———. 2007. History. www.honorthechief.org/history.html (accessed November 2, 2007).

Mastro, Dana E., and Elizabeth Behm-Morawitz. 2005. Latino representation on primetime television. *Journalism and Mass Communication Quarterly 82*, 110–130.

Mastro, Dana E., and Susannah R. Stern. 2003. Representations of race in television commercials: A content analysis of prime-time advertising. *Journal of Broadcasting and Electronic Media 47*, 638–647.

Moscovici, Serge. 1988. Notes toward a description of social representations. *European Journal of Social Psychology 18*, 211–250.

———. 1984. The phenomena of social representations. In *Social Representations*, ed. Robert M. Farr and Serge Moscovici, 18–77. Cambridge, England: Cambridge University Press.

———. 1973/1988. Preface to *Health and Illness: A Social Psychological Analysis*, by Claudine Herzlich. London: Academic Press.

Price, S. L. 2002. The Indian wars. *Sports Illustrated*, March 4, 66–72.

Steele, Claude M. 1997. A threat in the air: How stereotypes shape intellectual identity and performance. *American Psychologist 52*, 613–629.

Steele, Claude M., and Joshua Aronson. 1995. Stereotype and the intellectual performance of African Americans. *Journal of Personality and Social Psychology 69*, 797–811.

Steele, Claude M., Steven J. Spencer, and Joshua Aronson. 2002. Contending with group image: The psychology of stereotype and social identity threat. In *Advances in Experimental Social Psychology*, Vol. 34, ed. Mark. P. Zanna, 379–440. San Diego, CA: Academic Press.

U.S. Census Bureau. 2000. Census 2000 Gateway. http://www.census.gov/main/www/cen2000.html (accessed October 10, 2005).

U.S. Commission on Civil Rights. 2001. Statement of the U.S. Commission on Civil Rights on the use of Native American images and nicknames as sports symbols. April 13, 2001. www.usccr.gov/press/archives/2001/041601st.htm (accessed December 20, 2006).

DISCUSSION QUESTIONS

1. Close your eyes and think of a Native American mascot that you recognize. Now describe the image of Native Americans that this symbol projects. What does it suggest about Native American cultures? If you knew nothing else about Native Americans, what does the symbol imply?

2. Were sports teams to adopt stereotyped mascots about other groups (Jewish people, women, Latinos/as, or others), what might be the public reaction? Why have some not reacted negatively to Native American mascots?

12

Race, Civil Rights, and Immigration Law After September 11, 2001

The Targeting of Arabs and Muslims

SUSAN M. AKRAM AND KEVIN R. JOHNSON

The events of September 11, 2001, dramatically changed the lives of many citizens and noncitizens in the United States, but the federal response has meant dramatic changes for noncitizens who are Arabs and Muslims. Regardless of their nations of origin, many find that they are suspects and can be targets of hate crimes by Americans who see them as terrorists. These noncitizens are also subjected to many federal legal actions that not only communicate their perceived lack of loyalty, but put them at risk for incarceration.

INTRODUCTION

Although only time will tell, September 11, 2001, promises to be a watershed in the history of the United States. After the tragic events of that day, including the hijacking of four commercial airliners for use as weapons of mass destruction, America went to "war" on many fronts, including but not limited to military action in Afghanistan.

As needed and expected, heightened security measures and an intense criminal investigation followed. Almost immediately after the tragedy, Arabs and Muslims, as well as those "appearing" to be Arab or Muslim, were subject to crude forms of racial profiling.... Immediately after September 11, hate crimes against Arabs, Muslims, and others rose precipitously. In Arizona, a U.S. citizen claiming vengeance for his country killed a Sikh immigrant from India based on the mistaken belief that this turban-wearing, bearded man was "Arab."

Supporters and critics alike saw the federal government as "pushing the envelope" in restricting civil liberties in the name of national security.... The federal government has acted more swiftly and uniformly than the states ever

SOURCE: Akram, Susan M., and Kevin R. Johnson. 2002. "Race, Civil Rights, and Immigration Law after September 11, 2001: The Targeting of Arabs and Muslims." *Annual Survey of American Law* 58(3): 295–356.

could, with harsh consequences to the Arab and Muslim community in the United States. That the reaction was federal in nature—and thus national in scope as well as uniform in design and impact, and faced precious few legal constraints—increased the severity of the impacts.

... The Immediate Impacts

The federal government responded with ferocity to the events of September 11. Hundreds of Arab and Muslim noncitizens were rounded up as "material witnesses" in the ongoing investigation of the terrorism or detained on relatively minor immigration violations. The dragnet provoked criticism as a poor law enforcement technique as well as a major intrusion on fundamental civil liberties. Congress swiftly passed the USA PATRIOT Act, which, among other things, allowed the government to detain suspected noncitizen "terrorists" for up to a week without charges, and bolstered federal law enforcement surveillance powers over citizens and noncitizens associated with "terrorism." President Bush issued a military order allowing alleged noncitizen terrorists, including those arrested in the United States, to be tried in military courts while guaranteed few rights....

To the extent that the U.S. responses to September 11 can be characterized as regulating immigration, existing case law affords considerable leeway to the political branches of the federal government. The Supreme Court has upheld immigration laws discriminating against noncitizens on the basis of race, national origin, and political affiliation that would patently violate the Constitution if the rights of citizens were at stake. The so-called "plenary power" doctrine creates a constitutional immunity from judicial scrutiny of substantive immigration judgments of Congress and the Executive Branch. The doctrine thus allows the federal government, through the immigration laws, to lash out at any group considered undesirable. Such authority increases exponentially when, as in the case of international terrorism, perceived foreign relations and national security matters are at issue. When immigration law and its enforcement rests primarily in the hands of the federal government, uniform, national civil rights deprivations may result.

The laws supporting much of the immigration and civil rights incursions of the "war on terrorism," including the plenary power doctrine, have been subjected to sustained scholarly criticism. In important ways, contemporary immigration law ignores a constitutional revolution that occurred in the area of civil rights over the latter half of the twentieth century. Nonetheless, much of this body of law—or, more accurately, the perceived immunity from any legal constraints—has guided the Bush administration's domestic responses to the legitimate concerns with terrorism.

The Dragnet   The events of September 11, 2001 understandably provoked an immediate federal governmental response. Heightened security measures were the first order of the day. Within a matter of weeks, the U.S. government arrested and detained in the neighborhood of 1000 people as part of the Justice Department's investigation into the September 11 attacks. The mass dragnet of men from many nations, with the largest numbers from Pakistan and Egypt, apparently failed to produce any direct links to the terrorists acts; about 100 were charged with minor crimes and another 500 were held in custody on immigration-related matters, such as having overstayed their temporary nonimmigrant visas. Attorney General John Ashcroft admitted that minor immigration charges would be used to hold noncitizens while the criminal investigation continues.

Information remains sketchy about the persons detained by the U.S. government because the Attorney General has refused to release specific information about them, prompting criticism from U.S. Senator Russell Feingold. The Justice Department issued a rule barring disclosure of information about INS detainees held by state and local authorities, which survived a legal challenge. After September 11 the immigration courts began holding secret hearings in immigration cases involving Arab and Muslim noncitizens. In sum, the federal government's treatment of the detainees, and its treatment of Arab and Muslim noncitizens in immigration proceedings, was shrouded in secrecy....

The nature and conditions of the initial wave of mass arrests and detentions warrant consideration. Arab and Muslim detainees were held for weeks—in some instances, months—without charges filed against them and without being provided information about why federal authorities continued to detain them.... The dragnet did not end there. The Justice Department also sought to interview about 5000 men—almost all of them Arab or Muslim—between the ages of 18 and 33 who had arrived on nonimmigrant visas in the United States since January 1, 2000. There was *no* evidence that any of the 5000 had been involved in terrorist activities. Although technically "voluntary," the interviews with law enforcement authorities undoubtedly felt compulsory to many. Arab and Muslim fears of detention and deportation were reinforced by the November 2001 arrest of Mazen Al-Najjar, who had previously been held on secret evidence and released after the government failed to provide evidence that he was engaged in terrorist activity. In March 2002, Attorney General Ashcroft asked U.S. attorneys to interview another 3000 or so Arab and Muslim noncitizens. Around the time of this announcement, the federal government conducted raids on Arab and Muslim offices and homes in search of terrorist connections.

The questioning of noncitizen Arabs and Muslims could be expected to alienate those interviewed, as well as the communities of which they are a part....

The questions directed at the noncitizens suggested that the Arabs and Muslims were prone to disloyalty. One line of questioning was as follows: "You should ask the individual if he noticed anybody who reacted in a surprising or inappropriate way to the news of September 11th attacks. You should ask him how he felt when he heard the news." This tracks questions reportedly asked by federal investigators soon after the bombing....

By almost all accounts, Muslims perpetrated the terrorism of September 11. A few Arab and Muslim noncitizens in the United States might have information about terrorist networks.... Nonetheless, the dragnet directed at all Arabs and Muslims is contrary to fundamental notions of equality and the individualized suspicion ordinarily required for a stop under the Fourth Amendment. It exemplifies the excessive reliance on race in the criminal investigation, a frequent law enforcement problem, and shows how, once race (at least of nonwhites) enters the process, it can come to dominate an investigation. To target an entire minority group across the country for questioning is obviously over-inclusive. Over one million of persons of Arab ancestry in the United States, all of whom may feel threatened and under suspicion, cannot miss the message sent by the nature of the federal government's investigation.

In important ways, the September 11 dragnet carried out by the federal government resembles the Japanese internment during World War II although detention fortunately does not appear to be a current part of the U.S. government's strategy. National identity and loyalty are defined in part by "foreign" appearance, ambiguous as that may be. In some ways, the current treatment of Arabs and Muslims is more extralegal than the internment. No Executive Order authorizes the treatment of Arabs and Muslims; nor has there been a formal declaration of war. Moreover, nationality, which is more objective and easier to apply than religious and racial classifications, is not used as the exclusive basis for the measures. Rather, the scope of the investigation is broad and amorphous enough to potentially include all Arabs and Muslims, who may be natives of countries from around the world....

In the aftermath of September 11, the U.S. government arguably over-reacted and appeared to place little value on the liberty and equality interests of Arabs and Muslims. The response may be motivated in part by invidious hostility based on race and religion. With few legal constraints, the federal government adopted extreme action, with a largely symbolic impact in fighting terrorism, while having devastating impacts on Arabs and Muslims in the United States.

Moreover, the dragnet might prove to be a poor law enforcement technique. Racial profiling in criminal law enforcement has been criticized for alienating minority communities and making it more difficult to secure their much-needed cooperation in law enforcement. In a time when Arab and Muslim communities might be of assistance in investigating terrorism, they are being rounded up, humiliated, and discouraged from cooperating with law enforcement by fear of arrest, detention, and deportation.

Ultimately, such tactics suggest to noncitizens and citizens of Arab and Muslim ancestry in the United States that they are less than full members of U.S. society. The various efforts by the U.S. government, even while it claims not to discriminate against Arabs or Muslims, marginalize these communities. Consequently, the legal measures taken by the federal government reinforce deeply held negative stereotypes—foreign-ness and possibly disloyalty—about Arabs and Muslims.

DISCUSSION QUESTIONS

1. What do the recent experiences of Arabs and Muslims in the United States suggest about the social construction of race?

2. What is the impact of being racially profiled on Arabs and Muslims who are living in the United States?

13

Racism and Popular Culture

DANIELLE DIRKS AND JENNIFER C. MUELLER

The authors of this article show how popular culture has historically depicted racial hierarchies. They then show how contemporary popular culture constructs subordinated racial-ethnic groups as "other," thus perpetuating some of the same racial-ethnic stereotypes of the past, even though modified in a contemporary and global culture.

In 2002, the board game Ghettopoly was released, promising "playas" the amusement of "buying stolen properties, pimpin' hoes, building crack houses and projects, paying protection fees, and getting car jacked" (Ghettopoly, 2002). Invoking stereotypical images that implicitly implicate the cultural deficiency of African Americans, the game pieces included a pimp, a hoe, a machine gun, a 40-ounce malt liquor beverage, a marijuana leaf, a basketball, and a piece of crack rock. The game garnered significant positive attention, advertised as a great way to entertain and introduce "homies," coworkers, and children to "ghetto life." Yet, this game must be grasped beyond simple considerations of entertainment or play. Ghettopoly must be added to the wide array of popular culture productions that exist as contemporary reflections of the continual distortion and misappropriation of so-called blackness by dominant groups in the United States. In this paper, we seek to illustrate the many ways in which racist popular culture images persist today, and how their continued existence reflects a white thirst for blackness that seems unquenchable. We adopt the view that marks popular culture as pedagogical and, against the backdrop of this assumption, consider what the racial lessons are that we learn from popular culture.

The concept of race in American social life is a concept under constant contestation, giving it no single fixed meaning in defining racial boundaries, hierarchies, and images (Guerrero, 1993). Despite this fluidity, both historically and today, ideas about race have dictated notions about white superiority as much as they have about black inferiority. Although ideas about race are in their rawest forms fictions of our collective imagination, they have real and meaningful consequences—economic, psychological, and otherwise. Popular culture has had a centuries-old history of communicating racist representations of blackness in

SOURCE: Dirks, Danielle, and Jennifer C. Mueller. 2010. "Racism and Popular Culture."
Pp. 115–129 in *Handbook of the Sociology of Racial and Ethnic Relations*, edited by Hernán Vera and Joe R. Feagin. New York: Springer.

Western societies, giving it the power to distort, shape, and create reality, often blurring the lines between reality and fiction (Baudrillard, 1981, 1989; Pieterse, 1992). We argue that these productions do not exist without consequences—they permeate every aspect of our daily lives.

Popular culture has served as part of the ideological and material apparatus of social life for as long as it has existed. Most cultural theorists today disavow the polarities of popular culture as merely pure and innocent entertainment or as an uncontested instrument for executing top-down domination, adopting instead, as Kellner (1995) does, the model of media cultural texts as complex artifacts that embody social and political discourses. The power of popular culture lies in its ability to distort, shape, and produce reality, dictating the ways in which we think, feel, and operate in the social world (Kellner, 1995). And while popular culture certainly exists in many ways as a contested terrain in the sense that Kellner asserts, it has been frequently used hegemonically, as an effective peda-gogical tool of dominant classes in Western culture, supporting the lessons that keep structural inequalities safely in place (hooks, 1996).

As theorists like Kellner (1995) and Guerrero (1993) have asserted, this is the promise and predicament of popular culture. Contemporary media culture cer-tainly provides a form for the reproduction of power relations based in racism (and classism and sexism), yet its very fluidity and contestation provide some space and resource for struggle and resistance. This is the sole reason why challenging racist representations—in their various recycled and newer transformations—is crucial if we are truly, vigorously devoted to making social change a reality.

HISTORICAL BACKGROUND

In the United States, popular culture has assisted in the maintenance of a white supremacist racial hierarchy since its American inception. We provide a brief history of American popular culture's racist past to show that there is nothing creative about present-day images, ideas, or material goods manufactured by today's merchants of culture. Antiblack images are central to our historical anal-ysis because, as Guerrero (1993) has contended, "Blacks have been subordinated, marginalized, positioned, and devalued in every possible manner to glorify and relentlessly hold in place the white-dominated order and racial hierarchy of American society" (p. 2). This is certainly not to deny a long history of exploita-tion and domination for other groups in the United States, particularly among popular culture ideas and images; yet we see antiblack ideology and iconography as structurally embedded in every aspect of American social life—historically and today. In many respects, this ideology contains the racial "yardstick" by which other groups have been and continue to be measured, and elevated or devalued. We hope to show that contemporary popular cultural ideas and images are recycled products and remnants of dominant ideologies past—ideologies that exploited, distorted, and oppressed people of color historically and continue to do so today. As popular culture is constantly reinventing itself under the guise of innovation, a historical understanding of these ideas and images is crucial to deconstructing their continued existence today as simply reformations of such deeply rooted ideologies, rather than truly novel inventions.

Contemptible Collectibles

Although the sale of actual African Americans ended in 1865 with the official demise of the state-supported U.S. slavery system, the consumption of blackness through popular culture ideas, images, and material goods marked an easy, if figurative, transition in the postbellum South. From black-faced caricatures found on postcards, children's toys, and household items to 19th-century minstrelsy, these examples provide only a smattering of the racist iconography and ideology found throughout Western culture. As such, images of coons, pickaninnies, mammies, bucks, and Uncle Toms were born, to live out lives distorting the image of black Americans for centuries to come.

In the United States, popular racist stereotypes of the Jim Crow era easily became the faces of mass-produced lawn ornaments, kitchen items, postcards, and children's toys such as noisemakers, dolls, and costumes. Many of the material goods depicting black personas from this time, such as the mammy or pickaninny, have been mistakenly called "Black Americana," suggesting that these items come from the creative endeavors of black Americans themselves. However, this description is as incorrect as it is insulting, leading one author to more accurately describe them as "contemptible collectibles" (Turner, 1994). Perversely, these items have become immensely popular among collectors, with some originals of the era fetching several thousands of dollars apiece.

Manufacturers of everything from coffee, hair products, and detergents plastered the insidious iconography on virtually every type of household product available. Particularly prevalent was the image of the "coon," who, in addition to being depicted as unreliable, lazy, stupid, and child-like, was known for his "quaking," superstitious nature, making him an ideal target. Similar characterizations included the wide-eyed pickaninny and the image of the mammy (Skal, 2002). Mammy—the rotund, smiling, benevolent, uniformed black woman—is by far the most popularly disseminated contemptible collectible of all. Today she continues to happily oversee our pancakes and waffles as Aunt Jemima. For all of her popularity, no other image has been so historically identified as a fiction of white imagination than she. Social historians have pointed out that the existence of any "real" mammies in the antebellum South would have been very few and far between; her being overweight would be equally implausible given the severe rationing of food for slaves (Clinton, 1982; Turner, 1994). Yet the image of this obsequious and docile black woman has survived only to become immortalized through the mass production (and reproduction) of thousands of household and kitchen items made for "sufficiently demented homemakers" (Turner, 42)....

During early American history, popular culture reflected and supported an ideology that sought to romanticize conditions of slavery—particularly when its eradication came into focus. As people worked to dismantle the U.S. slavery system, the rise of dehumanizing images such as the contented Sambo and coon served to whitewash the depravity of plantation life and ease white consciences. These caricatures mirrored the prevailing belief that slaves were not human, therefore not deserving of full and free citizenship. Over time researchers have assigned additional functions to the continually expanding dehumanizing characterizations, suggesting, for instance, that they assuaged white male economic insecurity or created solidarity for the KKK by asserting the image of black male rapist (Gayle, 1976; Guerrero, 1993).

PRESENT-DAY REALITIES

Despite the advances made during the Civil Rights movement, we live in a post–Civil Rights era where social progress has been co-opted to help deny the existence of racism today. We view contemporary forms of racist popular culture as dangerous not only for the same reasons they were in the past, but also because we live in a slippery, self-congratulatory era where we can easily look back at popular images of the past with such disdain, that it temporarily blinds most from its subtle, yet equally egregious, forms today....

Yet racism in popular culture has not gone uncontested, and in recent years well-organized and successful protests have risen up in various forms against corporations, athletic organizations, and other purveyors of racialized popular media. However, for as many successful protests, decades-long battles continue today to end the dehumanizing portrayals of marginalized groups in the United States....

Contested Images

Corporate entities, in their push for profits, have misappropriated images of the racialized Other for as long as they have existed. Yet these images have not gone uncontested, and social organizing around these movements has been swift and well-organized, despite severe corporate foot dragging in recent decades. One example comes from Frito-Lay's 1967 introduction of the Frito Bandito—a greasy, pudgy character who would steal Anglos' Frito corn chips at gunpoint (Noriega, 2000). The company launched several commercials depicting the corporate mascot singing: "Ayiee, yie-yie-yieeee/I am dee Frito Bandito/I love Frito's Corn Chips/I love dem I do/I love Frito's Corn Chips/I take dem from you."

Chicano groups such as the National Mexican-American Anti-Defamation Committee and Involvement of Mexican-Americans in Gainful Endeavors organized and appealed to Frito-Lay on moral grounds to remove the negative image and replace it with a more positive one. In response, Frito-Lay "sanitized" the *bandito*, deciding to remove his gun and his gold tooth, making him less grimacing—an utter disregard for the moral pleas that the image was damaging to Mexican Americans. It was only after the threat of a class action anti-defamation lawsuit on behalf of the 6.1 million Mexican Americans in the United States at the time that Frito-Lay dropped the corporate mascot, after four years of immense profiteering (Carrillo, 2003; Noriega, 2000)....

Some of the most widely contested and long-standing controversies over dehumanizing and degrading images are those surrounding athletic team mascots. American Indians have been widely targeted with the naming of teams, such as the Washington Redskins, Cleveland Indians, and Atlanta Braves. Images of so-called Indianness are inaccurate and inappropriate cultural fictions of the white imagination that are disturbing on several levels. First, these images continue today despite decades-long fights over their use. Second, like blackface, they perpetuate a perverse means by which whites can "play Indian" during halftime spectacles (Deloria, 1998). Third, these images relegate Native Americans to the "mascot slot," denying them a meaningful sociopolitical identity in American public life (Strong, 2005; Trouillot, 1991). Overall, the continued existence of these racist representations—despite other images that have been resisted and retired—indicate that white America is so deeply invested in these cultural inventions, that they are unconcerned if the images bear any resemblance to reality as long as they can still "participate" in the mythologized dances, rituals, and movements they have come to love so dearly. Whites' resistance reflects an unjust sense of entitlement to "owning" these images, as well as their devotion to profit from the continued use of these racist representations.

Social movements against Native American mascot images remain some of the most visible and arguably most successful examples of American Indian activism and sociocultural resurgence, and over 1,000 mascot images have been retired as a result (King & Springwood, 2001). Much of this protest has invoked comparison among other marginalized groups, stating that groups such as the "Pittsburgh Negroes, the Kansas City Jews, and the San Diego Caucasians" would cause outrage, asking why these logos continue to exist for Native Americans (Strong, 2005, 81). Using this logic, a University of Northern Colorado intramural basketball team called themselves "The Fighting Whities," in protest of a local high school team, The Fighting Reds. In one year, they raised over $100,000 for scholarships for American Indians, selling clothing items with their name and mascot, a 1950s-style caricature of a middle-aged white man in a suit, bearing the phrase "Every thang's gonna be all white!" (Rosenberg, 2002).

Despite these successes, there is clear evidence that the critical evaluation and challenge of racist representations is more often the exception than the rule—both in real life as well as on screen....

Something Old, Something New, Something Borrowed, Something ...

Under critical historical examination, images of "blackness" found in popular culture today have shifted very little from their historical counterparts. Yet, as Patricia Hill Collins (2004) explains, "In modern America, where community institutions of all sorts have eroded, popular culture has increased in importance as a source of information and ideas" (p. 121). This is particularly problematic for black American youth, as popular culture has come to authoritatively fill the void where other institutions that could "help them navigate the challenges of social inequality" are beginning to disappear (p. 121).

Although whites have appropriated black popular culture throughout history, in recent decades it has reached new heights of global commodification—circulating problematic ideas about race, class, gender, and sexuality domestically and globally. Black women's roles in popular culture have been limited to mammies, matriarchs, jezebels, or welfare queens, yet we have seen these images being repackaged for contemporary consumption and global exportation (Collins, 2001). Contemporary hip hop portrays black women—lyrically and visually—as golddiggers and sexualized bitches who like to "get a freak on," an updated form of the jezebel (Collins, 2004).

Sexualized images of black men have also been repackaged for contemporary popular consumption as well, being touted as a way of life for many black American young men. bell hooks (2004) writes that, "Gangsta culture is the essence of patriarchal masculinity. Popular culture tells young black males that only the predator will survive" (p. 27). Today's *criminal-blackman* is not much different from the historical stereotype of bucks who are "always big, baadddd niggers, oversexed and savage, violent and frenzied as they lust for white flesh" (Bogle, 2000 [1973], 13; Russell, 2001). Currently, sexualized images of black femininity and black masculinity have become highly marketable yet remain historically rooted in an intersectional racialized sexism. "These controlling images are designed to make racism, sexism, poverty, and other forms of injustice appear to be natural, normal, and inevitable parts of everyday life" (Collins, 2001, 69). Such lessons are not only learned all too well domestically, but globally as well, with their continued popularity and exportation.

THE FUTURE OF RACISM AND POPULAR CULTURE

Pieterse (1992) tells us, "The racism that [has] developed is not an American or European one, but a Western one" (p. 9). With the global exportation of Western popular culture, it is no surprise that racist Western iconography and ideology have enjoyed immense popularity as well. The Hollywood film industry is a prime example of this problematic globalization of images, with U.S. studios controlling three-quarters of the distribution market outside the United States (Movie Revenues, 2006). When Disney's Uncle Remus tale, *Song of the South* (1946), was highly contested for its "this is how the niggers sing" jubilant portrayal of plantation life, its distribution was blocked in the United States after serious protest (Bernstein, 1996; Neupert, 2001; Schaffer, 1996; Vera & Gordon, 2003). However, the film was quickly made available for global distribution, making it the highest-grossing film in 1946 with $56.4 million in worldwide sales (World Wide Box Office, 2006)....

Global recycling of contested antiblack images and ideas has been found in numerous other examples. In 2003, the Bubble Sisters, an all-female quartet in Korea, made headlines when they used a "blackface gimmick" to gain popularity among pop music fans. Performing in black-face makeup, afros, grotesquely caricatured rubber lips, and dancing in pajamas, the group received airtime from several sources, including MTV Korea, leading to swift protests against the Bubble Sisters and their producers (Hodges, 2003). In response, Bubble Sister Seo Seung-hee explained the group "loved music by black people," and "we happened to have black makeup. With the makeup we felt good, natural, free and energized. In taking the real album cover photos, we finally decided to go for it" (KOCCA, 2003). Similar to other contemporary examples of people who have reported "accidentally" donning blackface, blackface appears to just spontaneously happen to people. After severe backlash, their manager reported, "To the 1 percent of people who were offended by this, we're really sorry ... we won't be performing with black faces" (Hodges, 2003).

In Japan, *Chikibura Sambo* (or "Little Black Sambo" in English), a children's book with a long history of controversy over its racial caricatures and stereotypes, was re-released in 2005, 17 years after Japanese booksellers agreed to pull it from shelves following a U.S.-led campaign against its racist imagery and language. Its contemporary re-release sold over 100,000 copies, making it a national bestseller in Japan....

With new technologies and the continued globalization of American popular culture, we can only imagine that these images will find their ways into more and more spaces—problematically defining blackness across the globe. As one study found with interviews of rural Taiwanese who had never traveled to the United States, they "knew" about race and black Americans in the United States from watching U.S. movies (Hsia, 1994). Like other immigrants who come to the United States, their exposure to U.S. movies undoubtedly shaped their stereotypical views and acceptance of racist ideas about black Americans. As popular culture's global audience grows, so do the lessons it provides about race and racism in the United States today. Without a critical resistance against these images, we can have no hope for racial equality in the United States or globally.

CONCLUSION

On any typical day, one could feasibly rise and dress in their Abercrombie and Fitch "Wok n Bowl" t-shirt, eat breakfast with Aunt Jemima. Get ready for lunch with, "Yo quiero Taco Bell!" Have dinner with Uncle Ben, before retiring to the television to watch the Indians, Redskins, or Braves (and don't forgot to throw down your "tomahawk chop" in [an] important moment of collective consciousness); After a leisurely game of "Ghettopoly" before heading to bed, you finally watch the late night news to get a daily dose of Arab and Muslim terrorists and criminal-blackmen bedtime stories.

As Noriega (2001) has argued, race in popular culture is in many ways a paradox—its representation has become regular in our media culture, while the profound ways it affects the real-life chances of individuals and groups remain hidden. And indeed, as Noriega notes, while popular media cannot be impli-cated as the "cause" of racism, neither does it offer a value-free medium for the exchange of ideas and information. The problem with the stranglehold popular culture has over dictating the way that the populace "knows" people of color is that for people who have very little real, interpersonal experience with indivi-duals from these groups, they can believe in an assentialist vision composed of every stereotype and myth promoted. In today's world of mass information, it is easy to see how the very ubiquity of such images makes keeping pace with them nearly impossible. As addressed above, this is the promise and predicament of popular culture. The deep need for a critical cultural studies is clear, one that seeks to understand the tools available, how they have been used in support of the dominant ideology, and how they might challenge such ideologies and offer counter-cultural solutions.

REFERENCES

Baudrillard, J. (1989). *America*. Translated by C. Turner. London: Verso.

Baudrillard, J. (1981). *Simulacra and Simulation*. Ann Arbor, MI: University of Michigan Press.

Bernstein, M. (1996). Nostalgia, ambivalence, irony: *Song of the South* and race relations in 1946 Atlanta. *Film History*, 8, 219–236.

Bogle, D. (2005 [1973]). *Toms, Coons, Mulattoes, Mammies, and Bucks: An Interpretive History of Blacks in American Films*. New York: Continuum.

Carrillo, K. J. (2003). Highly offensive: Karen Juanita Carrillo examines the ongoing currency of racist curios. Retrieved November 6, 2005 from: http://www.ferris .edu/news/jimcrow/links/newslist/offensive.htm.

Clinton, C. (1982). *The Plantation Mistress: Woman's World in the Old South*. New York: Pantheon.

Collins, P. H. (2004). *Black Sexual Politics: African Americans, Gender, and the New Racism*. New York: Routledge.

Collins, P. H. (2001). *Black Feminist Thought: Knowledge, Consciousness, and the Politics of Empowerment*. New York: Routledge.

Deloria, P. J. (1998). *Playing Indian*. New Haven, CT: Yale University Press.

Gayle, A., Jr. (1976). *The Way of the New World, the Black Novel in America*. New York: Doubleday Anchor.

Ghettopoly. (2002). *Ghettopoly*. Retrieved October 7, 2005 from: http://www .ghettopoly.com.

Guerrero, E. (1993). *Framing Blackness: The African American Image in Film*. Philadelphia, PA: Temple University Press.

Hodges. M. (2003, February 27). Bubble rap: How not to package a band. *Digital Korea Herald*. Retrieved November 6, 2005 from: http://wk.koreaherald.co.kr/SITE/ data/html_dir/2003/02/27/200302270007.asp.

hooks, b. (2004). *We Real Cool: Black Men and Masculinity*. New York: Routledge.

Hsia, Hsia-Chian. 1994. "Imported Racism and Indegenous Biases: The Impact of U.S. Media on Taiwanese Images of African Americans." Paper presented at the Annual Meetings of the American Sociological Association, Los Angeles, CA.

Kellner, D. (1995). *Media Culture: Cultural Studies, Identity, and Politics between the Modern and the Postmodern*. New York: Routledge.

King, C. R. & Springwood, C. F. (2001). *Team Spirits: The Native American Mascots Controversy*. Lincoln, NE: University of Nebraska Press.

KOCCA (Korea Culture and Content Agency). (2003, February 25). Trouble bubbles up around pop group's look. Retrieved November 6, 2005 from: http://www.kocca.or.kr/ctnews/eng/SITE/data/html_dir/2003/02/25/200302250128.html.

Movie Revenues. (2006). Global film industry. Retrieved January 7, 2006 from: http://www.factbook.net/wbglobal_rev.htm.

Nelson, G. (2002, November 14). Collection on blackface raises larger race issues. *The Phoenix Online*. Retrieved June 1, 2004 from: http://www.sccs.swarthmore.edu/org/phoenix/2002/2002-11-07/news/12410.php.

Neupert, R. (2001). Trouble in watermelon land: George Pal and the Little Jasper cartoons. *Film Quarterly*, 55, 14–26.

Noriega, C. (2000). *Shot in America: Television, the State and the Rise of Chicano Cinema*. Minneapolis: University of Minnesota Press.

Noriega, C. (2001). *Race matters, media matters*. Viewing Race Project. Retrieved January 11, 2006 from: http://www.viewingrace.org/content.php?sec=essay&sub=1.

Pieterse, J. N. (1992). *White on Black: Images of Africa and Blacks in Western Popular Culture*. New Haven, CT: Yale University Press.

Rosenberg, M. (2002, June 10). Take that, paleface! *The Nation*. Retrieved January 6, 2006 from: http://www.thenation.com/doc/20020610/editors2.

Russell, K. K. (2001). *The Color of Crime: Racial Hoaxes, White Fear, Black Protectionism, Police Harassment, and Other Macroaggressions*. New York: New York University Press.

Schaffer, S. (1996). Disney and the imagineering of histories. *Postmodern Culture*, 6, 3. Retrieved January 7, 2006, from the ProjectMuse database.

Skal, D. J. (2002). *Death Makes a Holiday: A Cultural History of Halloween*. New York: Bloomsbury.

Strong, P. T. (2004). The mascot slot: Cultural citizenship, political correctness, and pseudo-Indian sports symbols. *Journal of Sport & Social Issues*, 28, 79–87.

Trouillot, M. R. (1991). Anthropology and the savage slot: The poetics and politics of otherness. In R. G. Fox (ed.), *Recapturing Anthropology: Working in the Present*. Santa Fe, NM: School of American Research Press.

Turner, P. A. (1994). *Ceramic Uncles and Celluloid Mammies: Black Images and Their Influence on Culture*. New York: Anchor.

Vera, H. & Gordon, A. M. (2003). *Screen Saviors: Hollywood Fictions of Whiteness*. Lanham, MD: Rowman & Littlefield.

World Wide Box Office. (2006). *Song of the South* worldwide sales. Retrieved January 6, 2006 from: http://worldwideboxoffice.com.

DISCUSSION QUESTIONS

1. Think about the films you have seen recently or films you watched as a child. What racial stereotypes appear? How does this shape your understanding of popular culture and its impact on racial inequality?

2. Why do Dirks and Mueller think that popular culture has a "stranglehold" on how people think about race? Do you agree?

14

How Social Status Shapes Race

ANDREW M. PENNER AND ALIYA SAPERSTEIN

*By studying how researchers classified different people into "races," Andrew M.
Penner and Aliya Saperstein show the influence that certain stereotypes have on
how people perceive racial status. Stereotypes that Black people are criminals or
poor actually influenced whether someone would be classified as "Black" or
"White."*

Racial perceptions are fluid; how individuals perceive their own race and
... how they are perceived by others depends in part on their social position.
Using longitudinal data from a representative sample of Americans, we find that
individuals who are unemployed, incarcerated, or impoverished are more likely
to be seen and identify as black and less likely to be seen and identify as white,
regardless of how they were classified or identified previously. This is consistent
with the view that race is not a fixed individual attribute, but rather a changeable
marker of status.

Since at least the 19th century, the dominant understanding of race has been
that racial divisions are rooted in biological differences between human populations.... For the past 50 years or more, social scientists have challenged that
notion, claiming that races are instead created through social processes and subject to economic and political calculation.... However, even in disciplines where
race is viewed as socially defined, most empirical studies continue to treat race as
a fixed attribute of a particular individual.... We examine two conceptions of
race—how individuals are racially classified by others and how they identify
themselves—and find that both change over time. Further, we show that this
temporal variation is related to the individuals' social position: People who are
unemployed, incarcerated, or impoverished are more likely to be classified and
identify as black, and less likely to be classified and identify as white, regardless of
how they were classified or identified previously. This study is the first to examine changes in racial classification using a representative longitudinal sample, and
our findings suggest that race is not a fixed characteristic, but rather a flexible
marker of social status.

To examine changes over time in racial classification and self-identification,
we analyze data from the National Longitudinal Survey of Youth (NLSY),

SOURCE: Penner, Andrew M., and Aliya Saperstein. 2008. "How Social Status Shapes Race."
Proceedings of the National Academy of Sciences 105(50): 19628–19630.

which contains multiple measures of interviewer-classified and self-identified race over a twenty-year period. In each survey year between 1979 and 1998, NLSY interviewers were instructed to record their assessment of whether respondents were "White," "Black," or "Other" at the end of the interview. Respondents also self-reported their race in 2 years: In 1979 they were asked for their "origin or descent," and in 2002 they were asked whether they were of Hispanic origin and the "race or races" they considered themselves to be.

RESULTS

We begin by examining changes in racial classification, an often overlooked aspect of race that is nevertheless important because discrimination presumably rests on how people are perceived by others.... Twenty percent of the 12,686 individuals in the sample experienced at least one change in how they were racially classified by interviewers over the 19-year period. This degree of fluidity is surprising because the United States is typically characterized as having uniquely rigid racial boundaries.... Yet, the variation is clearly illustrated by the respondents' racial classification histories: If we represent being classified as white, black or other in a given year by the letters w, b, and o respectively, we see that some people are consistently classified over time ... or have only one discrepant classification ..., whereas other people vary considerably over time ... or experience a shift in their racial classification at some point.... It is possible that these changes could be the result of coding mistakes made by the interviewers; for example, where interviewers meant to record "White" but mistakenly recorded "Black." However, we find that changes in respondents' gender classification, which was also recorded by interviewers at the end of the survey, occur in just 0.27% of the cases. The much higher percentage of changes in racial classification from year-to-year (6%) suggests that the variation cannot be attributed to coding mistakes alone.

To assess whether these changes in racial classification are related to differences in social status, we focus our analysis on the likelihood of being classified by an interviewer as white ... or black.... We find that individuals who were classified as white in the previous year are less likely to be seen as white if they are currently incarcerated, unemployed, or have household incomes below the poverty line. For example, among respondents who were classified as white in the previous year ... 96% of nonincarcerated respondents are classified as white the following year, whereas only 90% of incarcerated respondents are still seen as white.

In contrast, respondents who were classified as black in the previous year are more likely to be seen as black in the current year if they are incarcerated, unemployed, or have incomes below the poverty line.... Although the differences in the likelihood of being classified as black are not as large as they are for being classified as white, all of the differences in racial classification between high- and low-status individuals are statistically significant....

… Among respondents who identified as white in 1979 …, we find 97% of never-impoverished respondents identified as white in 2002, whereas just 93% of respondents who experienced poverty (between 1979 and 2002) still identified as white. These results underscore not only that racial self-identification can be fluid …, but also that changes in identification are related to the respondents' social position.

DISCUSSION

The variation over time in racial classification and identification is at odds with the view that race is an attribute of individuals that is fixed at birth, and thus predates subsequent life outcomes, such as income or incarceration. Instead, we might think of individuals as having competing propensities for being classified into or identifying with different racial groups. Changes in these propensities likely reflect, in part, the imprecision of dividing continuous human variation into a few discrete categories …, as well as contextual variation in the location of racial divisions.… However, our findings also support the idea that racial propensities can be altered by changes in social position, much as a change in diet or stress level can alter a person's propensity to die of heart disease as opposed to cancer. This suggests that racial stereotypes can become self-fulfilling prophesies: although black Americans are overrepresented among the poor, the unemployed and the incarcerated, people who are poor, unemployed or incarcerated are also more likely to be seen and identify as black and less likely to be seen and identify as white. Thus, not only does race shape social status, but social status shapes race.…

DISCUSSION QUESTIONS

1.   This article suggests that a person's social status can influence how others perceive her or him. Explain how the authors reached this conclusion through their research method.

2.   The results of Penner and Saperstein's study were based on a national survey. How might their results also be found in other day-to-day interactions, for example, when interacting with people of color or different social class statuses? In different social settings? What does this thinking teach you about the social construction of race?

Student Exercises

1. For one week, keep a written log of every time the subject of race comes up in the conversations you hear around you. Make note of what people said and the tone in which they said it. You should also note, if possible, the age and race of the person making the comments. At the end of the week, review your log and answer the following questions:

    a. What evidence of racial prejudice did you find?

    b. Is the prejudice you observed related to racial discrimination? If so, how? If not, why?

    c. What do your observations reveal to you about the everyday reality of racism?

2. Identify a particular form of media that interests you—film, television, magazines, or books, for example—and design a research plan that will examine some aspect of the images you find of a racial-ethnic group. Narrow your topic so it won't be overly general. For example, if you choose films, pick only those nominated for the Academy Award for Best Film in a given year, or if you choose television, look only at prime-time situation comedies. Alternatively, you could examine images of women of color in top fashion magazines or watch Saturday morning children's cartoons to see how people of color are portrayed.

    Once you have narrowed your topic, design a systematic way to catalog your observations, such as counting the number of times people of color are represented in the medium you select, listing the type of characters portrayed by Asian men, or comparing the portrayal of White men and men of color in women's fashion magazines. What do your observations tell you about the representation of race in the form you chose? If you were to design your project to study such images as seen now and in the past, what might you expect to find? What impact do you think the images you found have on the beliefs of different racial-ethnic groups?

PART IV

Why Can't We Just Get Along?

Racial Identity and Interactions

ELIZABETH HIGGINBOTHAM AND
MARGARET L. ANDERSEN

This book shows how race and ethnicity are part of the social structure of society. As such, you might think they are "out there"—but they are also in us and in our relationships with other people and groups. As Peter Berger (1963) once pointed out, people live in society but society also lives in people. Similarly, in the United States people live within a system of race and ethnic relations, but race and ethnic relations also live within us. How society has organized race and ethnicity is reflected in our identities and in our relationships with others. As society becomes more racially and ethnically diverse, people's identities become more complex. At the same time, multiracial identities and the possibilities for interracial relationships increase.

Identity means the self-definition of a person or group, but it is not free-floating or devoid of society. *Identity is anchored in a social context:* We define ourselves in relationship to the social structures that surround us. Moreover, identities are multidimensional and thus include many of the social spaces we occupy. At any given time, some identities may be more salient than others—age as you grow older, gender if you are confronted with a sexist experience, race as you confront the realities of a racially stratified society, and so forth.

Racial identity is learned early in life, although those in the dominant group may take it for granted. People of color likely learn explicit lessons about racial identity early on, as parents prepare them for living in a society in which their racial status will make them vulnerable to harm. As Beverly Tatum (1997) shows, forming a positive racial identity—for both dominant and subordinate groups—means having to grapple with the realities of race. For people of color, this can mean surrounding themselves with others of their group, such as sitting together in the school cafeteria or joining an African American fraternity, even though they may then be blamed for "self-segregating" by Whites who do not understand or appreciate the support this affiliation can provide.

The formation of racial identity is especially complex when multiple races are involved. The children of biracial couples may define themselves as being of two or more races. Thus, a child born to a White parent and a Black parent may identify as Black, but appear White to others, and then identify as biracial or multiracial. As our society becomes more diverse, multiracial identities are becoming increasingly common. Racial and ethnic identities can also be complex because we have so many immigrants from nations where race and ethnicity may be constructed differently than in the United States. Such complex identities hold out the possibility that the rigid thinking about race that has prevailed for so long might break down.

The articles in this section each explore different dimensions of racial identity. We open with the views of psychologist Beverly Tatum ("Why Are the Black Kids Sitting Together?," interviewed by John O'Neil), whose work is well known for explaining the formation of racial identity. Tatum explains how racial identity emerges within the context of racial inequality. She challenges us to think about what it means when people of color, especially young people, choose to be among people like themselves. Although people of color get accused of "self-segregating" (even while White people who do the same thing are rarely so accused), Tatum shows the affirmation that such behavior provides.

It is important to remember, though, that racial identity is not just acquired by people of color. White people are generally not considered to have a racial identity, because they are not assumed to "have race" and may see themselves as the norm by which others are distinguished. Because White people are the dominant group, their identity has been considered transparent, taken for granted, and not marked as are the identities of racial and ethnic minorities.

White people actually do have a racial identity, but it is often not salient until they encounter experiences wherein that identity is brought to light. Thus, White college students may confront their own racial experiences for the first time when they interact with students of color on campus. Tim Wise ("White Like Me: Reflections on Race from a Privileged Son") shows that being "White" involves certain common experiences, even when the particular experiences of any given White person will vary because of other social factors, such as class, age, gender, and so forth. Recognizing "Whiteness" as race is not meant to essentialize race—that is, make it a fixed, unchanging, somehow "natural" phenomenon. Quite the contrary; it shows how racial identity is fluid and changing in society, but emerges in the context of racial inequality.

How we perceive ourselves and others in terms of race also shapes the relationships groups and people have with each other. Such interactions can be positive or negative, but in a segregated society, cross-race interaction of a positive sort can be rare or hard to achieve. Without regular interaction across different racial-ethnic groups, misunderstandings abound, and people—even those with good intentions—can reproduce exclusionary or hurtful practices. One example of this is what has come to be called *racial microaggression*—examined in the article by Derald Wing Sue et al. ("Racial Microaggressions in Everyday Life"). Racial microaggressions are not the same as old-fashioned, blatant hatred or bigotry, but these interactions can be just as hurtful to people of color—or people in other groups who might experience similar verbal or behavioral insensitivities, such as gendered microaggressions or homophobic microaggressions. Racial microaggressions can be verbal slights or insults, inadvertent forms of exclusion, or even just plain misunderstandings. In whatever way they are experienced, racial microaggressions reverberate in the experiences of people of color who may experience them on a daily basis, leaving them feeling injured, offended, angry, or all of the above.

Taken together, the three articles in this section illustrate the many ways that everyday behavior is patterned by racial and ethnic inequalities. Keep in mind that such patterns are not fixed and, with education and consciousness of how behaviors and attitudes shape individuals and society, we can work toward social change.

REFERENCES

Berger, Peter L. 1963. *Invitation to Sociology: A Humanist Perspective.* New York: Doubleday-Anchor.

Tatum, Beverly Daniel. 1997. *Why Are All the Black Kids Sitting Together in the Cafeteria? And Other Conversations about Race.* New York: Basic.

FACE THE FACTS: HOW WELL DO RACIAL-ETHNIC GROUPS GET ALONG?

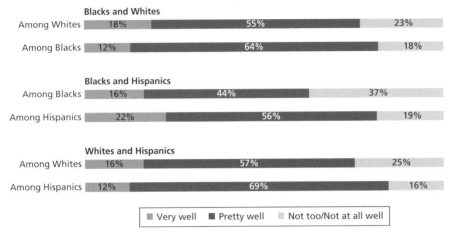

Percent saying these groups get along...

Blacks and Whites
Among Whites: 18% | 55% | 23%
Among Blacks: 12% | 64% | 18%

Blacks and Hispanics
Among Blacks: 16% | 44% | 37%
Among Hispanics: 22% | 56% | 19%

Whites and Hispanics
Among Whites: 16% | 57% | 25%
Among Hispanics: 12% | 69% | 16%

■ Very well   ■ Pretty well   ▨ Not too/Not at all well

SOURCE: Pew Research Social & Demographic Trends. 2014 (August 22, 2013). "King's Dream Remains an Elusive Goal: Many Americans See Racial Disparities." Washington, DC: Pew Research Center. www.pewsocialtrends.org

Think about It: What differences do you see in how different groups perceive how well these groups get along? Whose point of view tends to prevail in terms of shaping social policy? If you were to replicate this question by asking students from different groups on your campus, would you see similar results?

15

Why Are the Black Kids Sitting Together?

A Conversation with Beverly Daniel Tatum

BEVERLY TATUM AND JOHN O'NEIL

Beverly Tatum, well-known educational leader, tackles the often-asked question here of why Black students (and, by implication, other students of color) tend to stick together when they are in predominantly White settings, such as a high school or college campus. Rather than blaming the students, as many do, for supposed "self-segregating," she interprets this behavior as a matter of social support and racial identity formation. Her work has been very influential in antiracism education.

You call your recent book, Why Are All the Black Kids Sitting Together in the Cafeteria? *What's the significance of the title?*

The question is one I'm asked over and over again when I do a workshop on racial identity at a racially mixed school. Educators notice that kids often group themselves with others of the same race, for reasons I'll explain later. But people have other concerns as well. How do we talk to young children about race? How do we address these issues with our colleagues? How do you even engage in conversations about such hot topics as affirmative action without alienating one another?

Why do we have such problems discussing racial issues? Is it because we don't really understand one another's experiences?

I think that's part of it. It's interesting to watch people's reactions when they are really forced to experience being in the minority. One of the exercises that I ask white students and educators to engage in is to create a situation in which they will be in the minority, for a short period of time. A common choice, for example, is to attend a black church on a Sunday morning. Another is to go to a place where you know there's going to be a large Spanish-speaking population.

Usually, whites are very nervous and anxious about doing this. Some are even unwilling to do it alone, so they find a partner to go with, which is fine. But it's just interesting to me how fearful people are about this kind of

SOURCE: Tatum, Beverly, and John O'Neil. 1997. "Why Are the Black Kids Sitting Together? A Conversation with Beverly Daniel Tatum." *Educational Leadership* 55 (December): 12–15.

experience. When they come back, they often report how welcomed they felt, what a positive experience it was. But I do point out to them how worried they were about their own discomfort. And I hope that they develop a greater sense of understanding of how a person of color might feel in an environment that is predominantly white.

Some people suggest that race relations among kids are much improved, compared to our generation or our parents'. What do you see?

Young children do interact across racial lines fairly comfortably at the elementary school grade level. If you visit racially mixed schools at the elementary level, you will see kids interacting in the lunch room and on the playground. To the extent that neighborhoods are segregated, their interracial friendships might be limited. But you see much more cross-racial interaction at the elementary level than you do at the junior high or high school level.

Why?

I think the answer has to do with the child's transition into adolescence. Adolescents are searching for identity; they're asking questions like: Who am I in the world? How does the world see me? How do I see others? What will I be in the future? and Who will love me? All those questions of identity are percolating during that time period. And particularly for adolescents of color, these questions cannot be answered without also asking: Who am I ethnically? Who am I racially? and What meaning does this have for how people view me and interact with me?

Can you give an example of how these issues might emerge in students of color?

Sure. Imagine a 7-year-old black boy who everybody thinks is cute. So he's used to the world responding to him in a certain way: Look at that cute kid!

Now imagine that same kid at 15. He's six feet tall, and people don't think he's cute anymore. They think he's dangerous or a potential criminal; maybe people are now crossing the street to avoid him. So the way he sees himself perceived by others is very different at 15 than at seven. And that 15-year-old has to start figuring out what this means. Why are people crossing the street when I walk down it? Why am I being followed around by the security guards at the mall? And as that young person is trying to make sense of his experience, in all likelihood he is going to seek out and try to connect with other people who are having similar experiences.

As a result, even the young person who has grown up in a multiracial community and had a racially mixed group of friends tends to start to pull away from his non-black friends, his white friends in particular. This happens, in part, because their white friends are not having the same experiences; they're not having the same encounters with racism. And, unfortunately, many white youth don't have an understanding of how racism operates in our society, so they're not able to respond in ways that would be helpful.

The example I used was of a black boy, but a similar process unfolds among black girls or children of another race or ethnic heritage.

What about white students? What are they experiencing during this time of self-identification?

They can be confused and hurt by some of the changes. For example, it's not uncommon for a white student in my college class on the psychology of racism to say: "You know, I had a really close black or Latino friend in elementary school, and when we went to junior high she didn't want to hang around with me anymore." The student reporting the story usually is quite confused about that; it's often a very hurtful experience.

The observation I make is that, again, many white students are oblivious to the power of racism and the way that it's operating in society. And so when their friends are starting to have encounters with racism, they don't necessarily know how to respond. An example from my book is when a teacher makes a racist remark to a young black female. Afterward, her white friend comes up to her and says, "You seem upset, what's the problem?" So she explains what upset her to her friend, and the friend says, quite innocently, "Gee, Mr. Smith is such a nice guy, I can't imagine he would intentionally hurt you. He's not a racist, you know."

So white students might discount it because they can't identify with it?

Exactly. They can't identify, and also in many situations people who try to comfort often end up invalidating the person's feelings, by saying things like, "Oh, come on, it wasn't that bad." What happens is that you withdraw from the conversation. The feeling is "Well, you don't get it, so I'll find somebody who does." It would have been a very different response, however, if this young white student understood stereotypes and the reality of racism in society and told her friend, "You know, that was a really offensive thing he said."

Should teachers or principals be concerned when students self segregate? Should they actively seek to integrate the groups?

During "downtime" like lunch or recess, students should be able to relax with their friends, regardless of whether or not those friendship groups are of the same race or ethnic groups. However, it is important to create opportunities for young people to have positive interactions across group lines in school. So structuring racially mixed work groups—for example, by using cooperative learning strategies in the classroom—can be a very positive thing to do.

Similarly, intentionally working to recruit diverse members of the student body to participate in extracurricular events is worthwhile. We need to take advantage of every opportunity we have to bring young people together where they can work cooperatively as equals toward a common goal. Sports teams are a good example of the kind of mutually cooperative environment where young people often develop strong connections across racial lines, and we should look to create more such opportunities in schools. Unfortunately, school policies like tracking (which tends to sort kids along racial lines) impede rather than facilitate such opportunities.

You've talked with students of color who attend integrated schools but find themselves isolated in honors or advanced classes. What are those students experiencing in terms of their identity?

Even in racially mixed schools, it is very common for young people of color, particularly black and Latino students, to find that the upper-level courses have very few students of color in them. And, of course, honors chemistry is only offered during a certain time period in the day, which means you might also be taking English and other courses with the same kids. And so those students in honors chemistry or advanced algebra may find that their black or Latino peers accuse them of "trying to be white" because they're hanging around with all white kids. So to the extent that you're frequently in the company of white students, and your black classmates who are in the lower tracks see you as somehow separating yourself from them—it's a hard place to be.

What can educators do to support the healthy development of kids as they work through these issues?

I think students of color really need to see themselves reflected in positive ways in the curriculum. And that probably sounds very obvious, but the fact is that too often they don't see themselves reflected in the curriculum.

When and how they see themselves reflected in the curriculum is so important, though. To use African Americans as an example: Most schools teach about slavery, and for many black students that's a point of real discomfort. Their experience of that is that the teacher's talking about slavery and all the white kids in the class are looking at us, to see what our reaction is. I'm certainly not suggesting we shouldn't teach about slavery, but I think it's important to teach it in an empowering way. Teachers need to focus on resistance to victimization. Students of color need to see themselves represented not just as victims but as agents of their own empowerment. And there are lots of ways to do that. You can talk about Sojourner Truth, you can talk about Harriet Tubman, Frederick Douglass, and so on.

At the same time, I think white children need to be helped to understand how racism operates. Inevitably, when you talk about racism in a predominantly white society, you generate feelings of discomfort and often guilt among white people because they might feel that you're saying that white people are bad. What do we do? In these discussions, we need to include examples of white people as agents of change. Teach students about the abolitionists. Teach students about Virginia Foster Durr, who was so active during the Civil Rights movement. All children need to learn about those white folks who worked against oppression. Unfortunately, many white students don't have that information.

Some people have suggested that the school curriculum be heavily focused on cultural heritage; that black students need an Afrocentric curriculum, and so on. What's your perspective?

It's important to have as diverse a curriculum as possible because all students need to be able to view things through multiple perspectives.

A high school teacher told me recently that the young white men in her English classes were reluctant to read about somebody's experience other than their own. For example, she had the class read *House on Mango Street* (Cisneros 1994), a book about a young Chicana adolescent coming of age in Chicago. These young men were complaining: "What does this have to do with me? I can't identify with this experience." But, at the same time, they never wonder why the Latino students in the class have to read Ernest Hemingway. We need to help them develop that understanding. All of us need to develop a sense of multiple perspectives, regardless of the composition of our classrooms.

The teaching ranks are predominantly white, even though the student population is becoming increasingly diverse. What does this mean for efforts to increase racial understanding?

It makes it harder, but it's not impossible. We should be working very hard to increase the diversity of the teaching pool, and many teacher education programs are trying to do that. Still, we need to recognize that it's going to be a long time before the teaching population reflects the classroom population. So it's really important for white teachers to recognize that it is possible for them to become culturally sensitive and to be proactive in an antiracist way. Many white educators have grown up in predominantly white communities, attended mostly white schools, and may have had limited experiences with people of color, and that is a potential barrier. But what that means is that people need to expand their experiences.

Dreamkeepers: Successful Teachers of African-American Children, by Gloria Ladson-Billings (1994) is a great resource. I often encourage educators I work with to read that because she profiles several teachers, some of them white. Those are teachers who probably didn't grow up in neighborhoods or communities where they had a lot of interaction with people of color, but, one way or another, they have really been able to establish great teaching relationships with kids of color. So it certainly can be done.

You train teachers to work on issues involving race in their schools. What kinds of things do they learn?

For a number of years, I've taught a professional development course called *Effective Anti-Racist Classroom Practices for All Students*. It's basically designed to help teachers recognize what racism is, how it operates in schools, and what the impact of that is personally and professionally. So the focus is not just the impact of racism on the racial identity development of the students but also on the teachers. I've found that teachers who have not reflected on their own racial identity development find it very difficult to understand why young people are reflecting on theirs. So it's important to engage in self-reflection even as we're trying to better understand our students.

The course also looks at stereotypes, omissions, distortions, how those have been communicated in our culture and in our classroom, and then, what that means in terms of how we think about ourselves either as people of color or as whites. And, finally, what we can do about it. I talk in my book about racism as a sort of "smog." People who aren't aware of it can unwittingly perpetuate a cycle of oppression. If you breathe that smog too long, you internalize these messages. We can't really interrupt that process until it becomes visible to us. That's the first step—making the process visible. And once it is visible, we can start to strategize about how we're going to interrupt it.

Many teachers have been caught short by a racist incident or comment in their classroom. It often happens suddenly, and the teacher may be at a loss for how to respond. How have you handled it?

Well, I've been teaching a course on the psychology of racism since 1980, so I feel like I've probably heard it all.

It is a difficult situation, because you want the classroom to be a safe place, where students can say anything, knowing that only by opening up will they get feedback about their comments and learn another perspective. At the same time, you want the classroom to be a safe place for someone who may be victimized by a comment.

One time, a white student in my class made a very offensive remark about Puerto Ricans being responsible for crimes. Well, one of the things I've learned is that it really helps to validate somebody's comment initially, even if it is outrageous. So I said something like: "You know, I'm sure there are many people who feel that way, and if you've been victimized by a crime, that's a very difficult experience. At the same time, I think it's important to say that not all Puerto Ricans are car thieves." From there, you can move into how making such statements can reinforce stereotypes.

It must be hard to make it a teachable moment for everyone in the class.

Absolutely. One time I was observing somebody else's teaching when there was a similar kind of incident—a student made comments that the teacher thought were inappropriate, but she didn't know how to respond. So she didn't respond to them. After the class, we talked, and she said she felt terrible—she knew she should have done something, but she didn't know what to do. And we talked about what the choices might have been.

Even though she felt badly about how she handled it, those moments can be revisited. So in this particular case, the teacher opened her next class by saying: "You know, in our last class something happened that really bothered me, and I didn't say anything. I didn't say anything because I wasn't sure what to say, but in my silence I colluded with what was being said. So I would really like to talk about it now." And she brought the class back to the incident, and it was not an easy conversation. But I think it really deepened the students' understanding—both of the teacher herself and of how racism operates, because it showed how even well-intentioned people may unwittingly contribute to perpetuating the problem.

Although integrated schools have been a goal for decades, current statistics show a growth in schools that are nearly all-black or all-Hispanic. What do you see as the likely impact of the trend toward even more racially identifiable schools?

It's a very difficult issue from a number of perspectives. The continuing pattern of white flight is one of the main reasons that schools resegregate. A lot of money is put into a busing plan, and then white families leave the school. So now many people are asking questions about whether it's a good idea to spend all that money transporting kids instead of just using it to improve the neighborhood school, regardless of who attends it.

Many parents of color experience this as a double-edged sword. They're offended by the notion that children of color can only learn when they are in classes with white kids. They know there is nothing magical about sitting next to a white kid in class. On the other hand, the reality of school funding is that schools with more white students receive more financial support.

And so the question that I hear people asking now is: Can separate ever be equal? That's one I don't have the answer to!

… Real progress is being made in starting conversations at the local community level. Many grass-roots organizations are encouraging this kind of dialogue.

For example, an organization in Connecticut has a program of Study Circles. They actually have a guidebook for facilitating conversations about race. Using the guides, people come together and begin to discuss the questions together to improve their understanding. Also, many houses of worship encourage cross-group dialogue, whether it means interfaith dialogue or cross-racial dialogue.

It's sometimes frustrating for people who have been doing this work for years, because it may seem like there's talk, talk, talk and it doesn't go anywhere. However, I do think that when you engage in open and honest dialogue, you start to recognize the other person's point of view, and that helps you see where your action might be needed most. So if people engage in dialogue with the understanding that dialogue is supposed to lead somewhere, it can be a very useful thing to do.

We can't afford to forget the institutional nature of racism. And so it's not just about personal prejudices, though obviously we want to examine those. We can't just aspire to be prejudice-free. We need to examine how racism persists in our institutions so we don't perpetuate it.

REFERENCES

Cisneros, S. (1994). *House on Mango Street*. New York: Random House.

Ladson-Billings, G. (1994). *Dreamkeepers: Successful Teachers of African-American Children*. San Francisco, Calif.: Jossey-Bass.

DISCUSSION QUESTIONS

1. How does Tatum's analysis link the everyday behavior of minority students to the social structures in which they live?

2. What lessons are there in Tatum's discussion for reducing racial prejudice?

3. What are the implications of Tatum's argument for educating teachers about race?

16

White Like Me

Reflections on Race from a Privileged Son

TIM WISE

Tim Wise, like many White scholars, is questioning his racial identity and the privileges that accompany this status. He provides a way for dominant groups to think about how they perceive themselves and the system of racial inequality.

It's a question no one likes to hear, seeing as how it typically portends an assumption on the part of the questioner that something is terribly wrong, something that defies logic and calls for an explanation.

It's the kind of query one might get from former classmates on the occasion of one's twenty-year high school reunion: "Dear God, what the hell happened to you?" Generally, people don't ask this question of those whom they consider to have dramatically improved themselves in some way, be it physical, emotional, or professional. Instead, it is more often asked of those considered to be seriously damaged, as if the only possible answer would be, "Well, I was dropped on my head as a baby," to which the questioner would then reply, "Aha, I see."

So whenever I'm asked this, I naturally recoil for a moment, assuming that the persons inquiring "what happened" likely want an answer only in order to avoid, at whatever cost, having it (whatever "it" may be) happen to them. In my case, however, I'm usually lucky. Most of the persons who ask me "what happened" seem to be asking less for reasons of passing judgment than for reasons of confusion. They appear truly perplexed about how I turned out the way I did, especially when it comes to my views on the matter of race.

As a white man, born and reared in a society that has always bestowed upon me advantages that it has just as deliberately withheld from people of color, I am not expected to think the way I do. I am not supposed to speak against and agitate in opposition to racism and institutionalized white supremacy. Indeed, for people of color, it is often shocking to see white people even thinking about race, let alone challenging racism. After all, we don't have to spend much time contemplating the subject if we'd rather not, and historically white folks have made something of a pastime out of ignoring racism, or at least refusing to call it out as a social problem to be remedied.

SOURCE: Wise, Tim. 2008. *White Like Me*. Brooklyn, NY: Soft Skull Press.

But for me, and for the white folks whom I admire in history, ignoring race and racism has never been an option. Even when it would have been easier to turn away, there were too many forces, to say nothing of circumstances, pulling me back, compelling me to look at the matter, square in the face—in *my* face, truth be told.

Although white Americans often think we've had few first-hand experiences with race, because most of us are so isolated from people of color in our day-to-day lives, the reality is that this isolation *is* our experience with race. We are all experiencing race, because from the beginning of our lives we have been living in a racialized society, where the color of our skin means something socially, even while it remains largely a matter of biological and genetic irrelevance. Race may be a scientific fiction—and given the almost complete genetic overlap between persons of the various so-called races, it appears to be just that—but it is a social fact that none of us can escape no matter how much or how little we may speak of it. Just as there were no actual witches in Salem in 1692, and yet anti-witch persecution was frighteningly real, so too race can be a falsehood even as racism continues to destroy lives, to maim, to kill, and, on the flipside, to advantage those who are rarely its targets.

A few words about terminology: When I speak of "whites" of "white folks," I am referring to those persons, typically of European descent, who are able, by virtue of skin color or perhaps national origin and culture, to be perceived as "white," as members of the dominant group. I do not consider the white race to be a real thing, in biological terms, as modern science pretty well establishes that there are no truly distinct races, genetically speaking, within the human species. But the white race certainly has meaning in social terms. And it is in that social sense that I use the concept here.

As it turns out, this last point is more important than you might think. Almost immediately upon publication, [my book's] first edition came under fire from various white supremacists and neo-Nazis, who launched a fairly concerted effort to discredit it, and me as its author. They sought to do this by jamming the review boards at Amazon.com with harsh critiques, none of which discussed the content—in all likelihood none of them had actually read the book—but which amounted, instead, to ad hominem attacks against me as a Jew. As several explained, being Jewish disqualifies me from being white, or writing about my experience as a white person, since Jews are, to them, a distinct race of evildoers that seeks to eradicate Aryan stock from the face of the earth.

On the one hand (and ignoring for a second the Hitler-friendly histrionics) of course, it is absurd to think that uniquely "Jewish genes" render Jews separate from "real" whites, despite our recent European ancestry. And it's even more ridiculous to think that such genes from one-fourth of one's family, as with mine, on my paternal grandfather's line, can cancel out the three quarters Anglo-Celtic contribution made by the rest of my ancestors. But in truth, the argument is completely irrelevant, given how I am using the concept of whiteness here. Even if there were something biologically distinct about Jews, this would hardly alter the fact that most Jews, especially in the United States, are sufficiently light skinned and assimilated so as to be fully functional as whites in the eyes of authority. This wasn't always the case but it is inarguably such now. American Jews are, by and large, able to reap the benefits of whiteness and white racial privilege, vis-à-vis people of color, in spite of our Jewishness, whether viewed in racial or cultural terms. My "claiming to be white," as one detractor put it, was not an attempt on my part to join the cool kids. I wasn't trying to fool anyone.

Whiteness is more about how you're likely to be viewed and treated in a white supremacist society than it is about what you *are*, in any meaningful sense. This is why even some very light-skinned folks of color have been able to access white privilege over the years by passing as white or being misperceived as white, much to their benefit. Whiteness is, however much clichéd the saying may be, largely a social construct....

... As for the concept of privilege, here, too, clarification is in order. I am not claiming, nor do I believe, that all whites are wealthy and powerful. We live not only in a racialized society, but also in a class system, a patriarchal system, and one of straight supremacy/heterosexism, able-bodied supremacy, and Christian hegemony. These other forms of privilege, and the oppression experienced by those who can't manage to access them, mediate, but never fully eradicate, something like white privilege. So I realize that, socially rich whites are more powerful than poor ones, white men are more powerful than white women, able-bodied whites are more powerful than those with disabilities, and straight whites are more powerful than gay, lesbian, bisexual or transgendered whites.

But despite the fact that white privilege plays out differently for different folks, depending on these other identities, the fact remains that when all other factors are equal, whiteness matters and carries great advantage. So, for example, although whites are often poor, their poverty does not alter the fact that, relative to poor and working-class persons of color, they typically have a leg up. In fact, studies suggest that working-class whites are typically better off in terms of assets and net worth than even middle-class blacks with far higher incomes, due to past familial advantages. No one privilege system trumps all others every time, but no matter the ways in which individual whites may face obstacles on the basis of nonracial factors, our race continues to elevate us over similarly situated persons of color.

The notion of privilege is a relative concept as well as an absolute one, a point that is often misunderstood. This is why I can refer to myself as a "privileged son," despite coming from a family that was not even close to wealthy. In relative terms, compared to persons of color, whites receive certain head starts and advantages, none of which are canceled out because of factors like class, gender, or sexual orientation. Likewise, heterosexuals receive privileges relative to LGBT folks, none of which are canceled out by the poverty that many straight people experience. So too, rich folks have certain privileges on the basis of their wealth, none of which vanish like mist just because some of those wealthy persons are disabled.

While few of us are located only in privileged groups, and even fewer are located only in marginalized or oppressed groups—we are all privileged in some ways and targets in others—the fact remains that our status as occasional targets does not obviate the need for us to address the ways in which we receive unjust advantages at the expense of others....

The first [lesson] is that to be white is to be "born to belonging." This is a term I first heard used by my friend and longtime antiracist white ally Mab Segrest, though she was using it in a different context. To be white is to be born into an environment where one's legitimacy is far less likely to be questioned than would be the legitimacy of a person of color, be it in terms of where one lives, where one works, or where one goes to school. To be white is, even more, to be born into a system that has been set up for the benefit of people like you (like us), and as such provides a head start to those who can claim membership in this, the dominant club.

Second, to be white means only that one will typically inherit certain advantages from the past, but also that one will continue to reap the benefits of ongoing racial privilege, which is itself the flipside of discrimination against persons of color. These privileges have both material components, such as better job opportunities, better schooling, and better housing availability, as well as psychological components, not the least of which is simply having one less thing to constantly worry about during the course of a day. To be white is to be free of the daily burden of constantly having to disprove negative stereotypes. It is to have one less thing to sweat, and in a competitive society such as ours, one less thing on your mind is no small boost.

Third …, in the face of these privileges, whether derived from past injustice handed down or present injustice still actively practiced, to be white is typically to be in profound denial about the existence of these advantages and their consequences. I say denial here, rather than ignorance, because the term *ignorance* implies an involuntary lack of knowledge, a purity, an innocence of sorts, that lets white Americans off the hook, even if only linguistically. The fact is, whites' refusal to engage the issues of race and privilege is due largely to a *willed* ignorance, a voluntary evasion of reality, not unlike the alcoholic or drug addict who refuses to face their illness. How else but as the result of willed ignorance can we understand polls taken, not today, but in the early sixties, which demonstrated that even then—at a time of blatant racism and legally accepted discrimination against black people— the vast majority of whites believed that everyone had equal opportunity?

The only way that one can be completely ignorant of the racial truth in the United States, whether in the sixties or today, is to make the deliberate choice to think about something else, to turn away, to close one's ears, shut one's eyes, and bury one's head in the proverbial sand.

Oh sure, there are young people, perhaps of high school age or even younger, who might truly be ignorant in the strictest sense of the word when it comes to issues of racism and privilege. But if so, this is only because their teachers, preachers, parents, and the mass media to which they are daily subjected have made the choice to lie to them, either directly or by omission. So even the ignorance of the young is willed, albeit by their elders, much to their own detriment.

Fourth, whites can choose to resist a system of racism and unjust privilege, but doing so is never easy. In fact, the fear of alienating friends and family, and the relative lack of role models from whom we can take direction, renders resistance rare, and even when practiced, often ineffective, however important it may be.

Learning how to develop our resistance muscles is of vital importance. Thinking about how and when to resist, and what to do (and not do) is critical.

Fifth, even when committed to resistance, and even while in the midst of practicing it, we sometimes collaborate with racism and reinforce racial domination and subordination. In other words, we must always be on guard against our own screwups and willing to confront our failings honestly.

Sixth, whites pay enormous costs in order to have access to the privileges that come from a system of racism—costs that are intensely personal and collective, and which should inspire us to fight racism *for our own sake.*

And finally, in the struggle against injustice, against racism, there is the possibility of redemption.

Belonging, privilege, denial, resistance, collaboration, loss, and redemption: the themes that define and delineate various aspects of the white experience. The trick is getting from privilege, collaboration, denial, and loss to resistance and redemption, so that we may begin to belong to a society more just and sustainable than what we have now.

DISCUSSION QUESTIONS

1. What does it mean to say that Whites receive the benefits of White privilege, even when they differ on the basis of social class or other social facts?

2. What costs does Wise identify as occurring for White people because of racism and unjust privilege?

17

Racial Microaggressions in Everyday Life

DERALD WING SUE, CHRISTINA M. CAPODILUPO, GINA C. TORINO, JENNIFER M.
BUCCERI, AISHA M. B. HOLDER, KEVIN L. NADAL, AND MARTA ESQUILIN

*This team of psychotherapists defines the concept of racial microaggressions—
behavioral practices that, even when subtle or covert, communicate the status of
"lesser" to people of color. This article examines the multiple ways that such
behaviors harm, insult, and invalidate people.*

... FORMS OF RACIAL MICROAGGRESSIONS

Racial microaggressions are brief and commonplace daily verbal, behavioral, and
environmental indignities, whether intentional or unintentional, that communicate hostile, derogatory, or negative racial slights and insults to the target
person or group. They are not limited to human encounters alone but may also
be environmental in nature, as when a person of color is exposed to an office
setting that unintentionally assails his or her racial identity (Gordon & Johnson,
2003; D. W. Sue, 2003). For example, one's racial identity can be minimized or
made insignificant through the sheer exclusion of decorations or literature that
represents various racial groups. Three forms of microaggressions can be identified: microassault, microinsult, and microinvalidation.

Microassault

A microassault is an explicit racial derogation characterized primarily by a verbal
or nonverbal attack meant to hurt the intended victim through name-calling,
avoidant behavior, or purposeful discriminatory actions. Referring to someone
as "colored" or "Oriental," using racial epithets, discouraging interracial interactions, deliberately serving a White patron before someone of color, and displaying a swastika are examples. Microassaults are most similar to what has been
called "old fashioned" racism conducted on an individual level. They are most
likely to be conscious and deliberate, although they are generally expressed in
limited "private" situations (micro) that allow the perpetrator some degree of

SOURCE: Wing Sue, Derald, Christina M. Capodilupo, Gina C. Torino, Jennifer M. Bucceri,
Aisha M. B. Holder, Kevin L. Nadal, and Marta Esquilin. 2007. "Racial Microaggressions in
Everyday Life." *American Psychologist* 62 (May–June): 271–286.

anonymity. In other words, people are likely to hold notions of minority inferiority privately and will only display them publicly when they (a) lose control or (b) feel relatively safe to engage in a microassault. Because we have chosen to analyze the unintentional and unconscious manifestations of microaggressions, microassaults are not the focus of our article. It is important to note, however, that individuals can also vary in the degree of conscious awareness they show in the use of the following two forms of microaggressions.

Microinsult

A microinsult is characterized by communications that convey rudeness and insensitivity and demean a person's racial heritage or identity. Microinsults represent subtle snubs, frequently unknown to the perpetrator, but clearly convey a hidden insulting message to the recipient of color. When a White employer tells a prospective candidate of color "I believe the most qualified person should get the job, regardless of race" or when an employee of color is asked "How did you get your job?", the underlying message from the perspective of the recipient may be twofold: (a) People of color are not qualified, and (b) as a minority group member, you must have obtained the position through some affirmative action or quota program and not because of ability. Such statements are not necessarily aggressions, but context is important. Hearing these statements frequently when used against affirmative action makes the recipient likely to experience them as aggressions. Microinsults can also occur nonverbally, as when a White teacher fails to acknowledge students of color in the classroom or when a White supervisor seems distracted during a conversation with a Black employee by avoiding eye contact or turning away (Hinton, 2004). In this case, the message conveyed to persons of color is that their contributions are unimportant.

Microinvalidation

Microinvalidations are characterized by communications that exclude, negate, or nullify the psychological thoughts, feelings, or experiential reality of a person of color. When Asian Americans (born and raised in the United States) are complimented for speaking good English or are repeatedly asked where they were born, the effect is to negate their U.S. American heritage and to convey that they are perpetual foreigners. When Blacks are told that "I don't see color" or "We are all human beings," the effect is to negate their experiences as racial/cultural beings (Helms, 1992). When a Latino couple is given poor service at a restaurant and shares their experience with White friends, only to be told "Don't be so oversensitive" or "Don't be so petty," the racial experience of the couple is being nullified and its importance is being diminished.

We have been able to identify nine categories of microaggressions with distinct themes: alien in one's own land, ascription of intelligence, color blindness, criminality/assumption of criminal status, denial of individual racism, myth of meritocracy, pathologizing cultural values/communication styles, second-class status, and environmental invalidation. Table 1 provides samples of comments

T A B L E 1   **Examples of Racial Microaggressions**

Theme	Microaggression	Message
Alien in own land	"Where are you from?"	You are not American.
When Asian Americans and Latino Americans are assumed to be foreign-born	"Where were you born?"	
	"You speak good English."	
	A person asking on Asian American to teach them words in their native language	You are a foreigner.
Ascription of intelligence	"You are a credit to your race."	People of color are generally not as intelligent as Whites.
Assigning intelligence to a person of color on the basis of their race	"You are so articulate."	It is unusual for someone of your race to be intelligent.
	Asking an Asian person to help with a math or science problem	All Asians are intelligent and good in math/sciences.
Color blindness	"When I look at you, I don't see color."	Denying a person of color's racial/ethnic experiences.
Statements that indicate that a White person does not want to acknowledge race	"America is a melting pot."	Assimilate/acculturate to the dominant culture.
	"There is only one race, the human race."	Denying the individual as a racial/cultural being.
Criminality/assumption of criminal status	A White man or woman clutching their purse or checking their wallet as a Black or Latino approaches or passes	You are a criminal.
A person of color is presumed to be dangerous, criminal, or deviant on the basis of their race	A store owner following a customer of color around the store	You are going to steal/ You are poor/ You do not belong.
	A White person waits to ride the next elevator when a person of color is on it	You are dangerous.
Denial of individual racism	"I'm not racist. I have several Black friends."	I am immune to racism because I have friends of color.
A statement made when Whites deny their racial biases	"As a woman, I know what you go through as a racial minority."	Your racial oppression is no different than my gender oppression. I can't be a racist. I'm like you.
Myth of meritocracy	"I believe the most qualified person should get the job."	People of color are given extra unfair benefits because of their race.
Statements which assert that race does not play a role in life successes		

(Continues)

Theme	Microaggression	Message
	"Everyone can succeed in this society, if they work hard enough."	People of color are lazy and/or incompetent and need to work harder.
Pathologizing cultural values/communication styles The notion that the values and communication styles of the dominant/White culture are ideal	Asking a Black person: "Why do you have to be so loud/animated? Just calm down."	Assimilate to dominant culture.
	To an Asian or Latino person: "Why are you so quiet? We want to know what you think. Be more verbal." "Speak up more."	
	Dismissing an individual who brings up race/culture in work/school setting	Leave your cultural baggage outside.
Second-class citizen Occurs when a White person is given preferential treatment as a consumer over a person of color	Person of color mistaken for a service worker	People of color are servants to Whites. They couldn't possibly occupy high-status positions.
	Having a taxi cab pass a person of color and pick up a White passenger	You are likely to cause trouble and/or travel to a dangerous neighborhood.
	Being ignored at a store counter as attention is given to the White customer behind you"	Whites are more valued customers than people of color.
	You people … "	You don't belong. You are a lesser being.
Environmental microaggressions Macro-level microaggressions, which are more apparent on systemic and environmental levels	A college or university with buildings that are all named after White heterosexual upper class males	You don't belong/You won't succeed here. There is only so far you can go.
	Television shows and movies that feature predominantly White people, without representation of people of color	You are an outsider/You don't exist.
	Overcrowding of public schools in communities of color	People of color don't/shouldn't value education.
	Overabundance of liquor stores in communities of color	People of color are deviant.

or situations that may potentially be classified as racial microaggressions and their accompanying hidden assumptions and messages....

The experience of a racial microaggression has major implications for both the perpetrator and the target person. It creates psychological dilemmas that unless adequately resolved lead to increased levels of racial anger, mistrust, and loss of self-esteem for persons of color; prevent White people from perceiving a different racial reality; and create impediments to harmonious race-relations (Spanierman & Heppner, 2004; Thompson & Neville, 1999).

THE INVISIBILITY AND DYNAMICS OF RACIAL MICROAGGRESSIONS

The following real-life incident illustrates the issues of invisibility and the disguised problematic dynamics of racial microaggressions.

> I [Derald Wing Sue, the senior author, an Asian American] recently traveled with an African American colleague on a plane flying from New York to Boston. The plane was a small "hopper" with a single row of seats on one side and double seats on the other. As the plane was only sparsely populated, we were told by the flight attendant (White) that we could sit anywhere, so we sat at the front, across the aisle from one another. This made it easy for us to converse and provided a larger comfortable space on a small plane for both of us. As the attendant was about to close the hatch, three White men in suits entered the plane, were informed they could sit anywhere, and promptly seated themselves in front of us. Just before take-off, the attendant proceeded to close all overhead compartments and seemed to scan the plane with her eyes. At that point she approached us, leaned over, interrupted our conversation, and asked if we would mind moving to the back of the plane. She indicated that she needed to distribute weight on the plane evenly.

> Both of us (passengers of color) had similar negative reactions. First, balancing the weight on the plane seemed reasonable, but why were we being singled out? After all, we had boarded first and the three White men were the last passengers to arrive. Why were they not being asked to move? Were we being singled out because of our race? Was this just a random event with no racial overtones? Were we being oversensitive and petty?

> Although we complied by moving to the back of the plane, both of us felt resentment, irritation, and anger. In light of our everyday racial experiences, we both came to the same conclusion: The flight attendant had treated us like second-class citizens because of our race. But this incident did not end there. While I kept telling myself to drop the matter, I could feel my blood pressure rising, heart beating faster, and face flush with anger. When the attendant walked back to make sure

our seat belts were fastened, I could not contain my anger any longer. Struggling to control myself, I said to her in a forced calm voice: "Did you know that you asked two passengers of color to step to the rear of the 'bus'"? For a few seconds she said nothing but looked at me with a horrified expression. Then she said in a righteously indignant tone, "Well, I have never been accused of that! How dare you? I don't see color! I only asked you to move to balance the plane. Anyway, I was only trying to give you more space and greater privacy."

Attempts to explain my perceptions and feelings only generated greater defensiveness from her. For every allegation I made, she seemed to have a rational reason for her actions. Finally, she broke off the conversation and refused to talk about the incident any longer. Were it not for my colleague who validated my experiential reality, I would have left that encounter wondering whether I was correct or incorrect in my perceptions. Nevertheless, for the rest of the flight, I stewed over the incident and it left a sour taste in my mouth.

The power of racial microaggressions lies in their invisibility to the perpetrator and, oftentimes, the recipient (D. W. Sue, 2005). Most White Americans experience themselves as good, moral, and decent human beings who believe in equality and democracy. Thus, they find it difficult to believe that they possess biased racial attitudes and may engage in behaviors that are discriminatory (D. W. Sue, 2004). Microaggressive acts can usually be explained away by seemingly nonbiased and valid reasons. For the recipient of a microaggression, however, there is always the nagging question of whether it really happened (Crocker & Major, 1989). It is difficult to identify a microaggression, especially when other explanations seem plausible. Many people of color describe a vague feeling that they have been attacked, that they have been disrespected, or that something is not right (Franklin, 2004; Reid & Radhakrishnan, 2003). In some respects, people of color may find an overt and obvious racist act easier to handle than microaggressions that seem vague or disguised (Solórzano et al., 2000). The above incident reveals how microaggressions operate to create psychological dilemmas for both the White perpetrator and the person of color. Four such dilemmas are particularly noteworthy for everyone to understand.

Dilemma 1: Clash of Racial Realities

The question we pose is this: Did the flight attendant engage in a microaggression or did the senior author and his colleague simply misinterpret the action? Studies indicate that the racial perception of people of color differ markedly from those of Whites (Jones, 1997; Harris Poll commissioned by the National Conference of Christians and Jews, 1992). In most cases, White Americans tend to believe that minorities are doing better in life, that discrimination is on the decline, that racism is no longer a significant factor in the lives of people of color, and that equality has been achieved. More important, the majority of Whites do not view themselves as racist or capable of racist behavior.

Minorities, on the other hand, perceive Whites as (a) racially insensitive, (b) unwilling to share their position and wealth, (c) believing they are superior, (d) needing to control everything, and (e) treating them poorly because of their race. People of color believe these attributes are reenacted everyday in their interpersonal interactions with Whites, oftentimes in the form of microaggressions (Solórzano et al., 2000). For example, it was found that 96% of African Americans reported experiencing racial discrimination in a one-year period (Klonoff & Landrine, 1999), and many incidents involved being mistaken for a service worker, being ignored, given poor service, treated rudely, or experiencing strangers acting fearful or intimidated when around them (Sellers & Shelton, 2003).

Dilemma 2: The Invisibility of Unintentional Expressions of Bias

The interaction between the senior author and the flight attendant convinced him that she was sincere in her belief that she had acted in good faith without racial bias. Her actions and their meaning were invisible to her. It was clear that she was stunned that anyone would accuse her of such despicable actions. After all, in her mind, she acted with only the best of intentions: to distribute the weight evenly on the plane for safety reasons and to give two passengers greater privacy and space. She felt betrayed that her good intentions were being questioned. Yet considerable empirical evidence exists showing that racial microaggressions become automatic because of cultural conditioning and that they may become connected neurologically with the processing of emotions that surround prejudice (Abelson et al., 1998). Several investigators have found, for example, that law enforcement officers in laboratory experiments will fire their guns more often at Black criminal suspects than White ones (Plant & Peruche, 2005), and Afrocentric features tend to result in longer prison terms (Blair, Judd, & Chapleau, 2004). In all cases, these law enforcement officials had no conscious awareness that they responded differently on the basis of race.

Herein lies a major dilemma. How does one prove that a microaggression has occurred? What makes our belief that the flight attendant acted in a biased manner any more plausible than her conscious belief that it was generated for another reason? If she did act out of hidden and unconscious bias, how do we make her aware of it? Social psychological research tends to confirm the existence of unconscious racial biases in well-intentioned Whites, that nearly everyone born and raised in the United States inherits the racial biases of the society, and that the most accurate assessment about whether racist acts have occurred in a particular situation is most likely to be made by those most disempowered rather than by those who enjoy the privileges of power (Jones, 1997; Keltner & Robinson, 1996). According to these findings, microaggressions (a) tend to be subtle, indirect, and unintentional, (b) are most likely to emerge not when a behavior would look prejudicial, but when other rationales can be offered for prejudicial behavior, and (c) occur when Whites pretend not to notice differences, thereby justifying that "color" was not involved in the actions taken. Color blindness is a major form of microinvalidation because it denies the racial and experiential reality of people of color and provides an excuse to White people to claim that they are not prejudiced (Helms, 1992; Neville, Lilly, Duran, Lee, & Browne, 2000). The flight attendant, for example, did not realize that her "not seeing color" invalidated both passengers' racial identity and experiential reality.

Dilemma 3: Perceived Minimal Harm of Racial Microaggressions

In most cases, when individuals are confronted with their microaggressive acts (as in the case of the flight attendant), the perpetrator usually believes that the victim has overreacted and is being overly sensitive and/or petty. After all, even if it was an innocent racial blunder, microaggressions are believed to have minimal negative impact. People of color are told not to overreact and to simply "let it go." Usually, Whites consider microaggressive incidents to be minor, and people of color are encouraged (oftentimes by people of color as well) to not waste time or effort on them.

It is clear that old-fashioned racism unfairly disadvantages people of color and that it contributes to stress, depression, shame, and anger in its victims (Jones, 1997). But evidence also supports the detrimental impact of more subtle forms of racism (Chakraborty & McKenzie, 2002; Clark, Anderson, Clark, & Williams, 1999). For example, in a survey of studies examining racism and mental health, researchers found a positive association between happiness and life satisfaction, self-esteem, mastery of control, hypertension, and discrimination (Williams, Neighbors, & Jackson, 2003). Many of the types of everyday racism identified by Williams and colleagues (Williams & Collins, 1995; Williams, Lavizzo-Mourey, & Warren, 1994) provide strong support for the idea that racial microaggressions are not minimally harmful. One study specifically examined microaggressions in the experiences of African Americans and found that the cumulative effects can be quite devastating (Solórzano et al., 2000). The researchers reported that experience with microaggressions resulted in a negative racial climate and emotions of self-doubt, frustration, and isolation on the part of victims. As indicated in the incident above, the senior author experienced considerable emotional turmoil that lasted for the entire flight. When one considers that people of color are exposed continually to microaggressions and that their effects are cumulative, it becomes easier to understand the psychological toll they may take on recipients' well-being.

We submit that covert racism in the form of microaggressions also has a dramatic and detrimental impact on people of color. Although microaggressions may be seemingly innocuous and insignificant, their effects can be quite dramatic (Steele, Spencer, & Aronson, 2002). D. W. Sue believes that "this contemporary form of racism is many times over more problematic, damaging, and injurious to persons of color than overt racist acts" (D. W. Sue, 2003, p. 48). It has been noted that the cumulative effects of racial microaggressions may theoretically result in "diminished mortality, augmented morbidity and flattened confidence" (Pierce, 1995, p. 281). It is important to study and acknowledge this form of racism in society because without documentation and analysis to better understand microaggressions, the threats that they pose and the assaults that they justify can be easily ignored or downplayed (Solórzano et al., 2000). D. W. Sue (2005) has referred to this phenomenon as "a conspiracy of silence."

Dilemma 4: The Catch-22 of Responding to Microaggressions

When a microaggression occurs, the victim is usually placed in a catch-22. The immediate reaction might be a series of questions: Did what I think happened, really happen? Was this a deliberate act or an unintentional slight? How should I respond? Sit and stew on it or confront the person? If I bring the topic up, how do I prove it? Is it really worth the effort? Should I just drop the matter? These questions in one form or another have been a common, if not a universal, reaction of persons of color who experience an attributional ambiguity (Crocker & Major, 1989).

First, the person must determine whether a microaggression has occurred. In that respect, people of color rely heavily on experiential reality that is contextual in nature and involves life experiences from a variety of situations. When the flight attendant asked the senior author and his colleague to move, it was not the first time that similar requests and situations had occurred for both. In their experience, these incidents were nonrandom events (Ridley, 2005), and their perception was that the only similarity "connecting the dots" to each and every one of these incidents was the color of their skin. In other words, the situation on the plane was only one of many similar incidents with identical outcomes. Yet the flight attendant and most White Americans do not share these multiple experiences, and they evaluate their own behaviors in the moment through a singular event (Dovidio & Gaertner, 2000). Thus, they fail to see a pattern of bias, are defended by a belief in their own morality, and can in good conscience deny that they discriminated (D. W. Sue, 2005).

Second, how one reacts to a microaggression may have differential effects, not only on the perpetrator but on the person of color as well. Deciding to do nothing by sitting on one's anger is one response that occurs frequently in people of color. This response occurs because persons of color may be (a) unable to determine whether a microaggression has occurred, (b) at a loss for how to respond, (c) fearful of the consequences, (d) rationalizing that "it won't do any good anyway," or (e) engaging in self-deception through denial ("It didn't happen."). Although these explanations for nonresponse may hold some validity for the person of color, we submit that not doing anything has the potential to result in psychological harm. It may mean a denial of one's experiential reality, dealing with a loss of integrity, or experiencing pent-up anger and frustration likely to take psychological and physical tolls.

Third, responding with anger and striking back (perhaps a normal and healthy reaction) is likely to engender negative consequences for persons of color as well. They are likely to be accused of being racially oversensitive or paranoid or told that their emotional outbursts confirm stereotypes about minorities. In the case of Black males, for example, protesting may lend credence to the belief that they are hostile, angry, impulsive, and prone to violence (Jones, 1997). In this case, the person of color might feel better after venting, but the outcome results in greater hostility by Whites toward minorities. Further, while the person of color may feel better in the immediate moment by relieving pent-up emotions, the reality is that the general situation has not been changed. In essence, the catch-22 means you are "damned if you do, and damned if you don't." What is lacking is research that points to adaptive ways of handling microaggressions by people of color and suggestions of how to increase the awareness and sensitivity of Whites to microaggressions so that they accept responsibility for their behaviors and for changing them (Solórzano et al., 2000)....

CONCLUSION

... Racial microaggressions are potentially present whenever human interactions involve participants who differ in race and culture (teaching, supervising, training, administering, evaluating, etc.). We have purposely chosen to concentrate on racial microaggressions, but it is important to acknowledge other types of microaggressions as well. Gender, sexual orientation, and disability microaggressions may have equally powerful and potentially detrimental effects on women, gay, lesbian, bisexual, and transgender individuals, and disability groups. Further, racial microaggressions are not limited to White–Black, White–Latino, or White–Person of Color interactions. Interethnic racial microaggressions occur between people of color as well.... No racial/ethnic group is immune from inheriting the racial biases of the society (D. W. Sue, 2003)....

REFERENCES

Abelson, R. P., Dasgupta, N., Park, J., & Banaji, M. R. (1998). Perceptions of the collective other. *Personality and Social Psychology Review, 2,* 243–250.

Blair, I. V., Judd, C. M., & Chapleau, K. M. (2004). The influence of afrocentric facial features in criminal sentencing. *Psychological Science, 15,* 674–679.

Chakraborty, A., & McKenzie, K. (2002). Does racial discrimination cause mental illness? *British Journal of Psychiatry, 180,* 475–477.

Clark, R., Anderson, N. B., Clark, V. R., & Williams, D. R. (1999). Racism as a stressor for African Americans. *American Psychologist, 54,* 805–816.

Crocker, J., & Major, B. (1989). Social stigma and self-esteem: The self-protective properties of stigma. *Psychological Review, 96,* 608–630.

Dovidio, J. F., & Gaertner, S. L. (2000). Aversive racism and selective decisions: 1989–1999. *Psychological Science, 11,* 315–319.

Franklin, A. J. (2004). *From brotherhood to manhood: How Black men rescue their relationships and dreams from the invisibility syndrome.* Hoboken, NJ: Wiley.

Gordon, J., & Johnson, M. (2003). Race, speech, and hostile educational environment: What color is free speech? *Journal of Social Philosophy, 34,* 414–436.

Helms, J. E. (1992). *A race is a nice thing to have: A guide to being a white person or understanding the white persons in your life.* Topeka, KS: Content Communications.

Helms, J. E., & Cook, D. (1999). *Using race and culture in counseling and psychotheraphy: Theory and process.* Needham Heights, MA: Allyn & Bacon.

Hinton, E. L. (2004, March/April). Microinequities: When small slights lead to huge problems in the workplace. *DiversityInc.* (Available at http://www.magazine.org/content/files/Microinequities.pdf)

Jones, J. M. (1997). *Prejudice and racism* (2nd ed.). Washington, DC: McGraw-Hill.

Keltner, D., & Robinson, R. J. (1996). Extremism, power, and imagined basis of social conflict. *Current Directions in Psychological Science, 5,* 101–105.

Klonoff, E. A., & Landrine, H. (1999). Cross-validation of the Schedule of Racist Events. *Journal of Black Psychology, 25,* 231–254.

National Conference of Christians and Jews. (1992). *Taking America's pulse: A summary report of the National Conference Survey on Inter-Group Relations.* New York: Author. (Available at http://eric.ed.gov/ERlCDocs/data/ericdocs2/content_storage_01/0000000b/80/23/84/59.pdf)

Nelson, T. D. (2006). *The psychology of prejudice.* Boston: Pearson.

Neville, H. A., Lilly, R. L., Duran, G., Lee, R., & Browne, L. (2000). Construction and initial validation of the Color Blind Racial Attitudes Scale (COBRAS). *Journal of Counseling Psychology, 47,* 59–70.

Pierce, C. (1995). Stress analogs of racism and sexism: Terrorism, torture, and disaster. In C. Willie, P. Rieker, B. Kramer, & B. Brown (Eds.), *Mental health, racism, and sexism* (pp. 277–293). Pittsburgh, PA: University of Pittsburgh Press.

Plant, E. A., & Peruche, B. M. (2005). The consequences of race for police officers' responses to criminal suspects. *Psychological Science, 16,* 180–183.

Reid, L. D., & Radhakrishnan, P. (2003). Race matters: The relations between race and general campus climate. *Cultural Diversity and Ethnic Minority Psychology, 9,* 263–275.

Ridley, C. R. (2005). *Overcoming unintentional racism in counseling and therapy* (2nd ed.). Thousand Oaks, CA: Sage.

Sellers, R. M., & Shelton, J. N. (2003). The role of racial identity in perceived racial discrimination. *Journal of Personality and Social Psychology, 84,* 1070–1092.

Solórzano, D., Ceja, M., & Yosso, T. (2000, Winter). Critical race theory, racial microaggressions, and campus racial climate: The experiences of African American college students. *Journal of Negro Education, 69,* 60–73.

Spanierman, L. B., & Heppner, M. J. (2004). Psychosocial Costs of Racism to Whites Scale (PCRW): Construction and initial validation. *Journal of Counseling Psychology, 51,* 249–262.

Steele, C. M., Spencer, S. J., & Aronson, J. (2002). Contending with group image: The psychology of stereotype and social identity threat. In M. Zanna (Ed.), *Advances in experimental social psychology* (Vol. 23, pp. 379–440). New York: Academic Press.

Sue, D. W. (2003). *Overcoming our racism: The journey to liberation.* San Francisco: Jossey-Bass.

Sue, D. W. (2004). Whiteness and ethnocentric monoculturalism: Making the "invisible" visible. *American Psychologist, 59,* 759–769.

Sue, D. W. (2005). Racism and the conspiracy of silence. *Counseling Psychologist, 33,* 100–114.

Thompson, C. E., & Neville, H. A. (1999). Racism, mental health, and mental health practice. *Counseling Psychologist, 27,* 155–223.

Williams, D. R., & Collins, C. (1995). U.S. socioeconomic and racial differences in health: Patterns and explanations. *Annual Review of Sociology, 21,* 349–386.

Williams, D. R., Lavizzo-Mourey, R., & Warren, R. C. (1994). The concept of race and health status in America. *Public Health Reports, 109,* 26–41.

Williams, D. R., Neighbors, H. W., & Jackson, J. S. (2003). Racial/ethnic discrimination and health: Findings from community studies. *American Journal of Public Health, 93,* 200–208.

DISCUSSION QUESTIONS

1. What are some of the possible interventions that can reduce the harm and hurt that racial microaggressions produce?

2. Racial microaggressions are typically invisible to members of the dominant group, perhaps leading them to think that racism is no longer a "big deal." Using Table 1 in the article, identify times when you might have witnessed or experienced the microaggression identified there.

Student Exercises

1. Think back to the first time you remember recognizing your own racial identity. What were the circumstances? What did you learn? Now ask the same question of someone whose race is different from your own. How do the two experiences compare and contrast? How do the answers illustrate how racial identity is formed in different contexts and with different meanings depending on the group's experience?

2. Having read the interview with Beverly Tatum, observe patterns of seating in public spaces on your campus (such as in student centers, eating areas, and classrooms). What patterns do you see and how would you interpret them based on Tatum's arguments?

PART V

Intersecting Inequalities
Race, Class, and Gender

ELIZABETH HIGGINBOTHAM
AND MARGARET L. ANDERSEN

As important as race is, it does not stand alone in shaping individual and group experiences. People are not solely members of a racial or ethnic group, but also have a gender and a social class position, along with other social attributes. **Gender** refers to the culturally and socially structured relationship between men and women. Gender is constructed as both a social location and an identity, meaning that what men and women do, how they relate to each other, and how they think about themselves and others is shaped by the society and the interaction of race with gender and social class. Like race, gender is a social construction, even though people often think (erroneously) that gender is simply a function of one's biological makeup.

Social class, or class, refers to the where people stand in relationship to the economic systems of society. In part, class refers to the material assets people have relative to others (wages, salaries, wealth, and so forth). Class, however, has a subjective dimension as well because, like race and gender, class involves systems of power as well as the social judgments that people make about others. Social class means people have different life chances, some having advantages; others, disadvantages. Like race and gender, class is based in social structural conditions.

Race, gender, and class can be thought of separately, but they intersect in how they shape people's life chances. Developing an inclusive perspective means understanding how race, class, and gender intersect and overlap in shaping people's lives (Andersen and Collins 2015). You might experience one as more salient (or important) at a given point in time, but, all told, race, class, and gender simultaneously influence your lived experience. As an example, Asian American women, regardless of their social class, will likely encounter sexism, but the particular forms sexism takes will also be shaped by the fact of their Asian American status. At the same time, a working-class Asian American woman may be stereotyped, but not in the same way as an Asian American woman of higher class status—who may be stereotyped as a "model minority" or a "tiger mom!"

Understanding the intersections of race, class, and gender means that you cannot see all members of a given group as the same. For example, because of the interaction of gender and race, the cultural images that are produced of young African American men differ from those of young African American women. As these young men and women move about society, their reception by others is often based on racial and gendered stereotypes. Furthermore, the interaction of gender and race means that, depending on your social location, different groups, by virtue of their race, gender, and class, will likely travel different streets, attend different schools, and get different jobs. Looking at all of these dimensions—race, class, and gender—is necessary for understanding the specific challenges people face and the resources they have. As you will see in these readings, the complex pattern of advantages and disadvantages is important both at the individual level and at the organizational and institutional levels.

Part V opens with a work by one of the pioneers in this intersectional approach in the social sciences, Patricia Hill Collins ("Toward a New Vision: Race, Class, and Gender as Categories of Analysis and Connection"). In her classic article, Collins identifies how using these three dimensions of oppression explains the structural basis of domination and subordination. Collins offers an analysis that explores the institutional, symbolic, and individual dimensions of race, class, and gender, arguing that people sit within a *matrix of domination*.

Collins's work cautions us about using a singular lens to understand people's experience, even though in some studies, people rely on only one such dimension. Thus, in women's studies, people sometimes only analyze gender; in race studies, attention is sometimes solely on racial differences; and in social class studies, people may only highlight class hierarchies.

In her article, Yen Le Espiritu ("Theorizing Race, Gender, and Class") calls for greater integration of these perspectives. She provides a quick history of these scholarly fields, making it clear that failure to have an intersectional analysis makes it difficult to build those coalitions for social change. The racial, gender, and class complexity of the United States means we must move toward this direction of analysis to understand each other and our society.

Adia Harvey Wingfield's research ("Racializing the Glass Escalator: Reconsidering Men's Experiences with Women's Work") shows us the importance of not generalizing from the experience of just one group when making claims about what people experience. By studying African American men who work as nurses, Wingfield demonstrates that assumptions about men's advancement in traditionally female occupations do not hold for all men, specifically African American men. Because of their social location in both a race and class and gender system, African American men experience particular forms of exclusion—experiences not like those of White men in the same work setting. Her research reminds us of the importance of keeping all three factors—race, class, and gender—in our thinking when observing and thinking about the experiences of different groups in society.

REFERENCE

Andersen, Margaret L., and Patricia Hill Collins, eds. 2015. *Race, Class, and Gender: An Anthology*, 9th ed. Belmont, CA: Cengage.

FACE THE FACTS: THE RACE-GENDER WAGE GAP

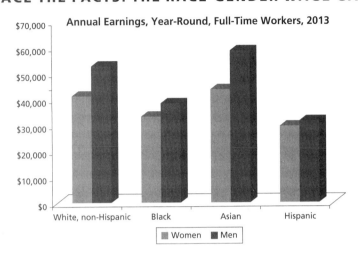

Annual Earnings, Year-Round, Full-Time Workers, 2013

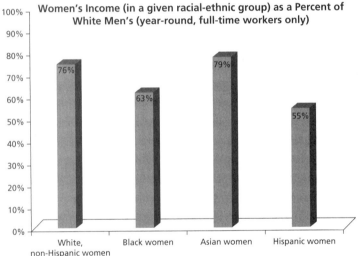

Women's Income (in a given racial-ethnic group) as a Percent of White Men's (year-round, full-time workers only)

SOURCE: U.S. Census Bureau. 2013. *Person Detailed Income Tables, PINC01.* Washington, DC: U.S. Department of Commerce. www.census.gov

Think about It: You have likely heard that the income gap between women and men is about 73 percent, but, as the top figure shows, when you add race to the picture, the gap between women and men within different racial–ethnic groups is different. Furthermore, when you compare women in different groups to the earnings of White men, you also see a different picture (see the bottom figure). Where are the gaps the largest? The smallest? What does this demonstrate about the intersection of race, gender, and class?

18

Toward a New Vision

Race, Class, and Gender as Categories of Analysis and Connection

PATRICIA HILL COLLINS

People are often focused on one dimension of oppression—the one they feel is most important in their lives, yet we are all positioned with regard to entire sets of advantages and disadvantages. Collins pushes readers to think about new ways of thinking that emerge when we explore the connection of institutional, symbolic, and individual dimensions of oppression.

> The true focus of revolutionary change is never merely the
> oppressive situations which we seek to escape, but the piece of
> the oppressor which is planted deep within each of us.
> —AUDRE LORDE, *SISTER OUTSIDER*, 123

A udre Lorde's statement raises a troublesome issue for scholars and activists working for social change. While many of us have little difficulty assessing our own victimization within some major system of oppression, whether it be by race, social class, religion, sexual orientation, ethnicity, age or gender, we typically fail to see how our thoughts and actions uphold someone else's subordination. Thus, white feminists routinely point with confidence to their oppression as women but resist seeing how much their white skin privileges them. African-Americans who possess eloquent analyses of racism often persist in viewing poor white women as symbols of white power. The radical left fares little better. "If only people of color and women could see their true class interests," they argue, "class solidarity would eliminate racism and sexism." In essence, each group identifies the type of oppression with which it feels most comfortable as being fundamental and classifies all other types as being of lesser importance.

Oppression is full of such contradictions. Errors in political judgment that we make concerning how we teach our courses, what we tell our children, and

SOURCE: From Collins, Patricia Hill. 1993. "Toward a New Vision: Race, Class, and Gender as Categories of Analysis and Connection." *Race, Gender, & Class* 1 (October): 25-46. Reprinted by permission of the author.

which organizations are worthy of our time, talents and financial support flow smoothly from errors in theoretical analysis about the nature of oppression and activism. Once we realize that there are few pure victims or oppressors, and that each one of us derives varying amounts of penalty and privilege from the multiple systems of oppression that frame our lives, then we will be in a position to see the need for new ways of thought and action.

To get at that "piece of the oppressor which is planted deep within each of us," we need at least two things. First, we need new visions of what oppression is, new categories of analysis that are inclusive of race, class, and gender as distinctive yet interlocking structures of oppression. Adhering to a stance of comparing and ranking oppressions—the proverbial, "I'm more oppressed than you"—locks us all into a dangerous dance of competing for attention, resources, and theoretical supremacy. Instead, I suggest that we examine our different experiences within the more fundamental relationship of domination and subordination. To focus on the particular arrangements that race or class or gender takes in our time and place without seeing these structures as sometimes parallel and sometimes interlocking dimensions of the more fundamental relationship of domination and subordination may temporarily ease our consciences. But while such thinking may lead to short-term social reforms, it is simply inadequate for the task of bringing about long-term social transformation.

While race, class and gender as categories of analysis are essential in helping us understand the structural bases of domination and subordination, new ways of thinking that are not accompanied by new ways of acting offer incomplete prospects for change. To get at that "piece of the oppressor which is planted deep within each of us," we also need to change our daily behavior....

In order to move toward new visions of what oppression is, I think that we need to ask new questions. How are relationships of domination and subordination structured and maintained in the American political economy? How do race, class and gender function as parallel and interlocking systems that shape this basic relationship of domination and subordination? Questions such as these promise to move us away from futile theoretical struggles concerned with ranking oppressions and towards analyses that assume race, class and gender are all present in any given setting, even if one appears more visible and salient than the others. Our task becomes redefined as one of reconceptualizing oppression by uncovering the connections among race, class and gender as categories of analysis.

1. THE INSTITUTIONAL DIMENSION
OF OPPRESSION

Sandra Harding's contention that gender oppression is structured along three main dimensions—the institutional, the symbolic and the individual—offers a useful model for a more comprehensive analysis encompassing race, class and gender oppression (Harding 1986). Systemic relationships of domination and subordination structured through social institutions such as schools, businesses, hospitals, the workplace and government agencies represent the institutional dimension of oppression. Racism, sexism and elitism all have concrete institutional locations. Even though the workings of the institutional dimension of oppression are often obscured with ideologies claiming equality of opportunity, in actuality, race, class and gender place Asian-American women, Native American men, White men, African-American women and other groups in distinct institutional niches with varying degrees of penalty and privilege.

Even though I realize that many … would not share this assumption, let us assume that the institutions of American society discriminate, whether by design or by accident. While many of us are familiar with how race, gender and class operate separately to structure inequality, I want to focus on how these three systems interlock in structuring the institutional dimension of oppression. To get at the interlocking nature of race, class and gender, I want you to think about the antebellum plantation as a guiding metaphor for a variety of American social institutions. Even though slavery is typically analyzed as a racist institution, and occasionally as a class institution, I suggest that slavery was a race, class, gender specific institution. Removing any one piece from our analysis diminishes our understanding of the true nature of relations of domination and subordination under slavery.

Slavery was a profoundly patriarchal institution. It rested on the dual tenets of White male authority and White male property, a joining of the political and the economic within the institution of the family. Heterosexism was assumed and all Whites were expected to marry. Control over affluent White women's sexuality remained key to slavery's survival because property was to be passed on to the legitimate heirs of the slave owner. Ensuring affluent White women's virginity and chastity was deeply intertwined with maintenance of property relations.

Under slavery, we see varying levels of institutional protection given to affluent White women, working class and poor White women and enslaved African women. Poor White women enjoyed few of the protections held out to their upper class sisters. Moreover, the devalued status of Black women was key in keeping all White women in their assigned places. Controlling Black women's fertility was also key to the continuation of slavery, for children born to slave mothers themselves were slaves.

African-American women shared the devalued status of chattel with their husbands, fathers and sons. Racism stripped Blacks as a group of legal rights, education and control over their own persons. African-Americans could be whipped, branded, sold, or killed, not because they were poor, or because they were women, but because they were Black. Racism ensured that Blacks would continue to serve Whites and suffer economic exploitation at the hands of all Whites.

So we have a very interesting chain of command on the plantation—the affluent White master as the reigning patriarch, his White wife helpmate to serve him, help him manage his property and bring up his heirs, his faithful servants whose production and reproduction were tied to the requirements of the capitalist political economy and largely propertyless, working class White men and women watching from afar. In essence, the foundations for the contemporary roles of elite White women, poor Black women, working class White men and a series of other groups can be seen in stark relief in this fundamental

American social institution. While Blacks experienced the most harsh treatment under slavery, and thus made slavery clearly visible as a racist institution, race, class and gender interlocked in structuring slavery's systemic organization of domination and subordination.

Even today, the plantation remains a compelling metaphor for institutional oppression. Certainly the actual conditions of oppression are not as severe now as they were then. To argue, as some do, that things have not changed all that much denigrates the achievements of those who struggled for social change before us. But the basic relationships among Black men, Black women, elite White women, elite White men, working class White men and working class White women as groups remain essentially intact.…

2. THE SYMBOLIC DIMENSION OF OPPRESSION

Widespread, societally sanctioned ideologies used to justify relations of domination and subordination comprise the symbolic dimension of oppression. Central to this process is the use of stereotypical or controlling images of diverse race, class and gender groups. In order to assess the power of this dimension of oppression, I want you to make a list, either on paper or in your head, of "masculine" and "feminine" characteristics. If your list is anything like that compiled by most people, it reflects some variation of the following:

Masculine	*Feminine*
aggressive	passive
leader	follower
rational	emotional
strong	weak
intellectual	physical

Not only does this list reflect either/or dichotomous thinking and the need to rank both sides of the dichotomy, but ask yourself exactly which men and women you had in mind when compiling these characteristics. This list applies almost exclusively to middle class White men and women. The allegedly "masculine" qualities that you probably listed are only acceptable when exhibited by elite White men, or when used by Black and Hispanic men against each other or against women of color. Aggressive Black and Hispanic men are seen as dangerous, not powerful, and are often penalized when they exhibit any of the allegedly "masculine" characteristics. Working class and poor White men fare slightly better and are also denied the allegedly "masculine" symbols of leadership, intellectual competence, and human rationality. Women of color and working class and poor White women are also not represented on this list, for they have never had the luxury of being "ladies." What appear to be universal categories representing all men and women instead are unmasked as being applicable to only a small group.

It is important to see how the symbolic images applied to different race, class and gender groups interact in maintaining systems of domination and subordination. If I were to ask you to repeat the same assignment, only this time, by making separate lists for Black men, Black women, Hispanic women and Hispanic men, I suspect that your gender symbolism would be quite different. In comparing all of the lists, you might begin to see the interdependence of symbols applied to all groups. For example, the elevated images of White womanhood need devalued images of Black womanhood in order to maintain credibility.

While the above exercise reveals the interlocking nature of race, class and gender in structuring the symbolic dimension of oppression, part of its importance lies in demonstrating how race, class and gender pervade a wide range of what appears to be universal language. Attending to diversity in our scholarship, in our teaching, and in our daily lives provides a new angle of vision on interpretations of reality thought to be natural, normal and "true." Moreover, viewing images of masculinity and femininity as universal gender symbolism, rather than as symbolic images that are race, class and gender specific, renders the experiences of people of color and of nonprivileged White women and men invisible. One way to dehumanize an individual or group is to deny the reality of their experiences. So when we refuse to deal with race or class because they do not appear to be directly relevant to gender, we are actually becoming part of someone else's problem.

Assuming that everyone is affected differently by the same interlocking set of symbolic images allows us to move forward toward new analyses. Women of color and White women have different relations to White male authority and this difference explains the distinct gender symbolism applied to both groups. Black women encounter controlling images such as the mammy, the matriarch, the mule and the whore, that encourage others to reject us as fully human people. Ironically, the negative nature of these images simultaneously encourages us to reject them. In contrast, White women are offered seductive images, those that promise to reward them for supporting the status quo. And yet seductive images can be equally controlling....

Both sets of images stimulate particular political stances. By broadening the analysis beyond the confines of race, we can see the varying levels of rejection and seduction available to each of us due to our race, class and gender identity. Each of us lives with an allotted portion of institutional privilege and penalty, and with varying levels of rejection and seduction inherent in the symbolic images applied to us. This is the context in which we make our choices. Taken together, the institutional and symbolic dimensions of oppression create a structural backdrop against which all of us live our lives.

3. THE INDIVIDUAL DIMENSION OF OPPRESSION

Whether we benefit or not, we all live within institutions that reproduce race, class and gender oppression. Even if we never have any contact with members of other race, class and gender groups, we all encounter images of these groups and are exposed to the symbolic meanings attached to those images. On this dimension of oppression, our individual biographies vary tremendously. As a result of our institutional and symbolic statuses, all of our choices become political acts.

Each of us must come to terms with the multiple ways in which race, class and gender as categories of analysis frame our individual biographies. I have lived my entire life as an African-American woman from a working class family and this basic fact has had a profound impact on my personal biography. Imagine how different your life might be if you had been born Black, or White, or poor, or of a different race/class/gender group than the one with which you are most familiar. The institutional treatment you would have received and the symbolic meanings attached to your very existence might differ dramatically from what you now consider to be natural, normal and part of everyday life. You might be the same, but your personal biography might have been quite different.

I believe that each of us carries around the cumulative effect of our lives within multiple structures of oppression. If you want to see how much you have been affected by this whole thing, I ask you one simple question—who are your close friends? Who are the people with whom you can share your hopes, dreams, vulnerabilities, fears and victories? Do they look like you? If they are all the same, circumstance may be the cause. For the first seven years of my life I saw only low income Black people. My friends from those years reflected the composition of my community. But now that I am an adult, can the defense of circumstance explain the patterns of people that I trust as my friends and collea-gues? When given other alternatives, if my friends and colleagues reflect the homogeneity of one race, class and gender group, then these categories of analysis have indeed become barriers to connection.

I am not suggesting that people are doomed to follow the paths laid out for them by race, class and gender as categories of analysis. While these three structures certainly frame my opportunity structure, I as an individual always have the choice of accepting things as they are, or trying to change them. As Nikki Giovanni points out, "we've got to live in the real world. If we don't like the world we're living in, change it. And if we can't change it, we change ourselves. We can do something" (Tate 1983, 68). While a piece of the oppressor may be planted deep within each of us, we each have the choice of accepting that piece or challenging it as part of the "true focus of revolutionary change."

REFERENCES

Harding, Sandra. 1986. *The Science Question in Feminism*. Ithaca, New York: Cornell University Press.

Lorde, Audre. 1984. *Sister Outsider*. Trumansberg, New York: The Crossing Press.

Tate, Claudia, ed. 1983. *Black Women Writers at Work*. New York: Continuum.

DISCUSSION QUESTIONS

1. What does Collins mean by "the contradictions of oppression"?
2. Can you identify how the interlocking nature of race, class, and gender structure inequality in your college or workplace?
3. What role does the symbolic dimension of oppression play in perpetuating inequalities?

19

Theorizing Race, Gender, and Class

YEN LE ESPIRITU

Here, Yen Le Espiritu discusses how approaches to gender issues can limit the way that scholars explore social issues and also how they understand different groups. As social scientists look at race and social class as well as gender, they are better able to understand the work experiences of men and women of color as well as gender relations within racial communities.

In the 1970s, when "second-wave" feminist scholars identified gender as a major social force, they pronounced that traditional scholarship represented the experiences of men as gender neutral, making it unnecessary to deal with women. With men's lives assumed to be the norm, women's experiences were subsumed under those of men, narrowly categorized, or excluded altogether. Responding to the omission of women, the feminist agenda at the time was to fill in gaps, correct sexist biases, and create new topics out of women's experiences. In the next stage, feminist thinking advanced from woman-centered analyses to providing a "gendered" understanding of sociological phenomena—one that traced the significance of gender organization and relations in all institutions and in shaping men's as well as women's lives. The feminist agenda shifted then from advocating the inclusion of women to transforming the basic conceptual and theoretical frameworks of the field.

In a 1985 statement on the treatment of gender in research, the American Sociological Association's Committees on the Status of Women in Sociology (1985) urged members of the profession "to explicitly acknowledge the social category of gender, and gender differences in power, at each step of the research process" (p. 1). But a gendered transformation of knowledge has not been achieved in sociology. In a review essay published in the mid-1980s, Judith Stacey and Barrie Thorne (1985) argued that although feminist scholarship has made important contributions to sociology in terms of uncovering and filling gaps in sociological knowledge, the gender paradigm, which positions gender as a prominent social category creating hierarchies of difference in society, remains a minority position within the discipline. As in anthropology and history, feminist sociology seems to have been "both co-opted and ghettoized, while the discipline as a whole and its dominant paradigms have proceeded relatively unchanged" (Stacey and Thorne, 1985, p. 302). The use of gender as a variable,

SOURCE: From Espiritu, Yen Le. 2008. *Asian American Women and Men: Labor, Laws and Love*, 2nd ed. Lanham, MD: Rowman & Littlefield.

conceptualized in terms of sex difference rather than as a central theoretical concept, is a prime example of the co-optation of feminist perspectives (Stacey and Thorne, 1985, p. 308). Six years later, Sylvia Pedraza (1991) reached the same conclusion: within sociology, "a truly gendered understanding of most topics has not been achieved" (p. 305). A 2007 assessment of the field remained bleak: although the quantity of work on gender in leading sociological journals in the twenty-first century has increased, "the critical turn for the discipline, one in which a structurally-oriented gender perspective would inform sociology as a whole, has yet to happen" (Ferree, Khan, and Morimoto, 2007, p. 473).

Although the concept of gender is invaluable, the gender process cannot be understood independently of class and race. Class and gender overlap when the culture of patriarchy enables capitalists to benefit from the exploitation of the labor of both men and women. Because patriarchy mandates that men serve as good financial providers, it obliges them to toil in the exploitative capitalist wage market. Correspondingly, the patriarchal assumption that women are not the breadwinners—and therefore can afford to work for less—allows employers to justify hiring women at lower wages and under poorer working conditions. On the other hand, in however limited a way, wage employment affords women a measure of economic and personal independence, thus strengthening their claims against patriarchal authority. This is just one of the many contradictions that occur in the interstices of the race/gender/class matrix.

Economic oppression in the United States is not only gendered but also racist. Historically, people of color in the United States have encountered institutionalized economic and cultural racism that has restricted their economic mobility. Due to their gender, race, and noncitizen status, immigrant women of color fare the worst because they are seen as being the most desperate for work at any wage. Within this racially based social order, there are no gender relations per se, only gender relations as constructed by and between races. Jane Gaines (1990, p. 198) suggested that insofar as the focus on gender keeps women from seeing other structures of oppression, it functions ideologically in the interests of the dominant group(s). That is, to conceptualize oppression only in terms of male dominance and female subordination is to obscure the centrality of classism, racism, and other forms of inequality in U.S. society (Stacey and Thorne, 1985, p. 311).

Women of color have charged that feminist theory falsely universalizes the category of "woman." As bell hooks (1984) observed, this gender analysis centers on the experiences of white, middle-class women and ignores the way women in different racial groups and social classes experience oppression. For women of color, gender is only part of a larger pattern of unequal social relations. In their daily lives, these women brave not only sexism but also the "entire system of racial and ethnic stratification that defines, stigmatizes, and controls the minority group as a whole" (Healey, 1995, p. 26). These systems of power render irrelevant the public/private distinction for many women of color. As Aida Hurtado (1989) reminded us,

> Women of color have not had the benefit of the economic conditions
> that underlie the public/private distinction.... Welfare programs and

policies have discouraged family life, sterilization programs have restricted production rights, government has drafted and armed disproportionate numbers of people of color to fight its wars overseas, and
locally, police forces in the criminal justice system arrest and incarcerate
disproportionate numbers of people of color. There is no such thing as a
private sphere for people of color except that which they manage to
create and protect in an otherwise hostile environment. (p. 849)

In this hostile environment, some women of color, in contrast to their white
counterparts, view unpaid domestic work—having children and maintaining
families—more as a form of resistance to racist oppression than as a form of
exploitation by men.

Women of color have also protested their marginalization in traditional feminist scholarship, charging that they have been added to feminist analysis only "as
an afterthought" (Gaines, 1990, p. 201). This tokenism is evident in the manner
in which race has been added to, rather than integrated into, traditional feminist
scholarship. Gaines (1990, p. 201) pointed out that although feminist anthologies
consistently include writings by women of color, the issue of race is conspicu-
ously absent from the rest of the volume. This practice suggests that white feminists view racism as an issue that affects only people of color and not as a system
that organizes and shapes the daily experiences of *all* people. In an anthology of
writings by radical women of color, Cherrie Moraga (1981) recorded her reaction to "dealing with white women": "I have felt so very dark: dark with anger,
with silence, with the feeling of being walked over" (p. xv). Similarly, Bettie
Luke Kan referred to racism when she explained why Chinese American women
rejected the National Organization of Women: "No matter how hard you fight to
reduce the sexism, when it's all done and over with, you still have the racism.
Because white women will be racist as easily as their male counterparts. And
white women continue to get preferential treatment over women of color"
(cited in Yung, 1986, p. 98). Recentering race, then, requires feminist scholars to
reshape the basic concepts and theories of their field and to acknowledge that the
experiences of white women and women of color are not just different but connected in systematic ways.

Bringing class and race into the study of gender also requires us to explicate
the positions that white men *and* white women occupy over men of color. Many
white feminists include all males, regardless of color and social class, into their
critique of sexist power relations. This argument assumes that *any* male in U.S.
society benefits from a patriarchal system designed to maintain the unequal relationship that exists between men and women. Pointing to the multiplicities of
men's lives, feminists of color have argued that, depending on their race and
class, men experience gender differently. Presenting race and gender as relation-
ally constructed, black feminists have referred instead to "racial patriarchy"—a
concept that calls attention to the white patriarch/master in American history
and his dominance over the black male as well as the black female (Gaines,
1990, p. 202). Providing yet another dimension to the intersections of race and
gender, Gaines (1990) pointed out that the "notion of patriarchy is most obtuse

when it disregards the position white women occupy over Black men as well as Black women" (p. 202). The failure of feminist scholarship to theorize the historically specific experiences of men of color makes it difficult for women of color to rally to the feminist cause without feeling divided or without being accused of betrayal. As Kimberlee Crenshaw (1989) observed, "Although patriarchy clearly operates within the Black community, ... the racial context in which Black women find themselves makes the creation of a political consciousness that is oppositional to Black men difficult" (p. 162). Along the same lines, King-Kok Cheung (1990) exhorted white scholars to acknowledge that, like female voices, "the voices of many men of color have been historically silenced or dismissed" (p. 246). David Eng (2001) reminded us that Asian Americanists have also neglected the topic of Asian American masculinities: in the last fifteen years, while the field of Asian American studies has produced substantial research on Asian American female identity, mother-daughter relations, and feminism, comparatively less critical attention has been paid to the formation of Asian American male subjectivity (p. 15).

REFERENCES

American Sociological Association, Committee on the Status of Women in Sociology. 1985. *The treatment of gender in research.* Washington, DC.

Cheung, K.-K. 1990. The woman warrior versus the Chinaman pacific: Must a Chinese American critic choose between feminism and heroism? In M. Hirsch and E. F. Keller (eds.), *Conflicts in feminism* (pp. 234–51). New York: Routledge.

Crenshaw, K. 1989. Demarginalizing the intersection of race and sex: A black feminist critique of antidiscrimination doctrine, feminist theory and antiracist politics. In *University of Chicago Legal Forum: Feminism in the law: Theory, practice, and criticism* (pp. 139–67). Chicago: University of Chicago Press.

Eng, D. L. 2001. *Racial castration: Managing masculinity in Asian America.* Durham, NC: Duke University Press.

Ferree, M. M., S. Khan, and S. A. Morimoto 2007. Assessing the feminist revolution: The presence and absence of gender in theory and practice. In C. Calhoun (ed.), *Sociology in America: A History* (pp. 438–79). Chicago: University of Chicago Press.

Gaines, J. 1990. White privilege and looking relations: Race and gender in feminist film theory. In P. Erens (ed.), *Issues in feminist film criticism* (pp. 197–214). Bloomington: Indiana University Press.

Healey, J. F. 1995. *Race, ethnicity, gender, and class: The sociology of group conflict and change.* Thousand Oaks, CA: Pine Forge.

hooks, b. 1984. *Feminist theory: From margin to center.* Boston: South End Press.

Hurtado, A. 1989. Relating to privilege: Seduction and rejection in the subordination of white women and women of color. *Signs: Journal of Women in Culture and Society,* 14, (pp. 833–55).

Moraga, C. 1981. Preface. In C. Moraga and G. Anzaldúa (eds.), *This bridge called my back: Writings by radical women of color* (pp. xiii–xix). Watertown, MA: Persephone.

Pedraza, S. 1991. Women and migration: The social consequences of gender. *Annual Review of Sociology*, 17, (pp. 303–25).

Stacey, J., and B. Thorne. 1985. The missing feminist revolution in sociology. *Social Problems*, 32, (pp. 301–16).

Yung, J. 1986. *Chinese women of America: A pictorial essay.* Seattle: University of Washington Press.

DISCUSSION QUESTIONS

1. What do scholars mean when they suggest that the category of woman has been universalized, and why is that a problem?

2. How does attention to race and class expand our understanding of gender and patriarchy?

3. How can we see both race and gender expressed in the work options of men and women of different racial groups?

20

Racializing the Glass Escalator

Reconsidering Men's Experiences
with Women's Work

ADIA HARVEY WINGFIELD

By locating her research within an intersectional analysis of race, class, and gender, Wingfield challenges assumptions about men riding a glass escalator to mobility in predominantly women's occupations. Her research on African American men who work as nurses shows the importance of understanding the simultaneous and overlapping influences of race, class, and gender.

Sociologists who study work have long noted that jobs are sex segregated and that this segregation creates different occupational experiences for men and women (Charles and Grusky 2004). Jobs predominantly filled by women often require "feminine" traits such as nurturing, caring, and empathy, a fact that means men confront perceptions that they are unsuited for the requirements of these jobs. Rather than having an adverse effect on their occupational experiences, however, these assumptions facilitate men's entry into better paying, higher status positions, creating what Williams (1995) labels a "glass escalator" effect.

The glass escalator model has been an influential paradigm in understanding the experiences of men who do women's work. Researchers have identified this process among men nurses, social workers, paralegals, and librarians and have cited its pervasiveness as evidence of men's consistent advantage in the workplace, such that even in jobs where men are numerical minorities they are likely to enjoy higher wages and faster promotions (Floge and Merrill 1986; Heikes 1991; Pierce 1995; Williams 1989, 1995). Most of these studies implicitly assume a racial homogenization of men workers in women's professions, but this supposition is problematic for several reasons. For one, minority men are not only present but are actually overrepresented in certain areas of reproductive work that have historically been dominated by white women (Duffy 2007). Thus, research that focuses primarily on white men in women's professions ignores a key segment of men who perform this type of labor. Second, and perhaps more important, conclusions based on the experiences of white men tend to overlook the ways that intersections of race and gender create different experiences for

SOURCE: From Wingfield, Adia Harvey. 2009. "Racializing the Glass Escalator: Reconsidering Men's Experiences with Women's Work." *Gender & Society* 23: 5–26.

different men. While extensive work has documented the fact that white men in women's professions encounter a glass escalator effect that aids their occupational mobility (for an exception, see Snyder and Green 2008), few studies, if any, have considered how this effect is a function not only of gendered advantage but of racial privilege as well.

... I examine the implications of race–gender intersections for minority men employed in a female-dominated, feminized occupation, specifically focusing on Black men in nursing. Their experiences doing "women's work" demonstrate that the glass escalator is a racialized as well as gendered concept....

The concept of the glass escalator provides an important and useful framework for addressing men's experiences in women's occupations, but so far research in this vein has neglected to examine whether the glass escalator is experienced among all men in an identical manner. Are the processes that facilitate a ride on the glass escalator available to minority men? Or does race intersect with gender to affect the extent to which the glass escalator offers men opportunities in women's professions? In the next section, I examine whether and how the mechanisms that facilitate a ride on the glass escalator might be unavailable to Black men in nursing....

RELATIONSHIPS WITH COLLEAGUES AND SUPERVISORS

... The congenial relationship with colleagues and gendered bonds with supervisors are crucial to riding the glass escalator. Women colleagues often take a primary role in casting these men into leadership or supervisory positions. In their study of men and women tokens in a hospital setting, Floge and Merrill (1986) cite cases where women nurses promoted men colleagues to the position of charge nurse, even when the job had already been assigned to a woman. In addition to these close ties with women colleagues, men are also able to capitalize on gendered bonds with (mostly men) supervisors in ways that engender upward mobility. Many men supervisors informally socialize with men worker in women's jobs and are thus able to trade on their personal friendships for upward mobility....

For Black men in nursing, however, gendered racism may limit the extent to which they establish bonds with their colleagues and supervisors. The concept of gendered racism suggests that racial stereotypes, images, and beliefs are grounded in gendered ideals (Collins 1990, 2004; Espiritu 2000; Essed 1991; Harvey Wingfield 2007). Gendered racist stereotypes of Black men in particular emphasize the dangerous, threatening attributes associated with Black men and Black masculinity, framing Black men as threats to white women, prone to criminal behavior, and especially violent....

... For Black men nurses, intersections of race and gender create a different experience with the mechanisms that facilitate white men's advancement in women's professions. Awkward or unfriendly interactions with colleagues, poor relationships with supervisors, perceptions that they are not suited for nursing, and an unwillingness to disassociate from "feminized" aspects of nursing constitute what I term *glass barriers* to riding the glass escalator.

RECEPTION FROM COLLEAGUES AND SUPERVISORS

When women welcome men into "their" professions, they often push men into leadership roles that ease their advancement into upper-level positions. Thus, a positive reaction from colleagues is critical to riding the glass escalator. Unlike white men nurses, however, Black men do not describe encountering a warm reception from women colleagues (Heikes 1991). Instead, the men I interviewed find that they often have unpleasant interactions with women coworkers who treat them rather coldly and attempt to keep them at bay. Chris is a 51-year-old oncology nurse who describes one white nurse's attempt to isolate him from other white women nurses as he attempted to get his instructions for that day's shift:

> She turned and ushered me to the door, and said for me to wait out here, a nurse will come out and give you your report. I stared at her hand on my arm, and then at her, and said, "Why? Where do you go to get your reports?" She said, "I get them in there." I said, "Right. Unhand me." I went right back in there, sat down, and started writing down my reports.

Kenny, a 47-year-old nurse with 23 years of nursing experience, describes a similarly and particularly painful experience he had in a previous job where he was the only Black person on staff:

> [The staff] had nothing to do with me, and they didn't even want me to sit at the same area where they were charting in to take a break. They wanted me to sit somewhere else.... They wouldn't even sit at a table with me! When I came and sat down, everybody got up and left.

... For these nurses, their masculinity is not a guarantee that they will be welcomed, much less pushed into leadership roles. As Ryan, a 37-year-old intensive care nurse says, "[Black men] have to go further to prove ourselves. This involves proving our capabilities, *proving to colleagues that you can lead*, be on the forefront" (emphasis added). The warm welcome and subsequent opportunities for leadership cannot be taken for granted. In contrast, these men describe great challenges in forming congenial relationships with coworkers who, they believe, do not truly want them there.

In addition, these men often describe tense, if not blatantly discriminatory, relationships with supervisors. While Williams (1995) suggests that men supervisors can be allies for men in women's professions by facilitating promotions and upward mobility, Black men nurses describe incidents of being overlooked by supervisors when it comes time for promotions. Ryan, who has worked at his current job for 11 years, believes that these barriers block upward mobility within the profession:

> The hardest part is dealing with people who don't understand minority nurses. People with their biases, who don't identify you as ripe for promotion. I know the policy and procedure, I'm familiar with past history. So you can't tell me I can't move forward if others did. [How did you deal with this?] By knowing the chain of command, who my supervisors were. Things were subtle. I just had to be better. I got this mostly from other nurses and supervisors. I was paid to deal with patients, so I could deal with [racism] from them. I'm not paid to deal with this from colleagues.

Kenny offers a similar example. Employed as an orthopedic nurse in a predominantly white environment, he describes great difficulty getting promoted, which he primarily attributes to racial biases:

> It's almost like you have to, um, take your ideas and give them to somebody else and then let them present them for you and you get no credit for it. I've applied for several promotions there and, you know, I didn't get them.... When you look around to the, um, the percentage of African Americans who are actually in executive leadership is almost zero percent. Because it's less than one percent of the total population of people that are in leadership, and it's almost like they'll go outside of the system just to try to find a Caucasian to fill a position. Not that I'm not qualified, because I've been master's prepared for 12 years and I'm working on my doctorate.

According to Ryan and Kenny, supervisors' racial biases mean limited opportunities for promotion and upward mobility. This interpretation is consistent with research that suggests that even with stellar performance and solid work histories, Black workers may receive mediocre evaluations from white supervisors that limit their advancement (Feagin 2006; Feagin and Sikes 1994). For Black men nurses, their race may signal to supervisors that they are unworthy of promotion and thus create a different experience with the glass escalator.

… Black men nurses do not speak of warm and congenial relationships with women nurses or see these relationships as facilitating a move into leadership roles. Nor do they suggest that they share gendered bonds with men supervisors that serve to ease their mobility into higher-status administrative jobs. In contrast, they sense that racial bias makes it difficult to develop ties with coworkers and makes superiors unwilling to promote them. Black men nurses thus experience this aspect of the glass escalator differently from their white men colleagues. They find that relationships with colleagues and supervisors stifle, rather than facilitate, their upward mobility.

PERCEPTIONS OF SUITABILITY

Like their white counterparts, Black men nurses also experience challenges from clients who are unaccustomed to seeing men in fields typically dominated by women. As with white men nurses, Black men encounter this in surprised or quizzical reactions from patients who seem to expect to be treated by white women nurses. Ray, a 36-year-old oncology nurse with 10 years of experience, states,

> Nursing, historically, has been a white female's job [so] being a Black male it's a weird position to be in…. I've, several times, gone into a room and a male patient, a white male patient has, you know, they'll say, "Where's the pretty nurse? Where's the pretty nurse? Where's the blonde nurse?."… "You don't have one. I'm the nurse."…

(White) men, by virtue of their masculinity, are assumed to be more competent and capable and thus better situated in (nonfeminized) jobs that are perceived to require greater skill and proficiency. Black men, in contrast, rarely encounter patients (or colleagues and supervisors) who immediately expect that they are doctors or administrators. Instead, many respondents find that even after displaying their credentials, sharing their nursing experience, and, in one case, dispensing care, they are still mistaken for janitors or service workers. Ray's experience is typical:

> I've even given patients their medicines, explained their care to them, and then they'll say to me, "Well, can you send the nurse in?"

Chris describes a somewhat similar encounter of being misidentified by a white woman patient:

> I come [to work] in my white uniform, that's what I wear—being a Black man, I know they won't look at me the same, so I dress the part—I said good evening, my name's Chris, and I'm going to be your nurse. She says to me, "Are you from housekeeping?" … I've had other cases. I've walked in and had a lady look at me and ask if I'm the janitor.

Chris recognizes that this patient is evoking racial stereotypes that Blacks are there to perform menial service work. He attempts to circumvent this very perception through careful self-presentation, wearing the white uniform to indicate his position as a nurse. His efforts, however, are nonetheless met with a racial stereotype that as a Black man he should be there to clean up rather than to provide medical care.

Black men in nursing encounter challenges from customers that reinforce the idea that men are not suited for a "feminized" profession such as nursing. However, these assumptions are racialized as well as gendered. Unlike white men nurses who are assumed to be doctors (see Williams 1992), Black men in nursing are quickly taken for janitors or housekeeping staff. These men do not simply describe a gendered process where perceptions and stereotypes about men serve to aid their mobility into higher-status jobs. More specifically, they describe interactions that are simultaneously raced *and* gendered in ways that reproduce stereotypes of Black men as best suited for certain blue-collar, unskilled labor.

These negative stereotypes can affect Black men nurses' efforts to treat patients as well. The men I interviewed find that masculinity does not automatically endow them with an aura of competency. In fact, they often describe inter-actions with white women patients that suggest that their race minimizes whatever assumptions of capability might accompany being men. They describe several cases in which white women patients completely refused treatment. Ray says,

> With older white women, it's tricky sometimes because they will come right out and tell you they don't want you to treat them, or can they see someone else.

Ray frames this as an issue specifically with older white women, though other nurses in the sample described similar issues with white women of all ages....

... These interactions do not send the message that Black men, because they are men, are too competent for nursing and really belong in higher-status jobs. Instead, these men face patients who mistake them for lower-status service work-ers and encounter white women patients (and their husbands) who simply refuse treatment or are visibly uncomfortable with the prospect. These interactions do not situate Black men nurses in a prime position for upward mobility. Rather, they suggest that the experience of Black men nurses with this particular mecha-nism of the glass escalator is the manifestation of the expectation that they should be in lower-status positions more appropriate to their race and gender....

CONCLUSIONS

Existing research on the glass escalator cannot explain these men's experiences. As men who do women's work, they should be channeled into positions as charge nurses or nursing administrators and should find themselves virtually pushed into the upper ranks of the nursing profession. But without exception, this is not the experience these Black men nurses describe. Instead of benefiting from the basic mechanisms of the glass escalator, they face tense relationships with colleagues, supervisors' biases in achieving promotion, patient stereotypes that inhibit care-giving, and a sense of comfort with some of the feminized aspects of their jobs. These "glass barriers" suggest that the glass escalator is a racialized concept as well as a gendered one. The main contribution of this study is the finding that race and gender intersect to determine which men will ride the glass escalator. The proposition that men who do women's work encounter undue opportunities and advantages appears to be unequivocally true only if the men in question are white....

REFERENCES

Charles, Maria, and David Grusky. 2004. *Occupational ghettos: The worldwide segregation of women and men.* Palo Alto, CA: Stanford University Press.

Collins, Patricia Hill. 1990. *Black feminist thought.* New York: Routledge.

——. 2004. *Black sexual politics.* New York: Routledge.

Duffy, Mignon. 2007. Doing the dirty work: Gender, race, and reproductive labor in historical perspective. *Gender & Society* 21:313–36.

Espiritu, Yen Le. 2000. *Asian American women and men: Labor, laws, and love.* Walnut Creek, CA: AltaMira.

Essed, Philomena. 1991. *Understanding everyday racism.* New York: Russell Sage.

Feagin, Joe. 2006. *Systemic racism.* New York: Routledge.

Feagin, Joe, and Melvin Sikes. 1994. *Living with racism.* Boston: Beacon Hill Press.

Floge, Liliane, and Deborah M. Merrill. 1986. Tokenism reconsidered: Male nurses and female physicians in a hospital setting. *Social Forces* 64: 925–47.

Harvey Wingfield, Adia. 2007. The modern mammy and the angry Black man: African American professionals' experiences with gendered racism in the workplace. *Race, Gender, and Class* 14 (2): 196–212.

Heikes, E. Joel. 1991. When men are the minority: The case of men in nursing. *Sociological Quarterly* 32:389–401.

Pierce, Jennifer. 1995. *Gender trials: Emotional lives in contemporary law firms.* Berkeley: University of California Press.

Snyder, Karrie Ann, and Adam Isaiah Green. 2008. Revisiting the glass escalator: The case of gender segregation in a female dominated occupation. *Social Problems* 55 (2): 271–99.

Williams, Christine. 1989. *Gender differences at work: Women and men in nontraditional occupations.* Berkeley: University of California Press.

——. 1992. The glass escalator: Hidden advantages for men in the "female" professions. *Social Problems* 39 (3): 253–67.

——. 1995. *Still a man's world: Men who do women's work.* Berkeley: University of California Press.

DISCUSSION QUESTIONS

1. What is meant by the "glass escalator?" How does Wingfield's intersectional perspective change prior understandings of how the glass escalator operates?

2. In what specific ways does African American men's race *and* gender shape their experiences in nursing? How would you describe the social class of this occupation?

Student Exercises

1. This part of the book argues for the importance of understanding race, class, and gender as overlapping in shaping people's experiences. Write a short biographical narrative introducing yourself to someone and explain how each of these three together have shaped your experience to date. Are there times when one seems more important than the other? Why?

2. Interview two or three women of color—or men of color—who work in an occupation where they are the numerical minority. How is their experience shaped by race, class, *and* gender? Are there patterns in this person's experience that are comparable to the experiences of the African American men in Wingfield's research?

PART VI

The Continuing Significance of Race

Racial Inequality in Social Institutions

ELIZABETH HIGGINBOTHAM AND MARGARET L. ANDERSEN

The history of discrimination still plagues us. The United States has made progress toward reducing inequality by changing laws that have blocked people's access to education and work opportunities. The passage of key civil rights legislation and important Supreme Court decisions have opened many doors to people of color but they have not dismantled **institutionalized racism**—that is, power and privilege based on race and resulting in discrimination in most areas of life. Discrimination shapes opportunities in key areas for people throughout the nation. Without intervention, discrimination will continue to promote inequalities among citizens and new immigrants.

In this section, we study some major social institutions, organized in five different sections (work, families and communities, housing and education, health care, and the criminal justice system). Each social institution is a site for shaping the everyday experience of all groups. As we will see, racial inequalities pervade these institutional structures, even with laws and policies in place that allegedly make them race-neutral. Simply put, institutionalized racism still operates today.

Section A

Work

Elizabeth Higginbotham and Margaret L. Andersen

We begin with an examination of race in the workplace. Labor exploitation has historically established a racialized labor system with people of color relegated only to certain areas of the labor force. In more recent decades, however, racial minorities work in a wider range of occupations than in the past. The **Civil Rights Act of 1964** outlawed discrimination based on race, creed (that is, religion), national origin, and sex whether in employment, education, or public accommodation (such as restaurants, buses, and trains). Because of this act and other antidiscrimination policies, many workplaces are more diverse, bringing people together. Some racial minorities have entered reasonably well-paid industrial and service jobs, as bus drivers and police officers, for example. The desegregation of education in many places has enabled racial minorities to move into more professional and managerial occupations, but as Adia Harvey Wingfield (in Part V) demonstrates in her study of Black male nurses, people of color still face the day-to-day mistreatment in workplaces, sometimes in the form of microaggressions.

Many of the disparities in employment opportunities can be linked directly to a legacy of exclusion. In their groundbreaking book *Black Wealth/White Wealth*, Melvin Oliver and Thomas Shapiro (2006) argue that exclusions from decent work and other social investments were historic barriers for racial minorities. The effects of those historic barriers persist over time. Because opportunities were denied to racial minorities for decades, even when jobs and opportunities were finally opened in the mid-1960s, racial minorities were still vulnerable. They continue to have a difficult path to advancement. Oliver and Shapiro call this phenomenon the **sedimentation of inequality**.

Efforts to end historic inequalities do not take place in a vacuum. The U.S. political economy has shifted from one based on industrial jobs to one based on service jobs, a process called **economic restructuring**. Technological changes mean that we need fewer unskilled and semiskilled workers in our industries, particularly within national borders. Economic restructuring produces jobs at the upper end of the workforce that require high levels of skill. Without higher education and training, workers are limited to low-wage employment, often in the growing personal services market. Given that racism is embedded in institutions, the options open to people at a time of economic restructuring can be influenced by discriminatory practices from the past. We can see this pattern in the location of new jobs, as they are not likely to be within the communities where many people of color, especially African Americans and Latinos/as live.

The sedimentation of inequality also helps you to understand various vulnerabilities in a shifting economy. People who lack social networks that can lead to jobs are at a disadvantage in a shrinking labor market. As global competition becomes more intense, some racial minority workers become vulnerable to new forms of job exclusion and blocked mobility. Other racial groups, such as better-educated Asian immigrants and Asian Americans, can find a niche in the new workplaces. The articles included here examine some of the dynamics that shape employment options.

William Julius Wilson ("Toward a Framework for Understanding Forces that Contribute to or Reinforce Racial Inequality") examines economic and political forces that can appear to be race neutral, but indirectly contribute to racial inequality. Technological innovations have revolutionized the workplace, and global trade has shifted job locations, often reshaping cities that were industrial hubs. This new landscape makes it difficult for African Americans with various levels of educational attainment to find work.

How do people navigate this shifting economic environment? Whether one can find a good job is, as the popular adage goes, often a matter of who you know. Deirdre A. Royster ("Race and the Invisible Hand: How White Networks Exclude Black Men from Blue-Collar Jobs") shows that racism trumps the "invisible hand" of the market by limiting the networks and connections that Black men have, leaving them less likely than White men to find employment, even in working-class jobs. Her research provides a rich account of the processes by which discrimination operates against Black men, as well as other groups who do not have access to White networks.

After decades of limited immigration, new policies and a new economy have made the United States a major destination for immigrants from around the globe, introducing new workplace challenges. Immigrants are settling in more regions of the country, including the South, Midwest, and rural areas, areas that have historically not been so diverse, but are now new gateway communities. How do immigrants confront a labor market marked by patterns of segregation and inequality? Chenoa A. Flippen ("Intersectionality at Work: Determinants of Labor Supply among Immigrant Latinas") elaborates on the factors in labor force participation for Latinas in the new gateway community of Durham, North Carolina. The women she studied face many levels of disadvantage: race, gender, social class, and immigration status. Flippen shows how the particular social location of women (meaning their immigration status, family structure, gender, and so forth) shapes their work opportunities.

Understanding the racial structures in the workplace and the problems they pose for people who are seeking work can help us think about the nature of future policy interventions. Employment is a major source of inequality in our time, creating comfortable lifestyles for many, marginality or poverty for others. As you read these articles, you might think about how we can build a sustainable workforce that is not based on exclusion and exploitation.

REFERENCE

Oliver, Melvin, and Thomas Shapiro. 2006. *Black Wealth/White Wealth: A New Perspective on Racial Inequality*, 2nd ed. New York: Routledge.

FACE THE FACTS: WAGES BY RACE AND GENDER IN OCCUPATIONAL CATEGORIES

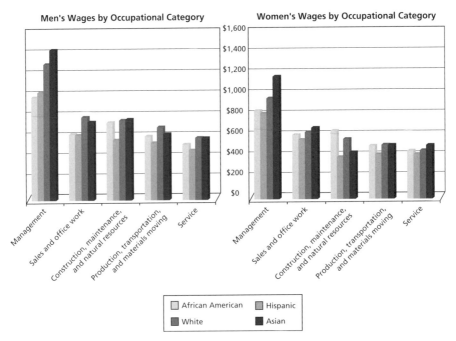

Men's Wages by Occupational Category   Women's Wages by Occupational Category

African American   Hispanic
White   Asian

SOURCE: Bureau of Labor Statistics. 2013. *Employment and Earnings.* Washington, DC: U.S. Department of Labor. www.bls.gov

Think about It:

1. What facts do you see reflected in these two graphs? What stands out to you?

2. Looking at the wages earned, can you identify some important differences across gender, even within racial groups?

Keep in mind that these five occupational categories are broad categories used by the Department of Labor to include a wide array of occupations. For example, "management" includes such things as CEOs of major companies, as well as small business owners and professional workers, such as teachers, social workers, nurses, and other professionals. With this in mind, what patterns of occupational segregation by race do you think influence the differences in wages for different groups? (You can look up more detailed information in Employment and Earnings on the Department of Labor website: www.dol.gov).

21

Toward a Framework for Understanding Forces that Contribute to or Reinforce Racial Inequality

WILLIAM JULIUS WILSON

We still have individuals who discriminate in hiring and promotion. These are individual acts, but there are structural processes that also affect racial inequality. Wilson highlights how economic and political changes, such as the shifts in the nature and location of jobs and government responses, shape the employment options for African Americans. The legacy of exclusion in education and employment continues to reverberate.

UNDERSTANDING THE IMPACT OF STRUCTURAL FORCES

Two types of structural forces contribute directly to racial group outcomes such as differences in poverty and employment rate: social acts and social processes. "Social acts" refers to the behavior of individuals within society. Examples of social acts are stereotyping; stigmatization; discrimination in hiring, job promotions, housing, and admission to educational institutions—as well as exclusion from unions, employers' associations, and clubs—when any of these are the act of an individual or group exercising power over others.

"Social processes" refers to the "machinery" of society that exists to promote ongoing relations among members of the larger group. Examples of social processes that contribute directly to racial group outcomes include laws, policies, and institutional practices that exclude people on the basis of race or ethnicity. These range from explicit arrangements such as Jim Crow segregation laws and voting restrictions to more subtle institutional processes, such as school tracking that purports to be academic but often reproduces traditional segregation, racial profiling by police that purports to be about public safety but focuses solely on minorities, and redlining by banks that purports to be about sound fiscal policy

SOURCE: From Wilson, William Julius. 2009. "Toward a Framework for Understanding Forces that Contribute to or Reinforce Racial Inequality." *Race and Social Problems* 1: 3–11.

but results in the exclusion of people of color from home ownership. In all of these cases, ideologies about group differences are embedded in organizational arrangements.

However, many social observers who are sensitive to and often outraged by the direct forces of racism, such as discrimination and segregation, have paid far less attention to those political and economic forces that *indirectly* contribute to racial inequality. I have in mind political actions that have an impact on racial group outcomes, even though they are not explicitly designed or publicly discussed as matters involving race, as well as impersonal economic forces that reinforce longstanding forms of racial inequality. These structural forces are classified as indirect because they are mediated by the racial groups' position in the system of social stratification (the extent to which the members of a group occupy positions of power, influence, privilege, and prestige). In other words, economic changes and political decisions may have a greater adverse impact on some groups than on others simply because the former are more vulnerable as a consequence of their position in the social stratification system....

Take, for instance, impersonal economic forces, which sharply increased joblessness and declining real wages among many poor African Americans in the last several decades. As with all other Americans, the economic fate of African Americans is inextricably connected with the structure and functioning of a much broader, globally influenced modern economy. In recent years, the growth and spread of new technologies and the growing internationalization of economic activity have changed the relative demand for different types of workers. The wedding of emerging technologies and international competition has eroded the basic institutions of the mass production system and eradicated related jobs in manufacturing in the United States. In the last several decades, almost all of the improvements in productivity have been associated with technology and human capital, thereby drastically reducing the importance of physical capital and natural resources. The changes in technology that are producing new jobs are making many others obsolete.

Although these trends tend to benefit highly educated or highly skilled workers, they have contributed to the growing threat of job displacement and eroding wages for unskilled workers. This development is particularly problematic for African Americans who have a much higher proportion of workers in low-skilled jobs than whites....

The workplace has been revolutionized by technological changes that range from mechanical development like robotics, to advances in information technology like computers and the internet. While even educated workers are struggling to keep pace with technological changes, lower-skilled workers with less education are falling behind with the increased use of information-based technologies and computers and face the growing threat of job displacement in certain industries. To illustrate, in 1962 the employment-to-population ratio—the percentage of adults who are employed—was 52.5% for those with less than a high school diploma, but by 1990 it had plummeted to 37.0%. By 2006 it rebounded slightly to 43.2%, possibly because of the influx of low-skilled Latino immigrants in low-wage service-sector jobs.

In the new global economy, highly educated, well-trained men and women are in demand, as illustrated most dramatically in the sharp differences in employment experiences among men. Compared to men with lower levels of education, college-educated men spend more time working, not less. The shift in the demand for labor is especially devastating for those low-skilled workers whose incorporation into the mainstream economy has been marginal or recent. Even before the economic restructuring of the nation's economy, low-skilled African Americans were at the end of the employment line, often the last to be hired and the first to be let go.

The computer revolution is a major reason for the shift in the demand for skilled workers. Even "unskilled" jobs such as fast food service require employees to work with computerized systems, even though they are not considered skilled workers. Whereas only one-quarter of U.S. workers directly used a computer on their jobs in 1984, by 2003 that figure had risen to more than half (56.1%) the workforce....

The shift in the United States away from low-skilled workers can also be related to the growing internationalization of economic activity, including increased trade with countries that have large numbers of low-skilled, low-wage workers. Two developments facilitated the growth in global economic activity: (1) advances in information and communication technologies, which enabled companies to shift work to areas around the world where wages for unskilled work are much lower than in the "first world;" and (2) the expansion of free trade, which reduced the price of imports and raised the output of export industries. But increasing imports that compete with labor-intensive industries (e.g., apparel, textile, toys, footwear, and some manufacturing) hurts unskilled labor.

Since the late 1960s international trade has accounted for an increasing share of the U.S. economy, and, beginning in the early 1980s, imports of manufactured goods from developing countries have soared. According to economic theory, the expansion of trade with countries that have a large proportion of relatively unskilled labor will result in downward pressure on the wages of low-skilled Americans because of the lower prices of the goods those foreign workers produce. Because of the concentration of low-skilled black workers in vulnerable labor-intensive industries (e.g., 40% of textile workers are African American even though blacks are only about 13% of the general population; this overrepresentation is typical in many low-skill industries), developments in international trade are likely to further exacerbate their declining labor market experiences.

Note that the sharp decline in the relative demand for low-skilled labor has had a more adverse effect on blacks than on whites in the United States because a substantially larger proportion of African Americans are unskilled. Indeed, the disproportionate percentage of unskilled African Americans is one of the legacies of historic racial subjugation. Black mobility in the economy was severely impeded by job discrimination as well as failing segregated public schools, where per-capita expenditures to educate African American children were far below amounts provided for white public schools. While the more educated and highly trained African Americans, like their counterparts among other racial groups, have very likely benefited from the shifts in labor demand, those with lesser skills have suffered....

The economic situation for many African Americans has now been further weakened because they tend not only to reside in communities that have higher jobless rates and lower employment growth—for example, places like Detroit or Philadelphia—but also they lack access to areas of higher employment growth....

The growing suburbanization of jobs means that labor markets today are mainly regional, and long commutes in automobiles are common among blue-collar as well as white-collar workers. For those who cannot afford to own, operate, and insure a private automobile, the commute between inner-city neighborhoods and suburban job locations becomes a Herculean task. For example, Boston welfare recipients found that only 14% of the entry-level jobs in the fast-growth areas of the Boston metropolitan region could be accessed via public transit in less than 1 hour. And in the Atlanta metropolitan area, fewer than one-half of the entry level jobs are located within a quarter mile of a public transit systems. To make matters worse, many inner-city residents lack information about suburban job opportunities. In the segregated inner-city ghettos, the breakdown of the infor-mal job information network magnifies the problems of *job spatial mismatch*—the notion that work and people are located in two different places.

Although racial discrimination and segregation exacerbate the labor-market problems of the low-skilled African Americans, many of these problems are cur-rently driven by shifts in the economy. Between 1947 and the early 1970s, all income groups in America experienced economic advancement. In fact, poor families enjoyed higher growth in annual real income than did other families. In the early 1970s, however, this pattern began to change. American families in higher income groups, especially those in the top 20%, continued to enjoy steady income gains (adjusted for inflation), while those in the lowest 40% expe-rienced declining or stagnating incomes. This growing disparity in income, which continued through the mid-1990s, was related to a slowdown in produc-tivity growth and the resulting downward pressure on wages....

More than any other group, low-skilled workers depend upon a strong eco-nomy, particularly a sustained tight labor market—that is, one in which there are ample jobs for all applicants. In a slack labor market—a labor market with high unemployment—employers can afford to be more selective in recruiting and granting promotions. With fewer jobs to award, they can inflate job require-ments, pursuing workers with college degrees, for example, in jobs that have tra-ditionally been associated with high school-level education. In such an economic climate, discrimination rises and disadvantaged minorities, especially those with low levels of literacy, suffer disproportionately.

Conversely, in a tight labor market job vacancies are numerous, unem-ployment is of short duration, and wages are higher. Moreover, in a tight labor market the labor force expands because increased job opportunities not only reduce unemployment but also draw in workers who previously dropped out of the labor force altogether during a slack labor market period. Thus, in a tight labor market the status of all workers—including disadvantaged minorities—improves....

Undoubtedly, if the robust economy could have been extended for several more years, rather than coming to an abrupt halt in 2001, joblessness and concentrated poverty in inner cities would have declined even more. Nonetheless, many people concerned about poverty and rising inequality have noted that productivity and economic growth are only part of the picture.

Thanks to the Clinton-era economic boom, in the latter 1990s there were signs that America's rising economic inequality that began in the early 1970s was finally in remission. Nonetheless, worrisome questions were raised by many observers at that time: Will this new economy eventually produce the sort of progress that prevailed in the two and a half decades prior to 1970—a pattern in which a rising tide did indeed lift all boats? Or would the government's social and economic policies prevent us from duplicating this prolonged pattern of broadly equal economic gains? In other words, the future of ordinary families, especially poor working families, depends a great deal on how the government decides to react to changes in the economy, and often this reaction has a profound effect on racial outcomes....

During Bill Clinton's 8 years in office, redistribution measures were taken to increase the minimum wage. But the George W. Bush administration halted increases in the minimum wage for nearly a decade, until the Democrats regained control of Congress in 2006 and voted to again increase the minimum wage in 2007. All of these political acts contributed to the decline in real wages experienced by the working poor. Because people of color are disproportionately represented among the working poor, these political acts have reinforced their position in the bottom rungs of the racial stratification ladder. In short, in terms of structural factors that contribute to racial inequality, there are indeed nonracial political forces that definitely have to be taken into account.

DISCUSSION QUESTIONS

1. Why does Wilson look beyond social acts of discrimination to explore the continuation of racial inequality?

2. Why are many African Americans unable to compete successfully in the new economy?

3. How can you apply Wilson's argument about the indirect effect of economic and political forces to understanding the prospects for different groups in today's economy?

22

Race and the Invisible Hand
How White Networks Exclude Black Men from Blue-Collar Jobs

DEIRDRE A. ROYSTER

The "invisible hand" refers to the fact that, who you know shapes employment options. By comparing the experiences of Black and White working-class men who graduated from the same vocational high school, Royster shows how race can shape Black men's access to job networks—making who you know a critical factor in their economic well-being.

In the late 1970s, [African American sociologist William Julius Wilson] published an extremely influential book, *The Declining Significance of Race*. In this book, Wilson argued that race was becoming less and less important in predicting the economic possibilities for well-educated African Americans. In other words, the black-led Civil Rights movement had been successful in removing many of the barriers that made it difficult, if not impossible, for well-trained blacks to gain access to appropriate educational and occupational opportunities. Wilson argued that this new pattern of much greater (but not perfect) access was unprecedented in the racial stratification system in the United States and that it would result in significant and lasting gains for African American families with significant educational attainment.

Recent research on the black middle class has only partially supported Wilson's optimistic prognosis. While blacks did experience significant educational and occupational gains during the 1970s, their upward trajectory appears to have tapered off in the 1980s and 1990s. Moreover, some blacks have found themselves tracked into minority-oriented community relations positions within the professional and managerial occupational sphere. Even more troubling are data indicating that the proportion of blacks who attend and graduate from college appears to be shrinking, with the inevitable result that fewer blacks will have the credentials and skills necessary to get the better jobs in the growing technical and professional occupational categories. Despite real concerns about the stability

SOURCE: From Royster, Deirdre A. 2003. *Race and the Invisible Hand: How White Networks Exclude Black Men from Blue-Collar Jobs*. Berkeley, CA: University of California Press, pp. 18–23, 58–59, 180–189. Copyright © 2003 by The Regents of the University of California. Reprinted by permission.

of the black middle class and some glitches in the workings of the professional labor market, no one doubts that a substantial portion of the black population now enjoys access to middle-class opportunities and amenities—including decent homes, educational facilities, public services, and most importantly, jobs—commensurate with their substantial education and job experience, or in economic terms, their endowment of human capital.

While other scholars were investigating his theories about the black middle class, Wilson became distressed about the pessimistic prospects of blacks who were both poorly educated and increasingly isolated in urban ghettos with high rates of poverty and unemployment. His main concern was that changing labor demands that increase opportunities for highly skilled workers have the potential of making unskilled black labor obsolete. According to Wilson's next two books, *The Truly Disadvantaged* and *When Work Disappears*, this group's inability to gain access to mobility-enhancing educational opportunities is exacerbated by the further problems of a deficiency of useful employment contacts, lack of reliable transportation, crowded and substandard housing options, a growing sense of frustration, and an image among urban employers that blacks are undesirable workers, not to mention the loss of manufacturing and other blue-collar jobs. These factors, and a host of others, contribute to the extraordinarily difficult and unique problems faced by the poorest inner-city blacks in attempting to advance economically. Wilson and hundreds of other scholars—even those who disagree with certain aspects of his thesis—argue that this group needs special assistance in order to overcome the obstacles they face.

If Wilson intended the *Declining Significance of Race* thesis (and its underclass corollary) to apply mainly to well-educated blacks and ghetto residents, then Wilson only explained the life chances of at most 30 to 40 percent of the black population.[1] The rest of the black population neither resides in socially and geographically isolated ghettoes, nor holds significant human capital, in the form of college degrees or professional work experience. Looking at five-year cohorts beginning at the turn of the century, Mare found that the cohort born between 1946 and 1950 reached a record high when 13 percent of its members managed to earn bachelor's degrees. Recent cohorts born during the Civil Rights era (1960s) have not reached the 13 percent high mark set by the first cohort to benefit from Civil Rights era victories. As a result, today the total percentage of African Americans age 25 and over who have four or more years of college is just under 14 percent.[2] According to demographer Reynolds Farley, while college attendance rates for white males (age 18–24) have rebounded from dips in the 1970s back to about 40 percent, black male rates have remained constant at about 30 percent since the 1960s. Figures like these suggest that Civil Rights era "victories" have not resulted in increasing percentages of blacks gaining access to college training. Instead, most blacks today attempt to establish careers with only modest educational credentials, just as earlier cohorts did. Thus the vast majority of blacks are neither extremely poor nor particularly well educated; most blacks would be considered lower-middle- or working-class and modestly educated. That is, most blacks (75 percent) lack bachelor's degrees but hold high school diplomas or GEDs; most blacks (92 percent) are working rather than unemployed; and most

(79 percent) work at jobs that are lower-white- or blue-collar rather than professional.[3] Given that modestly educated blacks make up the bulk of the black population, it is surprising that more attention has not been devoted to explicating the factors that influence their life chances.

Wilson's focus on the extremes within the black population, though understandable, points to a troubling under-specification in his thesis: it is unclear whether Wilson sees individuals with modest educational credentials—high school diplomas, GEDs, associate's degrees, or some college or other postsecondary training, but not the bachelor's degree—as co-beneficiaries of civil rights victories alongside more affluent blacks. The logic of his thesis implies that as long as they do not reside in socially and geographically isolated communities filled with poor and unemployed residents, from which industrial jobs have departed, then modestly educated blacks, like highly educated blacks, ought to do about as well as their white counterparts....

Because he argues that *past* racial discrimination created the ghetto poor, or underclass, while macro-economic changes—and not current racial discrimination—explain their current economic plight, Wilson's perspective implies that white attempts to exclude blacks are probably of little significance today. In addition, Wilson offers a geographic, rather than racial, explanation for whites' labor market advantages when he argues that because most poor whites live outside urban centers, they do not suffer the same sort of structural dislocation or labor obsolescence as black ghetto residents. If Wilson's reasoning holds, there is no reason to expect parity among the poorest blacks and whites in the United States without significant government intervention. Despite a conspicuous silence regarding the prospects for parity among modestly educated blacks and whites, Wilson's corpus of research and theory offers the most race and class integrative market approach available. First, Wilson specifies how supply and demand mechanisms work differently for blacks depending on their class status. Specifically, he argues that there is now a permanent and thriving pool (labor supply) of educationally competitive middle-class blacks, while simultaneously arguing that changes in the job structure in inner cities have disrupted the employment opportunities (labor demand) for poorer blacks. Second, Wilson argues that contemporary racial disparity results, by and large, from the structural difficulties faced by poor blacks rather than racial privileges enjoyed by (or racial discrimination practiced by) poor or more affluent whites. One of the questions guiding this study is whether the life chances of modestly educated whites and blacks are becoming more similar, as with blacks and whites who are well educated, or more divergent, as with blacks and whites on the bottom....

The fifty young men interviewed for this study may have been, in some ways, atypical. For example, none of them had dropped out of school, and all were extremely polite and articulate. I suspect that these men were among the easiest to contact because of high residential stability and well-maintained friendship networks. Their phone numbers had remained the same or they had kept in touch with friends since graduating from high school two to three years earlier. Because of these factors, I may have tapped into a sample of men who were more likely to be success stories than most would have been. Researchers call this sampling dilemma creaming, because the sample may reflect those who were most likely to rise to the top or be seen as the cream of the crop, rather than those of average or mixed potential. In this study, however, it may have been an advantage to have "creamed," since I wanted to compare black and white men with as much potential for success as possible. Moreover, men who are personable and who have stable residences and friendship networks might be most able to tap into institutional and personal contacts in their job searches—one of my main research queries....

While I don't think there were idiosyncratic differences among the black and white men I studied or between the men I found and their same-race peers that I didn't find, some of the positive attributes of my sample suggest that my findings may generalize only to young men who generally play by the rules. Of course, it isn't all that clear what proportion of young working-class males (black or white) try to play by the rules—maybe the vast majority try to do so. Nor is it clear by what full set of criteria my subjects did, in fact, play by the rules. My sample includes men who had brushes with the law as well as some who might be considered "goody two-shoes." In other words I'm not sure that the specific men with whom I spoke are atypical among working-class men, but I am willing to acknowledge that they may be. Perhaps what is most important to remember about the sample, who seemed to me to be pretty ordinary, "All-American" men, is my contention that this set of men ought to have similar levels of success in the blue-collar labor market—if, that is, we have finally reached a time when race doesn't matter....

One narrative, the achievement ideology, asserts that formal training, demonstrated ability, and appropriate personal traits will assure employment access and career mobility. The second narrative, the contacts ideology, emphasizes personal ties and affiliations as a mechanism for employment referrals, access, and mobility.... The achievement ideology has persistently dominated American understanding of occupational success, even though everyone, it seems, is willing to admit that "who you know" is at least as important as "what you know" in gaining access to opportunities in American society. All of the men in this study, for example, said that contacts were very important in establishing young men like themselves in careers. One offered a more nuanced explanation: "It's not [just] who you know, it's how they know you." That is, it is not simply knowing the right people that matters; it is sharing the right sort of bonds with the right people that influences what those people would be willing to do to assist you.

The black and white men in this study had more achievements in common than contacts. They were trained in the same school, in many of the same trades, and by the same instructors. They had formal access to the same job listing services and work-study programs. Instructors and students alike agreed, and records confirm, that in terms of vocational skills and performance, the blacks and whites in this sample were among the stronger students....

Black Glendale graduates trail behind their white peers. They are employed less often in the skilled trades, especially within the fields for which they have been trained; they earn less per hour; they hold lower status positions; they receive fewer promotions; and they experience more and longer periods of unemployment. No set of educational, skill, performance, or personal characteristics unique to either the black or white students differentiates them in ways that would explain the unequal outcomes.... Only their racial status and the way it situates them in racially exclusive networks during the school-work transition process adequately explain their divergent paths from seemingly equal beginnings.

In this study of one variant of the "who you know" versus "what you know" conundrum, it is manifestly and perpetually evident that racial dynamics are a key arbiter of employment outcome. Yet challenging the power of the achievement ideology in American society requires a careful exposition of how factors such as race throw a wrench into the presumption of meritocracy. In addition, the contacts ideology must be uniquely construed to take into account the significance of racially determined patterns of affiliation within a class, in this case the working class.

RACE, AN ARBITER OF EMPLOYMENT NETWORKS

Researchers have long argued that black males lack access to the types of personal contacts that white males appear to have in abundance.[4] I would argue that it's more than not having the right contacts. In terms of social networks, black men are at a disadvantage in terms of configuration, content, and operation, a disadvantage that is exacerbated in sectors with long traditions of racial exclusion, such as the blue-collar trades. Even when blacks and whites have access to some of the same connections, as in this study, care must be taken to examine exactly what transpires. For example, I noted that black and white males were assisted differently by the same white male teachers. If I had only asked students whether they considered their shop teachers contacts on whom they could rely, equal numbers of black and white males would have answered affirmatively. But this would have told us nothing about *how* white male teachers *chose to know and help* their black male students. The teachers chose to verbally encourage black students, while providing more active assistance to white students. I discovered a munificent flow of various forms of assistance, including vacancy information, referrals, direct job recruitment, formal and informal training, vouching behaviors, and leniency in supervision. For white students, this practice, which repeated neighborhood and community patterns within school walls, served to convert institutional ties (as teachers) into personal ones (as friends) that are intended to and do endure well beyond high school....

Even without teachers consciously discriminating, significant employment information and assistance remained racially privatized within this public school context. In that white male teachers provided a parallel or shadow transition system for white students that was not equally available to black students, segregated networks still governed the school-work transition at Glendale even though classrooms had long been desegregated.

The implications for black men are devastating. Despite having unprecedented access to the same preparatory institution as their white peers, black males could not effectively use the institutional connection to establish successful trade entry. Moreover, segregation in multiple social arenas, beyond schools, all but precluded the possibility of network overlaps among working-class black and white men. As a result, black men sought employment using a truncated, resource-impoverished network consisting of strong ties to other blacks (family, friends, and school officials) who like themselves lacked efficacious ties to employment.

Beyond school, matters were even worse. Without being aware of it, white males' descriptions of their experiences revealed a pattern of intergenerational intraracial assistance networks among young and older white men that assured even the worst young troublemaker a solid place within the blue-collar fold. The white men I studied were not in any way rugged individualists; rather they survived and thrived in rich, racially exclusive networks.

For the white men, neighborhood taverns, restaurants, and bars served as informal job placement centers where busboys were recruited to union apprentice programs, pizza delivery boys learned to be refrigeration specialists, and dishwashers studying drafting could work alongside master electricians then switch back to drafting if they wished. I learned of opportunities that kept coming, even when young men weren't particularly deserving. One young man had been able to hold onto his job after verbally abusing his boss. Another got a job installing burglar alarms after meeting the vice president of the company at a cookout—without ever having to reveal his prison record, which included a conviction for burglary.

Again and again, the white men I spoke with described opportunities that had landed in their laps, not as the result of outstanding achievements or personal characteristics, but rather as the result of the assistance of older white neighbors, brothers, family friends, teachers, uncles, fathers, and sometimes mothers, aunts, and girlfriends (and their families), all of whom overlooked the men's flaws. It never seemed to matter that the men were not A students, that they occasionally got into legal trouble, that they lied about work experiences from time to time, or that they engaged in horseplay on the job. All of this was expected, brushed off as typical "boys will be boys" behavior, and it was sometimes the source of laughter at the dining room table. In other words, there were no significant costs for white men associated with being young and inexperienced, somewhat immature, and undisciplined.

The sympathetic pleasure I felt at hearing stories of easy survival among working-class white men in an era of deindustrialization was only offset by the depressing stories I heard from the twenty-five black men. Their early employment experiences were dismal in comparison, providing a stark and disturbing contrast. Whereas white men can be thought of as the second-chance kids, black men's opportunities were so fragile that most could not have recovered from even the relatively insignificant mishaps that white men reported in passing.

Black men were rarely able to stay in the trades they studied, and they were far less likely than white men to start in one trade and later switch to a different one, landing on their feet. Once out of the skilled trade sphere, they sank to the low-skill service sector, usually retail or food services. The black men had numerous experiences of discrimination at the hands of older white male super-visors, who did not offer to help them and frequently denigrated them, using familiar racial epithets. The young black men I spoke with also had to be careful when using older black social contacts. More than one man indicated to me that, when being interviewed by a white person, the wisest course of action is to behave as if you don't know anyone who works at the plant, even if a current worker told you about the opening. These young black men, who had been on the labor market between two and three years, were becoming discouraged. While they had not yet left the labor force altogether, many (with the help of parents) had invested time and resources in training programs or college courses that they and their families hoped would open up new opportunities in or beyond the blue-collar skilled labor market. Many of the men had begun to lose the skills they had learned in high school; others, particularly those who'd had a spell or two of unemployment, showed signs of depression.

My systematic examination of the experiences of these fifty matched young men leads me to conclude that the blue-collar labor market does not function as a market in the classic sense. No pool of workers presents itself, offering sets of skills and work values that determine who gets matched with the most and least desir-able opportunities. Rather, older men who recruit, hire, and fire young workers choose those with whom they are comfortable or familiar. Visible hands trump the "invisible hand"—and norms of racial exclusivity passed down from genera-tion to generation in American cities continue to inhibit black men's entry into the better skilled jobs in the blue-collar sector.

Claims of meritocratic sorting in the blue-collar sector are simply false; equally false are claims that young black men are inadequately educated, inher-ently hostile, or too uninterested in hard work or skill mastery to be desirable workers. These sorts of claims seek to locate working-class black men's employ-ment difficulties in the men's alleged deficits—bad attitudes, shiftlessness, poor skills—rather than in the structures and procedures of worker selection that are typically under the direct control of older white men whose preferences, by cus-tom, do not reflect meritocratic criteria.

Few, if any, political pressures, laws, or policies provide sufficient incentives or sanctions to prevent such employers from arbitrarily excluding black workers or hiring them only for menial jobs for which they are vastly overqualified. More-over, in recent years, affirmative action policies that required that government contracts occasionally be awarded to black-owned firms or white-owned firms that consistently hire black workers have come under attack—eroding the paltry incentives for inclusion set forth during the Civil Rights era, nearly forty years ago. Indeed, there is far less pressure today than in the past for white-owned firms to hire black working men. And given persistent patterns of segregation—equivalent to an American apartheid, according to leading sociologists—there remain few incentives for white men to adopt young black men into informal, neighborhood-generated networks. As a result, occupational apartheid reigns in the sector that has always held the greatest potential for upward mobility, or just basic security, for modestly educated Americans.

IDEOLOGY AND THE DEFENSE OF RACIALIZED EMPLOYMENT NETWORKS

The public perception of the causes of black men's labor difficulties—namely, that the men themselves are to be blamed—contrasts with my findings. And my research is consistent with that of hundreds of social scientists who have demonstrated state-supported and informal patterns of racial exclusion in housing, education, labor markets, and even investment opportunities. Racism continues to limit the life chances of modestly educated black men.…

WHITE PRIVILEGE, BLACK ACCOMMODATION

How, then, do black males, if they wish to earn a living in the surviving trades, negotiate training and employment opportunities in which networks of gatekeepers remain committed to maintaining white privilege? The present research suggests that the options are few: either accommodation to the parameters of a racialized system or failure in establishing a successful trade career. The interviews revealed that forms of black accommodation begin early, as when young men avoided training in trades of interest because they were known to hold little promise for integration and advancement. For those who made such discoveries later, accommodation took the form of disengaging from specific trades, such as electrical construction, and pursuing whatever jobs became available. For some, the disengaging process involved the claim that they were never really committed to the original trade field, but I suspect that such claims merely served to soften the blow of almost inevitable career failure. For the determined, accommodation required suppressing anger at racially motivated insults and biased employment decisions in the majority-white trade settings. If this strategy wore thin, two difficult accommodations remained. The first involved finding a work setting—not necessarily within one's trade—in which the workplace culture was, if not actively receptive to black inclusion, at least neutral. The second involved finding ways to work in white-dominated fields without having to work beside whites.

A word needs to be said about a particularly troubling accommodative behavior adopted by the black men: not actively and persistently pursuing offers of assistance. It is not clear to what extent the black men were fully cognizant of the extent and potency of whites' informal networks or of the cultural norms governing their operation. But, while it is evident that the older white men who were network gatekeepers did not extend the same access and support to the black men, the black men may also have been less proactive in pursuing older white men who might have assisted them.

Generally, the white men in the study appear to have more actively followed up on offers of assistance. And although their careers developed much more smoothly than those of the black men, they were certainly not without the difficulties of not being hired, workplace dissatisfaction, competing vocational interests, and unemployment. Nevertheless, they returned, sometimes repeatedly, to contacts for further assistance. Certainly, demonstrating proactivity toward a typically racially exclusive white network would be especially problematic for black men.

Undoubtedly, black men's exclusion from white personal settings where easy informal contact is facilitated, like neighborhoods and family, contributes to black men's reluctance to pursue whites for assistance. In addition, black men's lack of personal familiarity with normative expectations among whites probably hampers their efforts to imitate their white peers' more forward network behaviors. Furthermore, any efforts by blacks to engage in such behaviors might not be similarly regarded as appropriate, and might instead be interpreted as aggressive, "uppity," or indicative of a feeling of entitlement. Finally, black men's early experiences of racial exclusion, bias, and hostility in the school and the workplace inform not only their assessment of employment prospects, but also their actual employment strategies. Given these complicated contingencies, perhaps the somewhat hesitant responses of black men are, on the whole, not unreasonable.

CONCLUSION

Black men have paid a great price for exclusion from blue-collar trades and the networks that supply those trades, but they have not paid it alone. The pain of black men's unemployment and underemployment spreads across black communities in a ripple effect. Less able to contribute financially to the care of children and parents, or to combine resources with black women or assist other men with work entry and "learning the ropes" on the job, black men withdraw from the support structures that they need and that they are needed to support emotionally as well as economically. The enduring power of segregated networks in the blue-collar trades is as responsible as segregated neighborhoods for the existence of extremely poor and isolated black communities and of the disproportionately black and male prison population—in fact, more so. While many black families live in stable communities that are mostly, if not entirely, black, the inability to find remunerative jobs that do not require expensive college training makes living decently anywhere extremely difficult. And the loss of manufacturing jobs cannot account for black men's underemployment in the remaining blue-collar fields—especially construction, auto mechanics, plumbing, computer repair, and carpentry....

My findings demonstrate that, without governmental initiatives that provide strong incentives for inclusion, white tradesmen will have no reason to open their networks to men of color. As a result, the work trajectories of white and black men who start out on an equal footing will continue to diverge into skilled and unskilled work paths because of business-as-usual patterns of exclusion. Although there are few precedents for intervening in the private sector, there are strong precedents for intervening in the public sector, where the tax dollars of majority and minority citizens must not be redistributed in ways that condone customs of exclusion.…

Without the government taking a lead, the young black men I studied—who played by the rules—are unlikely to ever reach their potential as skilled workers or to take their places as blue-collar entrepreneurs, as so many of their white peers are poised to do. This tragedy could have been averted. My hope is that it will be averted in the next generation.

NOTES

1. Haywood Horton, Beverlyn Lundy, Cedric Herring, and Melvin E. Thomas, "Lost in the Storm: The Sociology of the Black Working Class, 1850 to 1990," *American Sociological Review* 65, no. 1 (2000): 128–137.

2. Robert Mare, "Changes in Educational Attainment and School Enrollment," in *State of the Union*, ed. Reynolds Farley (New York: Russell Sage Foundation, 1995); Nancy Folbre, *The Field Guide to the U.S. Economy* (New York: New Press, 1999).

3. Ibid.

4. Richard Freeman and Harry J. Holzer, *The Black Youth Employment Crisis* (Chicago: University of Chicago Press, 1986); William Julius Wilson, *The Truly Disadvantaged* (Chicago: University of Chicago Press, 1987); Paul Osterman, *Getting Started: The Youth Labor Market* (Cambridge: Massachusetts Institute of Technology Press, 1980).

DISCUSSION QUESTIONS

1. Why is economic advancement difficult for inner-city Black men with modest levels of education?

2. What are the differences in how networks operated for Black men and White men in Royster's study? What are the implications of those outcomes for their futures?

3. Rather than just let market forces work as they will, why does Royster think government intervention is important?

23

Intersectionality at Work

Determinants of Labor Supply among Immigrant Latinas

CHENOA A. FLIPPEN

Now, as in other eras, employers hire immigrants in low-skill positions, but we need to understand this issue from the perspective of women. Immigrant women have particularly complex situations because of their family, gender, and immigrant status. In her research on Latinas in North Carolina, Flippen demonstrates the complexities of factors that push and pull women between work and family.

Latin American immigration to the United States has grown sharply in recent years, with the number of foreign-born Latinos topping 21 million in 2010. As their numbers have grown, the prospects for immigrants' economic incorporation have become an increasingly pressing concern in popular, policy, and academic circles alike. While Latino immigration has long been disproportionately male, a growing number of married and unmarried women have not only entered the country in recent decades, they have also increasingly joined the labor force as well. Indeed, with stagnating male wages in the lower segment of the labor market and the growth of female-headed households, women's employment has become a progressively more important determinant of family economic wellbeing.

In spite of the importance of these trends, we still know relatively little about the forces shaping employment patterns among immigrant Latinas. Most examinations of immigrant incorporation into the labor market tend to focus on men, and our understanding of the forces that are unique to Latinas remains poorly developed. An important and growing body of work on intersectionality emphasizes that women's outcomes are not a simple extension of the male experience, and that researchers must explicitly consider how the multiple dimensions of stratification are intertwined....

... This article examines the determinants of multiple facets of labor supply among immigrant Latinas in Durham, North Carolina, a new immigrant destination. I examine how women's position in three different institutional spheres structures their labor force participation: the larger political and legal structure, the economy, and the family. I argue that, to fully grasp the employment position

SOURCE: Flippen, Chenoa A. 2014. "Intersectionality at Work: Determinants of Labor Supply among Immigrant Latinas." *Gender & Society* 28 (June): 404–434.

of immigrant women, it is necessary to take the broadest view possible of labor force participation, considering both whether women work for pay and, among working women, the number of hours and weeks worked. It is also essential to consider multiple aspects of immigration status, economic position, and family structure.… I expect that not only will these multiple dimensions cumulatively describe women's market position, their interaction will be central to immigrant women's trajectories. For immigrant Latinas, it is particularly important to examine the nexus between family structure and other social positions, examining differences in the determinants of employment among married and single women and considering transnational dimensions of family life.… [I]mmigrant Latinas face multiple and intersecting disadvantages owing to their position as largely undocumented entrants into the United States, their concentration in low-wage and unstable occupations, and the particular demands associated with family responsibilities in immigrant Latino communities.…

IMMIGRANT WOMEN IN THE U.S. LABOR MARKET

In recent decades, the United States, like other postindustrial economies, has undergone dramatic transformation of its employment structure. The heightened emphasis on free trade, deregulation, and flexibility have contributed to rising inequality by skill, and to a substantial erosion of work conditions in the lower segment of the employment hierarchy. Nonstandard work arrangements, such as on-call work, temporary help agencies, subcontracting, and part-time employment have all grown dramatically, both across and within industries, to the serious detriment of wages and job quality (Kalleberg 2011).

The declining fortunes in the low-skilled labor market have heightened the demand for immigrant labor. In fact, the share of all workers in the low-skill labor market who were foreign-born rose from a mere 12 percent in 1980 (Enchautegui 1998) to 50 percent in 2010 (Bureau of Labor Statistics 2011). Within this larger trend, the demand for female immigrant labor has also grown sharply. Globalization contributed not only to the exodus of U.S. manufacturing employment to low-income countries but also to the downgrading of manufacturing that remained. In tandem with the explosive growth of service work and "caring" jobs in child and health care, this has dramatically increased the demand for low-skill immigrant women's labor (Myers and Cranford 1998).

As the concentration of immigrants in the low-wage market grows, the conditions prevalent in that sector will necessarily loom ever larger in their prospects for socioeconomic advancement. Human capital theory posits that the return to education, work experience, and skill raises the opportunity costs for nonwork for highly educated and experienced women. They should thus be more likely to work than their otherwise comparable peers. However, the unique characteristics of the low-wage labor market, particularly the segments where immigrant women are concentrated, may constrain the role of human capital in shaping employment decisions. Immigrant women are highly segregated in a handful of occupations and industries that are characterized by low wages, small firms, and subcontractors

(Catanzarite and Aguilera 2002), and even receive a lower return to skill than other workers (Blau and Kahn 2007)....

Immigrant incorporation is also profoundly shaped by the legal and political context of reception. In recent years, the consequences of lacking legal authorization to work have grown substantially. The 1986 Immigration Reform and Control Act (IRCA) initiated employer sanctions for the hiring of undocumented workers, and the 1996 Illegal Immigration Reform and Immigrant Responsibility Act (IIRIRA) heightened these sanctions and devoted considerably more resources to enforcement. These laws, and an increasingly harsh anti-immigrant sentiment more generally, have reverberated through the low-wage labor market, transforming industries that rely heavily on immigrant labor. In these sectors, the need to insulate employers from the risks associated with hiring undocumented workers has hastened the shift to subcontracting and other forms of nonstandard work arrangements, to the detriment of all who work in those fields (Massey and Bartley 2005). Indeed, in the increasingly bifurcated employment structure of more developed economies, it is increasingly nativity and legal status that determine the deployment of workers into the worst jobs (Hudson 2007); in 2005 it was estimated that fully 23 percent of all U.S. low-skill workers were undocumented (Capps, Fortuny, and Fix 2007).

While numerous studies have sought to identify the wage penalty associated with undocumented status (Donato et al. 2008; Flippen 2012; Hall, Greenman, and Farkas 2010), few, if any, have examined the impact of legal status on women's labor *supply*. Because IRCA and IIRIRA were ostensibly aimed at preventing the employment of unauthorized workers, we might expect undocumented women to work at lower rates than their legal resident peers. However, while U.S. immigration law and enforcement patterns have failed to prevent unauthorized employment, they have nevertheless indelibly shaped the pattern of incorporation of undocumented workers. The laws have, in effect, created a differentiated set of workers who face unique structural limitations....

THE FAMILY AND IMMIGRANT WOMEN'S EMPLOYMENT

The family is another social institution central to immigrant Latinas' employment and working hours. Economic theory views labor force decisions as part of the allocation of time between work and leisure. Because women bear disproportionate responsibility for family reproduction, their labor supply decisions are also shaped by the demand for nonmarket household labor. A wide body of research has shown that women with children are less likely to work than those without children (Cohen and Bianchi 1999; England, Garcia-Beaulieu, and Ross 2004). The relationship between marriage and women's employment is less uniform. On the one hand, if husbands encourage a traditional division of labor, marriage could undermine women's paid employment even over and above its association with childbearing; on the other hand, if husbands were to share in child rearing responsibilities, marriage could facilitate women's employment outside the home. While it is important to recognize the tremendous diversity in gender norms and attitudes both within Latin America and the United States, Latinos are generally thought to be more patriarchal in their family orientations than other groups. For instance, while marriage appears either unrelated or even slightly positively predictive of women's paid employment for non-Latino white and black women (Christopher 1996), the effect is negative among Latinas (Kahn and Whittington 1996; Read and Cohen 2007)....

Moreover, it is essential to recognize that the family does not shape women's employment decisions in a vacuum, but rather its influence interacts with women's position in other social fields, particularly larger economic and political structures. Immigrant women are concentrated in a handful of occupational niches in the United States, particularly in cleaning, food preparation, factories, child care, and laundry. The conflict between work and family is likely to differ among occupations; in some jobs (particularly child care) women are more likely to be able to bring their own children to work, while in others (such as independent house cleaning and laundry) work hours are more flexible and may end before school-age children are home from school. Different occupations also vary in the extent to which they conflict with sex stereotypes, with work in food preparation and child care potentially less challenging to patriarchy than work in factories or outside of immigrant niches. It is important to conceive of these decisions, whether or not to work and what kind of job to pursue, as jointly determined to assess the roles of human capital, legal status, and family constraints in jointly shaping employment outcomes.

It is likewise critical to consider how family structure interacts with legal status to shape immigrant women's employment. A burgeoning literature on the impact of migration on gender roles highlights the many obstacles that immigrant women face in translating their greater access to employment and wages into more egalitarian relations within the family. The precarious structural position of undocumented women *and* men in the U.S. economic and political system undermines women's ability to leverage economic gains into less patriarchal family arrangements (Deeb-Sossa and Bickham-Mendez 2008; Hondagneu-Sotelo 1994; Menjivar 2000; Parrado and Flippen 2005; Parrado, Flippen, and McQuiston 2005; Schmalzbauer 2009)....

... Many immigrant women from Latin America left children in the care of other relatives in their countries of origin. The growing global trend of migrant mothers is transforming the boundaries of motherhood in more traditional societies to increasingly include economic provision for children (Hondagneu-Sotelo 1997; Parreñas 2005)....

CONCLUSIONS

... [My r]esults demonstrate numerous ways in which Latino immigrant women's position in the overall economy constrain their labor supply. In the occupational niches in which immigrant women in Durham concentrate, there are few signs that human capital is a significant determinant of labor force participation. While older women are more likely to work, are more often engaged in the labor market full time, and average fewer weeks of nonwork over the course of the year than their younger counterparts, education has virtually no effect on employment probabilities or work effort among the employed. Among human capital and immigration characteristics, it is primarily English language ability and time in Durham that shape immigrant women's labor supply. Those with better English skills are more likely to work overall and to work full time, and are also more likely to procure jobs in child care and non-niche occupations, which entail more interaction with nonimmigrants. Women with longer tenures in Durham are also more likely to work full time, and are better able to move out of child care and into cleaning, which pays a higher hourly wage. However, while they also spent less time out of the labor market over the previous year, it is primarily because they spent fewer weeks attending to children rather than because they were more successful at avoiding idleness due to slack labor demand.

The disadvantage associated with immigrant Latinas' position in the legal system is also clearly evident. While undocumented women are no less likely to work overall than their legal resident peers, their employment opportunities are significantly curtailed in other ways. First, lack of documentation blocks women's entry into non-niche occupations, where pay is often higher and work tends to be more stable, instead pushing them into child care, where the opposite is true. Second, undocumented women also experience greater employment instability and time out of work during the previous year than their legal resident counterparts. Moreover, the impact on weeks without work is evident only for reasons relating to insufficient labor demand, and not for those pertaining to child rearing. Given the economic need of the families in which immigrant women are embedded, the numerous, sizeable impediments to stable, full-time employment are a cause for serious concern. While employment conditions improve with age and time in the local labor market, the effects are relatively modest and suggest sustained disadvantage over the course of women's lives.

Results also demonstrate the profound limits to employment posed by immigrant Latinas' family structure. Both marriage and child rearing exert independent negative effects on women's employment. Indeed, marriage itself seems to be as large an impediment, if not larger, to work as the presence of coresident children is. Married women and those with coresident children are not only less likely to work, they also work fewer hours per week and experience significant family-related inactivity over the course of the year. However, the effect of children on women's employment does not end with those who are living in the household. It is important to acknowledge transnational aspects of family and child rearing; women with minor children abroad are more likely to work than other women, even net of household structure in Durham. This finding adds to a growing literature on transnational families suggesting that migration is expanding the boundaries of motherhood to include not only caring work but also financial provision (Hondagneu-Sotelo 1997; Parreñas 2005). Taken together, these findings support the need to take a broad view of the family when considering immigrant women's work experiences. Both gender roles and family arrangements are profoundly altered through migration, and a deeper understanding of immigrant women's work requires greater attention to the nexus between the family and other structures.

While each of these domains taken separately provides valuable insight into the complex employment patterns of immigrant Latinas, results also indicate significant intersection between structural constraints. One of the most consequential interactions is between marital and legal status. Married women are not simply less likely to work than their unmarried counterparts; their labor force participation is also more sensitive to legal status. These patterns temper our view of the Latino (particularly Mexican) family as a patriarchal constraint on women's paid employment, and shifts the emphasis to the structural conditions affecting Latinas in the U.S. labor market....

There are also important interactions among immigrant Latinas' human capital and family characteristics and structural conditions in the low-wage labor market. While documentation does not impede work overall, it does channel women into child care and raises significant barriers to non-niche employment. Moreover, the conflict between work and family is not uniform across occupations. Child care and laundry work, in particular, stand out as posing fewer barriers to work for women who are married or with children. While most of the niches in which immigrant Latinas are concentrated could be considered "women's work," and thus represent more modest challenges to patriarchy than work in male-dominated fields, child care and laundry work are unique in their ability to accommodate young children at work and long hours of operation, respectively, which seem to mitigate the impact of family responsibilities on market work....

REFERENCES

Blau, Francine, and Lawrence Kahn. 2007. Gender and assimilation among Mexican Americans. In *Mexican immigration to the United States*, edited by George Borjas. Washington, DC: National Bureau of Economic Research.

Bureau of Labor Statistics. 2011. Foreign-born workers: Labor force characteristics— 2010. Economic News Release, http://www.bls.gov/news.release/archives/forbrn_ 05272011.htm.

Capps, Randy, Karina Fortuny, and Michael Fix. 2007. *Trends in the low-wage immigrant labor force, 2000-2005.* Washington, DC: Urban Institute.

Catanzarite, Lisa, and Michael Aguilera. 2002. Working with co-ethnics: Earnings penalties for Latino immigrants at Latino jobsites. *Social Problems* 49:101–27.

Christopher, Karen. 1996. Explaining the recent employment gap between black and white women. *Sociological Focus* 29:263–80.

Cohen, Philip, and Suzanne Bianchi. 1999. Marriage, children, and women's employment: What do we know? *Monthly Labor Review* 122:22–31.

Deeb-Sossa, Natalia, and Jennifer Bickham-Mendez. 2008. Enforcing borders in the Nuevo South: Gender and migration in Williamsburg, Virginia, and the Research Triangle, North Carolina. *Gender & Society* 22:613–38.

Donato, Katharine, Chizuko Wakabayashi, Shirin Hakimzadeh, and Amada Armenta. 2008. Shifts in the employment conditions of Mexican immigrant men and women: The effects of U.S. immigration policy. *Work and Occupations* 35:462–95.

Enchautegui, Maria. 1998. Low-skilled immigrants and the changing American labor market. *Population and Development Review* 24:811–24.

England, Paula, Carmen Garcia-Beaulieu, and Mary Ross. 2004. Women's employment among Blacks, Whites, and three groups of Latinas: Do more privileged women have higher employment? *Gender & Society* 18:494–509.

Flippen, Chenoa. 2012. Laboring underground: The employment patterns of Latino immigrant men in Durham, NC. *Social Problems* 59:21–42.

Hall, Matthew, Emily Greenman, and George Farkas. 2010. Legal status and wage disparities for Mexican immigrants. *Social Forces* 89:491–514.

Hondagneu-Sotelo, Pierrette. 1994. *Gendered transitions: Mexican experiences of immigration.* Berkeley: University of California Press.

Hondagneu-Sotelo, Pierrette. 1997. "I'm here but I'm there": The meanings of Latina transnational motherhood. *Gender & Society* 11:548–71.

Hudson, Kenneth. 2007. The new labor market segmentation: Labor market dualism in the new economy. *Social Science Research* 36:286–312.

Kahn, Joan, and Leslie Whittington. 1996. The labor supply of Latinas in the USA: Comparing labor force participation, wages, and hours worked with Anglo and Black women. *Population Research and Policy Review* 15:45–73.

Kalleberg, Arne. 2011. *Good jobs, bad jobs: The rise of polarized and precarious employment systems in the United States, 1970s-2000s.* New York: Russell Sage.

Massey, Douglas, and Katherine Bartley. 2005. The changing legal status distribution of immigrants: A caution. *International Migration Review* 39:469–84.

Menjivar, Cecilia. 2000. *Fragmented ties: Salvadoran immigrant networks in America.* Berkeley: University of California Press.

Myers, Dowell, and Cynthia Cranford. 1998. Temporal differentiation in the occupational mobility of immigrant and native-born Latina workers. *American Sociological Review* 63:68–93.

Parrado, Emilio, and Chenoa Flippen. 2005. Migration and gender among Mexican women. *American Sociological Review* 70:606–32.

Parrado, Emilio, Chenoa Flippen, and Chris McQuiston. 2005. Migration and relationship power among Mexican women. *Demography* 42:347–72.

Parreñas, Rhacel. 2005. *Children of global migration: Transnational families and gendered woes.* Stanford, CA: Stanford University Press.

Read, Jen'nan, and Philip Cohen. 2007. One size fits all? Explaining U.S.-born and immigrant women's employment across 12 ethnic groups. *Social Forces* 85:1713–34.

Schmalzbauer, Leah. 2009. Gender on a new frontier: Mexican migration in the rural mountain west. *Gender & Society* 23:747–67.

DISCUSSION QUESTIONS

1. What unique challenges do Latinas face in finding work in the new gateway community of Durham?

2. How does immigrant status influence employment options for Latinas?

3. What role do family characteristics play in the paths to employment for Latinas?

Section B

Families and Communities
Elizabeth Higginbotham and Margaret L. Andersen

In his theorizing on the differences between immigrant and colonized minorities, Robert Blauner (1972) identified how the lack of economic and political rights made African Americans, Asian Americans, Native Americans, and Latino groups vulnerable to many cultural assaults, including their ability to function as families. Families' ability to function well is linked to their available resources and to how well the state supports families, through public schools, health care, public safety, and other state-based assistance. Race also influences the opportunities for individuals to form families, provide for families, and live together as families. We might celebrate "family values," but in reality the resources available for family well-being are structured by race, social class, and citizenship status.

Most people marry within their own racial or ethnic groups, but in a diverse society people can be attracted to individuals from different backgrounds. With racial integration, this becomes more likely. Race has historically been a factor in people's abilities to form the families of their choice. One of the signs of a free society is the freedom of people to freely associate with others—as peers, friends, neighbors, lovers, marriage partners, or in any other relationship. It is hard to believe that for much of our history people have not had that opportunity. Racism has invaded the most intimate areas of people's lives. For many decades, thirty states had laws that prohibited White people from marrying someone of a different race. These **antimiscegenation laws** prohibited so-called race mixing. For example, the state of California passed a law in 1880 prohibiting any White person from marrying a "negro, mulatto, or Mongolian." This law was designed to prevent marriages between White people and Chinese immigrants (Takaki 1989). Specific laws against intermarriage varied from state to state. All southern states prohibited White people from marrying "Negroes," while some western states, like California, were also anti-Asian. Laws did not prohibit non-White groups from marrying each other. Thus, in Mississippi, the Chinese could and did marry Negroes, although neither group was allowed to marry White people.

In order to enforce antimiscegenation laws, states had to devise ways to define race. States varied in this practice as well. "Alabama and Arkansas defined anyone with one-drop of 'Negro' blood as Black; Florida had a one-eighth rule;" Other states varied in their racial definitions (Lopez 1996: 118). If someone wanted to marry a person of another race, they had to do so in a state that did not prohibit the union, but they risked having their marriage denied if they moved to a state where such arrangements were illegal.

Mildred Jeter (a Black woman) and Richard Loving (a White man), who had been married in the District of Columbia in 1958, had this experience. When they returned to their home state of Virginia, they were indicted and charged with violating Virginia's law banning interracial marriage. They were convicted and sentenced to one year in jail—a term they never served because the judge suspended the sentence on the condition that they leave Virginia. They challenged this law when they returned to Virginia five years later to appeal the decision through the courts. Their case, *Loving v. Virginia*, went all the way to the U.S. Supreme Court, which declared antimiscegenation laws unconstitutional. In 1967, the Supreme Court decided the case based on the argument that laws against intermarriage violated the Fourteenth Amendment, which states: "No State shall make or enforce any laws which shall abridge the privileges or immunities of citizens of the United States; nor shall any State deprive any person of life, liberty, or property, without due process of law; nor deny to any person within its jurisdiction the equal protection of the laws" (U.S. Constitution, Amendment 14, Section 1).

In this section, we explore families and communities, the intimate social institutions where we should be cared for and nurtured. We see how racism influences the challenges that families face. Recently, there has been an increase in the numbers of couples marrying outside their race or ethnicity, as about 15 percent of new marriages are interracial or interethnic (Passel, Wang, and Taylor 2010). The numbers are higher if we include cohabitation. People who seek partners from a race other than their own still face many challenges and tensions even in the context of great love.

Amy Steinbugler ("Loving across Racial Divides") interviewed mixed-race couples around the nation. Different from such couples in earlier years, these couples rarely face open hostility, yet they live in communities that reflect the residential segregation of our era. These couples report thinking about race all the time, as they negotiate racial profiling and other inequities in society. A couple might live in a residential area where one of them is comfortable, while the other is not and may even feel isolated.

Joe R. Feagin and Karyn D. McKinney ("The Family and Community: Costs of Racism") discuss the impact of racism on Black families and communities, exploring the real harm of racism to people and the places they live. People of color do actively resist such psychic and physical assaults, working as they can to build families and communities for social support.

Race, as well as other dimensions of inequality like social class, influences a family's resources, as well as how others in the society perceive them. There are many myths about the "absent Black father." It is true that there is a high rate of single-parent families headed by women, but this does not always mean that the father is absent. Fathers may be present but not necessarily as prescribed by dominant norms. Racism results in many social constraints on a people's ability to parent. Roberta L. Coles and Charles Green ("The Myth of the Missing Black Father") identify some of the unique challenges African American men face in the context of deindustrialization and the consequent lack of employment opportunities for Black men. They show that, contrary to dominant assumptions, Black men value fatherhood and fulfill parental roles both inside and outside of marriage.

Increasingly, the United States is made of people with mixed racial and ethnic heritages. Diversity will likely continue in that direction, particularly as the walls of segregation are shattered and people have more face-to-face interaction with people of different backgrounds. Dismantling all forms of segregation is important because with integration people are less dependent on media images for information about people. As people meet, they can connect across shared interests and passions.

REFERENCES

Blauner, Robert. 1972. *Racial Oppression in America*. New York: Harper & Row.

Lopez, Ian F. Haney. 1996. *White by Law: The Legal Construction of Race*. New York: New York University Press.

Passel, Jeffery, Wendy Wang, and Paul Taylor. 2010. "Marrying Out: One-in-Seven New U.S. Marriage is Interracial or Interethnic." Washington, D.C.: Pew Research Center Publication.

Takaki, Ronald. 1989. *Strangers from a Different Shore: A History of Asian Americans*. New York: Penguin.

FACE THE FACTS: MARITAL STATUS OF THE U.S. POPULATION, 2014

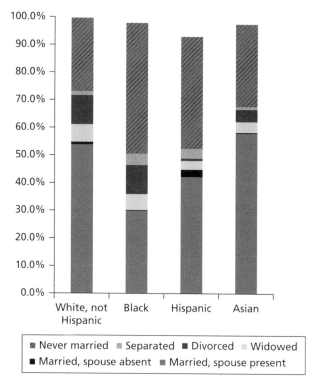

SOURCE: U.S. Census Bureau. 2014. America's Families and Living Arrangements: 2014 (Adults; over 15 years of age). Washington, DC: U.S. Department of Commerce. www.census.gov

Think about It: Diversity in families is a fact of life in the United States. What factors do you think influence the differences in marital status of the different groups you see here? (Note: These data were gathered at a time when same-sex marriages would not have been included in the data.)

24

Loving across Racial Divides

AMY STEINBUGLER

What are the daily challenges that mixed-race couples face, especially as they have to live in communities that are still highly segregated? What would be the consequences of facing a public that is not totally supportive of your choice of a life partner? Steinbugler's interviews with mixed-race couples give us a sense of how far we need to go to welcome all in the society.

Leslie acknowledges that African Americans have to "think about race all the time," but insists that there are unique racial issues that stem from being in an interracial relationship.

Sylvia and Leslie face challenges that are usually less overt than those black/ white couples would have confronted 50 years ago. There is little chance that they might lose their jobs, get kicked out of their church, or be denied housing simply because one of them is black and the other is white. While injustices like these still occur, when they do they are noteworthy. Undisguised discrimination against interracial couples is no longer typical. Nor do relationships like theirs inspire the raw disbelief that Sidney Poitier and Katharine Houghton famously elicited when they portrayed a young couple in the 1967 film, *Guess Who's Coming to Dinner.*

Still, Sylvia says, "We're always looking for spaces where we can be together as a couple, [be] validated, feel comfortable." Compared to straight interracial pairs, same-sex partners like Sylvia and Leslie can count on far fewer legal protections, and also are vulnerable to homophobia from coworkers, family members, and strangers on the street.

Interracial couples in the United States have always attracted public scrutiny. Until the mid-twentieth century, this attention was almost entirely negative. Legal sanctions prevented people of African, Asian and, sometimes, Native American descent from marrying whites, Black/white relationships, especially those between white women and black men, drew the harshest condemnation. Black communities treated such couples as disreputable; white communities often threatened, physically harmed, or ostracized them.

In recent years, interracial couples are more likely to encounter hope than censure, at least in terms of public discourse. Some observers liken current legal prohibitions against same-sex marriage to anti-miscegenation laws before the

SOURCE: From Steinbugler, Amy. 2014. "Loving across Racial Divides." *Contexts* 13 (Spring): 32–37.

1967 Supreme Court ruling in *Loving v. The State of Virginia*. Social commentators paint contemporary interracial marriage as a victory for equality and freedom. A 2001 *Time* magazine article celebrated interracial unions as representing an intimate "vanguard" who "work on narrowing the divisions between groups in America, one couple at a time."

More recently, a Pew Research Center report released in 2012 suggests a positive shift in public attitudes towards intermarriage. Forty-three percent of Americans now view the trend for more people of different races to marry each other as a change for the better. About two-thirds say it would be fine with them if a family member "married out" of their racial or ethnic group. Even black/white relationships, which have long elicited the fiercest disapproval and the strongest legal sanctions, are becoming more acceptable.

Despite the supposed acceptance of dating and marrying across racial lines, only a small percentage of people in the United States—according to the 2010 Census, less than 7 percent of all heterosexual married couples—actually do so. Among gay and lesbian couples, approximately 14 percent are interracial—about the same proportion as among heterosexual unmarried partners.

Low intermarriage rates notwithstanding, many people embrace the popular notion that Americans have truly become "colorblind." But racism is more than just a matter of prejudice; liberalizing racial attitudes coexist with the stubborn persistence of racism. For the past 12 years, I have studied couples who love across racial difference in Philadelphia, New York, and Washington, D.C. What I've found is that while hostility toward interracial pairs, like racism itself, has become more subtle, race continues to powerfully impact everyday life for Sylvia, Leslie, and the other 39 interracial couples I interviewed. Racism, manifested in neighborhood segregation and racial self-understandings, shapes everyday life, creeping up in the most ordinary circumstances, like walking through their neighborhood, or deciding where to get a drink.

RACED SPACES

Mary Chambers, a heterosexual woman of Afro-Caribbean descent, knew that her husband Neil was sometimes uncomfortable in their middle-class, majority-black neighborhood. The neighborhood, which includes many sprawling, three-story, Tudor houses, feels suburban, though it is located in a small city less than 30 miles from Manhattan. Mary thinks that Neil "would prefer to live in a community where he's more comfortable with the people, in a community where he can look around and see his own race. [One] with more white people." She finds this discouraging. "I wish that I could change his perspective on it and really make him see that the community we live in is valuable."

In the years Neil spent in this neighborhood, he became very aware of his own whiteness. It was impossible for him not to think about race in everyday social interactions. It was typical, he says, to go into the grocery store or the post office and be "the only white guy there." He continues, "It's not a bad thing, you know. It's not like I feel like I'm going to get mugged or, um, I'm going to get hurt. But you have to understand, it's kind of like when you deal with white people, they have their prejudice—they have what they're used to. And then when you deal with black people, it's really the same thing, you know?"

The residential racial segregation of blacks and whites has been slowly declining nationwide. But in northeastern cities like New York and Philadelphia, highly segregated neighborhoods remain the norm. (In New York, the racial composition of neighborhoods is so lopsided that 79 percent of blacks would have to move in order to achieve a balanced distribution—in which the percentage of blacks in every neighborhood mirrors their share of the city's total population.)

Neighborhoods that are black or white often pose problems for interracial couples because they set the stage for situations in which one partner feels uncomfortable or conspicuous. "No matter where we live," one white woman lamented, "one of us is not going to be in the right neighborhood." A black partner agreed, "As diverse as [New York] city is, to me it's still pretty segregated." This sense of belonging or not belonging is something interracial partners often brought up as we talked.

Neighborhood divisions are stressful when one partner feels conspicuous and has to look out for racial undercurrents in everyday social interactions. Such divisions create racial fatigue, though they affect black and white partners differently. For Neil Chambers and other whites in my study, being in the racial minority feels awkward because it happens so rarely. Noticing one's own whiteness is a new experience that can prompt an unsettled feeling. One of the taken-for-granted privileges of being white is the tendency to think of yourself not as a *white person*—*just* as a person. I asked one white woman who is married to a black man how often she thinks of herself as white. "I don't," she said. "Well, maybe if I were in an all-black environment, and I'm the only white person there. That's the only time."

Black interracial partners also noticed when they were among only a handful of blacks in a neighborhood or social gathering. But for these middle-class Americans, that experience was not uncommon. Many worked or had gone to school in majority-white environments. Compared to the whites in my study, black partners tended to be much more accustomed to being in the numerical minority. In contrast to the unease of white partners, who sometimes felt intimidated in black neighborhoods, the discomfort of black partners was linked to a history of violence against their racial group. It was one of many instances in which black and white partners perceived race very differently.

RACIAL ORIENTATIONS

Tamara is white, and Scott is black. When this 30-something unmarried straight couple decided to move in together, they needed to transport countless boxes of books and clothes from Tamara's place in Philadelphia to a nearby city. Tamara wanted Scott to drive the SUV she had borrowed from a friend. But like many other black men in the United States, Scott was concerned about racial profiling. He didn't want to get pulled over driving a borrowed car, especially given that his cell phone wasn't working properly.

Scott recalls telling Tamara, "I'd really rather you drive … because when—if [I] get pulled over … I can't dial [the woman who owns the car]. Now it's just me and some cop and he's probably going to treat you better than he's going to treat me." For Scott, this was a routine calculation. For Tamara, it didn't seem like a big deal. This reflects a broader disjuncture in how—and how often—each of them thinks about race. Scott, continuing the story of the move, tells me, "It's my job to consider that, whereas … I don't think [me getting pulled over] is automatically something that she considers—and you know what? It doesn't bother me that it isn't, because how could it be? Someone [who is] Black can really, really think, like could have their mind go to that."

For Scott, anticipating everyday acts of prejudice and discrimination is second nature, ever since his grandmother told him about the lynching of Emmett Till. But Tamara, whose whiteness has shielded her from being the target of racial animus, is only now learning to consider how the accumulation of a lifetime of racial experiences informs even small decisions, like who will drive an SUV full of books and clothes 30 minutes away.

Daniel, who is black, and Shawn, who is white, thought their racial orientations were very similar until they became the adoptive parents of two black boys. When Daniel and Shawn began to talk about their sons' future schooling, they soon discovered that they conceptualize racism differently. "We both share the same basic political views," Shawn explained. But Daniel, he believed, "subscribes to a kind of conspiracy theory that white America has banded together to exclude black America." Shawn sees the racism, he says, but "I don't see it as being organized in the same way. Because I'm white and nobody ever came to me and said, 'Hey, let's get together and do this thing to the black people.' So that's a difference of philosophy." Daniel sees "[racism] as institutionalized and I see it just as kind of widespread."

Daniel and Shawn's perspectives lead to important differences in dealing with racism. While Shawn has no intention of letting his sons get hurt, he is more comfortable taking a wait-and-see approach. Daniel feels more strongly about the need to be pro-active about racial discrimination at school. "I'm not looking for trouble," says Daniel. It's just that "I don't want to be asleep when that stuff happens."

For Daniel, there are limits to what a white person can understand about discrimination. Shawn, he says, "doesn't expect that kind of behavior and it's hard [for him] to believe that in fact that can happen." He "just doesn't believe that a teacher would look at an eight-year-old and actually treat one eight-year-old differently from another simply because of the color of their skin." While both men are deeply invested in protecting their sons from racism, their conflicting racial orientations to the subject remain unresolved.

Shawn's strategies for parenting black sons and Tamara's skepticism about racial profiling reflect the attitudes of many white Americans who question the scope and severity of contemporary racism. Surprisingly, white partners' intimacy with black people did not substantially challenge their racial perspectives, casting doubt on the notion that interracial partners represent an enlightened, "post-racial" vanguard.

STEREOTYPES—AND EXCEPTIONS

Gary, 54 and his wife Soonja, 58, met in Korea. They have been married for 20 years. Gary chose to marry a Korean woman because, he said, a "good" wife should be loyal and subservient, and to him, Soonja's race signifies these traits. Soonja, too, believes that her choice of husband may reflect upon the kind of person she is. She chafes at being associated with what Koreans regard as stereotypical "international marriages," temporary sexual relationships between local, uneducated Korean women and American military men.

Gary's whiteness and American citizenship did not hold any special appeal for Soonja or her family. But over time, part of what has made their marriage work is that Soonja believes there are distinct cultural differences that make American husbands better partners than Korean husbands. "I'm glad that I didn't marry a Korean, who ignores his wife, drinks a lot, and comes home late." Gary confides that many of his friends "actually say that they envy me because they understand that Asian women are very good wives and … good mothers."

Sociologist Kumiko Nemoto has researched marriages between whites and Asians in the United States. White men, according to Nemoto, commonly associate Asian and Asian American women with family and domesticity, Younger Asian women in interracial relationships, she found, are more likely to define themselves as egalitarian, ambitious, and aesthetically (as opposed to domestically) feminine, challenging these stereotypes.

Vivian, 25, who grew up in a Chinese family is attracted to Peter, 27, in part because she sees them as equals, He is intelligent, got good grades in college, and is economically mobile. "We have a mental connection," she says. Peter's professional ambitions and work ethic help Vivian think of herself as a modern Asian American woman. Peter, for his part, is proud to appreciate beauty that falls outside of normative white femininity: "[I'm] more attracted to ideas and people who are more exotic … I think Asian features are prettier than white features." He likes "darker-skinned women" and the "shapes of Asian eyes."

In the U.S. racial order, particular Asian groups (such as Japanese and Chinese) are positioned much closer to whites than blacks, Asian Americans certainly experience racial discrimination and the false presumptions embedded within the idea of the "model minority." Even so, some Asian groups are increasingly seen as what sociologist Eduardo Bonilla-Silva calls "honorary Whites." As the 2010 Census shows, intermarriages between Asian Americans and whites are far more common than those between blacks and whites.

Couples in my study also used racial-gender stereotypes about their partners to describe themselves. Still, some whites were careful to portray their partner as exceptional, rather than a typical example of their racial group. As Neil said of

Mary, who is of African American descent: "She's not someone who would curse or, you know, say anything that's inappropriate or off color. Not that she's a saint but—she has a certain background—she's not offensive to you. She's very pleasant."

What this and other examples suggest is that the prejudice interracial couples encounter from strangers is only one small part of how race shapes their everyday lives. Race is a social system that shapes neighborhoods, orientations, and identities, and plays a critical role in intimate relationships. Despite the gains of the Civil Rights movement and the historic election of the first black president, the racial categories we are assigned to at birth have tremendous material consequences for how our lives unfold. Racial inequalities affect the wealth which we have access to, the neighborhoods we live in, the type and amount of the healthcare we are able to get, and the quality of our children's schools. Even in this supposed post-racial moment, our position in the racial system shapes the way we see the world.

DISCUSSION QUESTIONS

1. Many surveys and opinion polls document that national attitudes toward interracial relationships have changed. How does Steinbugler's research show the continuing difficulties that mixed-race couples experience on a daily basis?

2. What dilemmas might you face raising a mixed-race child if you are a member of the majority group? How might this approach differ from the attitude of a parent who is a racial minority?

25

The Family and Community
Costs of Racism

JOE R. FEAGIN AND KARYN D. MCKINNEY

Racism takes a toll on people of color in multiple ways. Feagin and McKinney show how, even in the face of racist beliefs and actions that harm Black families and communities, those same institutions provide a defense against the harm that racism produces.

Most African Americans see their families and communities as primary defenses against the daily assaults of racism. It is in these families that most black children first learn how to cope with racism. Additionally, the local black community often operates as extended kin. As they become adults, most black Americans seek support from others in their communities when they face increasing numbers of racist incidents in a variety of societal settings, including places of employment. As our respondents often note, the damage of a racially hostile workplace does not end at the workplace door. A black individual's experience with racial animosity and mistreatment is personally painful at the moment it happens, and also can have a cumulative and negative impact on other individuals, on one's family, and on one's community.

African American families and communities are negatively affected in many ways by continuing white-on-black racism. The black family has faced physical, ideological, and material assaults from whites for nearly four centuries. Over these centuries, family and community have been closely linked for African Americans, and for that reason the various white assaults on black families often have a significant impact on the larger black communities....

THE LONG-TERM IDEOLOGICAL ASSAULT
ON THE BLACK FAMILY

Most social science researchers view African American families as distinctive in certain characteristics when compared to similar white families. What is disputed

SOURCE: Feagin, Joe R., and Karyn D. McKinney. 2003. "The Family and Community: Costs of Racism." In *The Many Costs of Racism*, by Joe R. Feagin and Karyn D. McKinney. Lanham, MD: Rowman and Littlefield.

is whether these differences are problematic. Rather than looking at African American families as simply *different* in certain ways from white families, most of the early literature and much recent literature addressing certain family forms found in black communities has described these family forms as more or less problematical or pathological....

Perhaps the best-known example of this pathological-family viewpoint is the 1965 report by Daniel Patrick Moynihan, *The Negro Family: The Case for National Action*. In this widely cited and influential government policy initiative, Moynihan writes about an allegedly distinctive pathology of black communities: "Obviously, not every instance of social pathology can be traced to the weakness of family structure.... Nonetheless, at the center of the tangle of pathology is the weakness of the family structure. Once or twice removed, it will be found to be the principal source of most of the aberrant, inadequate, or antisocial behavior, that did not establish, but now serves to perpetuate the cycle of poverty and deprivation."[1] In this report, white society is all but absolved of contemporary responsibility for poverty, deprivation, and family stresses in black communities. This labeling of urban black communities as a "tangle of pathology" is common-place and has been copied to the present day. According to many advocates of this pathology, the typical black family is matriarchal in character, has weak kinship ties, has "illegitimate" children, lacks mainstream values, and is present-oriented rather than future-oriented. This group of factors is sometimes referred to as a "culture of poverty."[2] Thus defined, it can seem inevitable and somehow right to whites that African Americans are impoverished....

A primary problem with the "culture of poverty" notion is that it confuses cause and result. While admitting that slavery and racism *began* the problems for the black family and community, those who adhere to this perspective believe that certain contemporary black family and community problems persist primarily because of an unhealthy black subculture. They do not recognize or admit that the original cause of stress on the black family, institutionalized racial oppression, is *still* the major and pervasive cause of much everyday stress for black families and communities today—and that the *results* of that stress include some of the adaptive characteristics often described in negative terms, or that are distorted in many analyses as supposed deviations from white family norms.

CONFRONTING AND COPING WITH RACISM: STRONG AND ADAPTIVE FAMILIES

Contrary to the common view of the pathology advocates, the African American family has long been characterized by strong kinship bonds and interaction. During slavery, one of the primary reasons that enslaved African Americans ran away was to return to their families. In fact, many who had escaped to freedom in the North again risked their lives to return South in the attempt to free their families. After more than two centuries of colonial and U.S. slavery, during Reconstruction many African American families that had been separated by slaveholders began to regroup and re-create themselves. Although most of these newly freed black parents were struggling to feed their children, they often added to their families' innumerable "fictive kin"—the extended family and non-kin children who were orphaned after slavery. Similar family restructuring, and the care of extended family and "fictive kin," were seen again during the Great Depression of the 1930s. These patterns of care and survival can still be seen in many African American families and communities today.

For several decades now, numerous scholars have shown that African American families and communities are remarkably adaptive and strong....

Contrary to the "culture of poverty" thesis and commonplace white stereotyping, most members of black families exhibit a strong orientation toward work. Many studies show that whites and blacks are similarly oriented in their attitudes toward work. Our data show that rather than working less hard than whites, most African Americans realize that they often must work *harder* than whites to achieve the same results. For example, a sheriff's deputy in one focus group put it this way: "And that's the same thing ... we were talking about on the energy. Burning so much energy trying to educate these people, that we qualify, you know? And I always said if you see a black doctor and a white doctor standing side by side, equal in status, that black man is *twice* as [good], because he had to work harder ... in every profession."

Another strength of the black family is its adaptability in regard to family roles. Contrary to the popular notion that the black family is usually "matriarchal," even when there is a husband present, scholars have found that African American families are often more egalitarian than are white families. Because they are not as patriarchal as the traditional white family, black families thus appear to many (especially white) outsiders as part of a "matriarchal" family system. Evidence of the adaptability of black families can also be seen when young women become mothers, in that the larger family is usually supportive. African American families are twice as likely as whites to take the children of premarital births into the homes of family or friends of the parents, and are *seven times less* likely than whites to put up these children for adoption.

Another adaptation to unfavorable circumstances is a strong achievement orientation.... Education is valued highly in the black family and community as one way to help overcome poverty and racism.

Another distinctive strength of African American families is their often strong religious orientation. This religious orientation contributes to cohesive black families, and has been used as a mechanism of survival for black communities from slavery through the 1960s civil rights movement to the present day....

BLACK FAMILIES: THE CHALLENGE OF DIFFICULT ECONOMIC CONDITIONS

Some of the most difficult problems faced by black families stem from outside economic forces. Some of these economic forces have, unintentionally, a differential impact on African Americans, while other economic problems are created by government or private sector policies that intentionally, or at least knowingly, have a negative impact on African Americans. In other work, Feagin has distinguished between direct and indirect institutionalized discrimination. Direct institutionalized discrimination is that discrimination intentionally built into organizations and institutions by whites in order to have a significant negative impact on people in subordinate racial groups, especially African Americans. Examples include the intentional exclusion, blatant or subtle, of African Americans from traditionally white jobs and neighborhoods, discrimination that can still be found in some sectors of U.S. society. Indirect institutionalized discrimination involves the differential and negative treatment by whites of subordinate racial groups without that treatment being intentional in the present moment. This is oppressive and damaging discrimination, nonetheless, because it carries into the present the impact of the often blatant discrimination of the past. For example, intentional discrimination in the past often creates limited resources for the descendants of those so discriminated against, and current generations of groups once severely subordinated have less inherited wealth and other socioeconomic resources than do current generations of whites. This means that, today, the average African American typically does not have the same economic and educational resources and, thus, opportunities as the average white American. For example, ... the average black family today has about 60 percent of the income of the average white family—and only about a tenth of the average white family's wealth....

In our focus groups and in informal discussions with a number of other African American husbands and wives, we have found that many people explain in some detail how the stress of discrimination at work places an added burden on their relationships with their spouses. When husbands or wives are under great pressure and stress from the racist incidents in their predominantly white workplaces, their energy for, or willingness to interact with, their spouses when they come home is affected. They may wish to be secluded from family relationships, and perhaps just watch television for a long time in order to unwind for a while from the stress of dealing with whites in the workplace or other social settings.... Many African Americans are aware of how this impacts the family, and they try to develop counter-strategies. In other cases, even these veterans of contending with racism do not realize fully just how that racism has negatively affected their intimate family relationships. In these cases, white racism has yet other negative, often energy-draining, effects on African American families.

For the family, another obvious consequence of workplace discrimination occurs when an important breadwinner is fired unfairly from his or her job because of discrimination, and then is not able to provide adequately for the family. Thus, one black man quoted in an earlier research study had worked for ten years as a school employee whose job it was to deal with the community. Although he was hired for his interpersonal skills and life experiences, after ten years on the job his employer decided to put some formal screening credentials into place for such community contact positions. Under this newly imposed standard, although he had personally moved up into the middle-class, his working-class demeanor and speech were no longer valued, and he was fired. At the time of his interview, he was preparing to sue to win back his job and was working a temporary job that paid a very low wage. He had also separated from his wife. In his interview, he spoke of the costs of this experience on his children:

> I honestly believed and felt that I was put on this earth to help to work with kids, I really did. And I was good, I was damn good at my job to have, let's say, less education than probably anybody at that school.... I *am* angry. I'm very angry. Because what they did, they didn't do because I was not performing my job. They did [it] because I was a black man and I spoke out on what I believe and felt. If I hadda been one of their little henchmen to say "yes, no" ... I would still be there.... They don't know what the hell they have put me through. I know I could have a job if it wasn't for them folks down there. They have shut off my livelihood. They don't know the suffering, not that I am absorbing, but my kids.... I don't know how many times I didn't have the money to pay the rent. They got food on their table when they come in, in the evening. Sometimes I don't have food here on my table for my kids.

Damage is done to African American families because of workplace mistreatment along racial or racial-class lines. This man cannot feed his children adequately. He also notes that he has had to take his daughter out of college, thereby jeopardizing her ability to be economically successful in the future. And he has separated from his wife. His family has been torn apart by the racist actions of whites in his workplace. We see here the powerful domino effects on families of racist actions in yet another U.S. workplace.

THE IMPACT OF RACISM: THE BLACK COMMUNITY

... The spin-off effects of animosity and mistreatment in employment settings can be seen in yet other areas of the lives of these African Americans. Another respondent in a focus group sadly noted the negative impact on participation in church activities:

> I have withdrawn from some of the things I was involved with at church that were very important to me, like dealing with the kids at church. Or we had an outreach ministry where we would go out into the low-income housing and we would share about our services, ... and I was just so drained, like [another respondent] said, if we are all so drained, and we stop doing that, then we lose our connection. But I, physically, by the time I got home at the end of the day, I was just so tired. I didn't even feel like giving back to my community; I didn't feel like doing anything. And so I withdrew from church activities, to the point where I just really was not contributing anything. And it was pulling all that energy; I was exhausted from dealing with what I had to at work. And then whatever little bit was left went to my family, so there was nothing there to give.

The considerable impact of workplace racism is graphically described here, for even church activities become a challenge for this person. What energy there is left after struggles at work with racism is reserved for the family. These economically successful African Americans can be important role models in their local communities, but only if they have the energy to participate actively in churches and other community organizations. These accounts of withdrawal from, or lack of energy for, community activities are worth considering in the light of the many accomplishments of African Americans in organizing to improve both their own communities, in a variety of churches and other important local organizations, and also in working together in national organizations to improve the larger U.S. society. In spite of the great strain, pain, and energy loss stemming from individual and institutional racism, the majority of African Americans manage to strive, endure, and succeed in raising families and building communities. Being tired from the daily struggle against racism in the workplace, schools, and public accommodations has not stopped many African Americans from continuing to organize, a point we will accent in the conclusion that follows.

CONCLUSION: CREATING COMMUNITIES OF RESISTANCE

Dealing with the family and community impact of racism is a constant challenge for African Americans. The question of how to develop effective community strategies for development or change has been a recurring topic of concern and analysis for African American intellectuals and other leaders....

It is evident in the words of numerous respondents that interpersonal relationships, including family relationships, are sometimes jeopardized because one or both partners does not have much energy left at the end of a racially stressful day. Yet most seem to manage to overcome these huge challenges most of the time. Indeed, from their discussions of the energy-draining aspects of discrimination, one might wonder how African Americans have developed vibrant community organizations and successful resistance movements over nearly four centuries now. Most African Americans persevere through the barriers and manage to overcome the constant "rain" of antiblack racism enough to stay centered in their life struggles and, remarkably, thrive....

These organized efforts have not only liberated African Americans from legal segregation, but also the *country as a whole* from these enormous barriers to human progress. All Americans are the current beneficiaries of organized African American resistance to racial oppression. There is greater freedom in the U.S. today than there would have been without the civil rights movement of the 1960s. This is as true today as it has been in the past. Racism is a destructive force that ultimately affects all residents of this country. As the English poet John Donne said long ago, "Any man's death diminishes me, because I am involved in mankind; and therefore never send to know for whom the bell tolls; it tolls for thee."

NOTES

1. Daniel P. Moynihan, *The Negro Family: The Case for National Action* (Washington, D.C.: U.S. Department of Labor, 1965), p. 30.

2. This term was first coined by Oscar Lewis to refer to Mexicans in Oscar Lewis, *The Children of Sanchez* (New York: Random House 1961). He later applied the term to Puerto Ricans. Lewis saw the "culture of poverty," at least in part, as a response to oppressive circumstances. It was Moynihan and others who later applied the term to black Americans and used it in more of a victim-blaming manner.

DISCUSSION QUESTIONS

1. What does it mean to say that the common view of African American families has been one of "pathology?" How do Feagin and McKinney evaluate this assumption?

2. In what specific ways do African American families adapt to the racism they encounter?

26

The Myth of the Missing Black Father

ROBERTA L. COLES AND CHARLES GREEN

*Why is there such a high number of mother-only families in the Black commu-
nity when this has not historically been the case? These authors identify how race
and social class barriers are important in shaping Black men's ability to fill an
important gendered role—being a father.*

The black male. A demographic. A sociological construct. A media caricature.
A crime statistic. Aside from rage or lust, he is seldom seen as an emotionally
embodied person. Rarely a father. Indeed, if one judged by popular and academic coverage, one might think the term "black fatherhood" an oxymoron. In
their parenting role, African American men are viewed as verbs but not nouns;
that is; it is frequently assumed that Black men *father* children but seldom *are*
fathers....

But this stereotype did not arise from thin air.... In 2000, only 16 percent of
African American households were married couples with children, the lowest of
all racial groups in America. On the other hand, 19 percent of Black households
were female-headed with children, the highest of all racial groups. From the perspective of children's living arrangements, shown in [Table 1] over 50 percent of
African American children lived in mother-only households in 2004, again the
highest of all racial groups. Although African American teens experienced the
largest decline in births of all racial groups in the 1990s, still in 2000, 68 percent
of all births to African American women were nonmarital, suggesting the pattern
of single-mother parenting may be sustained for some time into the future. This
statistic could easily lead observers to assume that the fathers are absent.

While it would be remiss to argue that there are not many absent black
fathers, absence is only one slice of the fatherhood pie and a smaller slice than is
normally thought. The problem with "absence," as is fairly well established now,
is that it's an ill defined pejorative concept usually denoting nonresidence with
the child, and it is sometimes *assumed* in cases where there is no legal marriage
to the mother. More importantly, absence connotes invisibility and noninvolve-
ment, which further investigation has proven to be exaggerated (as will be discussed below). Furthermore, statistics on children's living arrangements (Table 1)
also indicate that nearly 41 percent of black children live with their fathers, either
in a married or cohabiting couple household or with a single dad.

SOURCE: Coles, Roberta L., and Charles Green, eds. 2010. *The Myth of the Missing Black Father.*
New York: Columbia University Press.

T A B L E 1   **Living Arrangements of Black Children, 2004**

Living arrangements	Black children
Two parents	37.6%
Married	33.9
Unmarried	3.7
Mother-only	50.4
Father-only	3.3
Neither parent	8.8

SOURCE: Kreider (2008).

These African American family-structure trends are reflections of large-scale societal trends—historical, economic, and demographic—that have affected all American families over the past centuries. Transformations of the American society from an agricultural to an industrial economy and, more recently, from an industrial to a service economy entailed adjustments in the timing of marriage, family structure and the dynamics of family life. The transition from an industrial to a service economy has been accompanied by a movement of jobs out of cities; a decline in real wages for men; increased labor-force participation for women; a decline in fertility; postponement of marriage; and increases in divorce, non-marital births, and single-parent and nonfamily households.

These historical transformations of American society also led to changes in the expected and idealized roles of family members. According to Lamb (1986), during the agricultural era, fathers were expected to be the "moral teachers"; during industrialization, breadwinners and sex-role models; and during the service economy, nurturers. It is doubtful that these idealized roles were as discrete as implied.... It is likely that many men had trouble fulfilling these idealized roles despite the legal buttress of patriarchy, but it was surely difficult for African American men to fulfill these roles in the context of slavery, segregation, and, even today, more modern forms of discrimination. A comparison of the socio-economic status of black and white fathers illustrates some of the disadvantages black fathers must surmount to fulfill fathering expectations.... In 1999 only 33.4 percent of black fathers had attained at least a college education, compared to 68.5 percent of white fathers. In 1998, 25.5 percent of black fathers were un- or underemployed, while 17.4 percent of white fathers fell into that category. Nearly 23 percent of black fathers' income was half of the poverty threshold, while 15 percent of white fathers had incomes that low.

The historical transformations were experienced across racial groups but not to the same extent. The family forms of all racial groups in America have become more diverse, or at least recognition of the diversity of family structure has increased, but the proportions of family types vary across racial groups. Because African American employment was more highly concentrated in blue-collar jobs, recent economic restructuring had harsher implications for black communities and families. The higher and more concentrated poverty levels and greater income and wealth inequality—both among African Americans and between African Americans and whites—expose African American men, directly and indirectly, to continued lower life expectancy, higher mortality, and, hence, a skewed gender ratio that leaves black women outnumbering black men by the age of eighteen.

All of these societal and family-level trends affect black men's propensity to parent and their styles of parenting in ways we have yet to fully articulate. For instance, Americans in general have responded to these trends by postponing marriage by two to four years over the last few decades, but that trend is quite pronounced among African Americans, to the point that it is estimated that whereas 93 percent of whites born from 1960 through 1964 will eventually marry, only 64 percent of blacks born in the same period ever will.

… Wilson (1987) and others have suggested that black men's underemployment, along with black women's higher educational attainment in relation to black men may decrease both men's and women's desire to marry and may hinder some black men's efforts to be involved fathers. However, other research has found that even college-educated and employed black men have exhibited declines in marriage, and yet additional research points to attitudinal factors with black men desiring marriage less than white and Latino men.…

In conjunction with [an] increased amount of research and, in fact, frequently fueling the research, has been a proliferation of public and private programs and grants aimed at creating "responsible fatherhood." While many of the programs have been successful in educating men on how to be qualitatively better fathers, many have aimed primarily either at encouraging fathers to marry the mothers of their children or at securing child support. Marriage and child support are important aspects of family commitment, but marriage is no guarantee of attentive fathering, and garnished child support alone, particularly if it goes to the state and not to the mother and child, is hardly better parenting.…

Given the increased focus on fatherhood in scholarly and popular venues, what do we really know about black men and parenting? … African American fathers, when discussed at all, continue to be addressed predominantly under categories frequently associated with parenting from afar, as nonresident, nonmarital fathers. Even books specifically on black fathers concentrate almost exclusively on nonresident fathers.

… Studies on this ilk of fathers indicate that generally a large portion of nonresident fathers are literally absent from their children's lives or, if in contact, their involvement decreases substantially over time. A number of memoirs by black men and women, sons and daughters of literally absent fathers, attest to the painful experience that this can be for the offspring—both sons and daughters—of these physically or emotionally missing fathers. For instance, writing in his 1999 book *Becoming Dad: Black Men and the Journey to Fatherhood*, award-winning journalist Leonard Pitts wrote of his own father and others:

> He was one thing many other fathers were not: He was there. Present and accounted for every day. Emotionally absent, mind you. But there, at least, in body. I know so many men, so many black men, who cannot

say the same. So many men for whom the absence of father is a wound that never scabbed over.

… Although these anguished experiences are too common, they remain only one part, though often the more visible part, of the larger fatherhood picture. An increasing number of quantitative and qualitative studies find that of men who become fathers through nonmarital births, black men are least likely (when compared to white and Hispanic fathers) to marry or cohabit with the mother. But they were found to have the highest rates (estimates range from 20 percent to over 50 percent) of visitation or provision of some caretaking or in-kind support (more than formal child support). For instance, Carlson and McLanahan's (2002) figures indicated that only 37 percent of black nonmarital fathers were cohabiting with the child (compared to 66 percent of white fathers and 59 percent of Hispanic), but of those who weren't cohabiting, 44 percent of unmarried black fathers were visiting the child, compared to only 17 percent of white and 26 percent of Hispanic fathers. These studies also suggested that black nonresident fathers tend to maintain their level of involvement over time longer than do white and Hispanic nonresident fathers.

Sometimes social, fictive, or "other" fathers step in for or supplement nonresident biological fathers. Little research has been conducted on social fathers, but it is known they come in a wide variety: relatives, such as grandfathers and uncles; friends, romantic partners and new husbands of the mother, cohabiting or not; and community figures, such as teachers, coaches, or community-center staff. Although virtually impossible to capture clearly in census data, it is known that a high proportion of black men act as social fathers of one sort or another, yet few studies exist on this group of dads.

Jarrett, Roy, and Burton's (2002:234) review of qualitative studies of black fathers managed to capture the perspectives of a few low-income social fathers. One sixteen-year-old talked about his fatherlike relationship with the young daughter of a friend.

> Tiffany (a pseudonym) is not my baby, but she needs a father. To be with her, I work in the day care center at school during my lunch hour. I feed her, change her diapers, and play with her. I buy her clothes when I can because I don't make much money. I keep her sometimes. Her mother and her family appreciate what I do and Tiffany loves me too. Every time she sees me she reaches for me and smiles.

Fagan's (1998) study of low-income biological and stepfathers in two-parent homes found that the two types of fathers of black children were equally involved. Similarly, McLanahan and Sandefur (1994) found that, compared to those who live in single-parent homes, black male teens who lived with stepfathers were significantly less likely to drop out of school and black teen females were significantly less likely to become teen mothers. The authors speculated that the income, supervision, and role models that stepfathers provide may help compensate for communities with few resources and social control. Although they are often pictured as childless men, these social fathers may also be some other child's biological father, sometimes a nonresident father himself. Consequently, it is not easy and is certainly misleading to discuss fathers as if they come in discreet, nonoverlapping categories of biological or social.

A smaller amount of research has been conducted on black fathers in two-parent families, which are more likely to also be middle-class families. Allen (1981), looking at wives' reports, found black wives reported a higher level of father involvement in childrearing than did white wives. McAdoo (1988) and Bowman (1993) also concluded that black fathers are more involved than white fathers in childrearing....

Finally, and ironically, most *absent* in the literature on black fatherhood have been those fathers who are most *present*: black, single full-time fathers. About 6 percent of black households are male-headed, with no spouse present; about half of those contain children under eighteen years old. These men also may be biological or adoptive fathers, but little is known about them....

In sum, research on black fathers has been limited in quantity and has narrowly focused on nonmarital, nonresident fathers and only secondarily on dads in married-couple households. This oversight is not merely intentional, for black men are only about 6 percent of the U.S. population and obviously a smaller percent are fathers.... We want to adjust the public's visual lens from a zoom to a wide angle to view black fathers in a realistic landscape, to illustrate that they are quite varied in their living arrangements, marital status, and styles of parenting.

... We want to consider policies, tried or suggested, that impede or facilitate parenting on the part of black men. We seek to provide a forum in which black fathers in their full range of parenting take center stage. We feel that the timing ... is opportune, with the recent election of the first black president of the United States. Many African Americans are optimistic that President Barack Obama, who experienced the absence of his own father and is expressly committed to furthering involved fatherhood, will be able to significantly weaken the existing stereotype of the black father, both through his own public example and through facilitative policy.

REFERENCES

Allen, W. 1981. "Mom, Dads, and Boys: Race and Sex Differences in the Socialization of Male Children." In *Black Men*, ed. L. Gary, 99–114. Beverly Hills, Calif.: Sage.

Bowman, P. 1993. "The Impact of Economic Marginality on African-American Husbands and Fathers." In *Family Ethnicity*, ed. H. McAdoo, 120–137. Newbury Park, Calif.: Sage.

Carlson, M. J. and S. S. McLanahan. 2002. "Fragile Families, Father Involvement, and Public Policy." In *Handbook of Father Involvement: Multidisciplinary Perspectives*, ed. Catherine Tamis-LeMonda and Natasha Cabrera, 461–88. Mahwah, N.J.: Lawrence Erlbaum.

Fagan, J. 1998. "Correlates of Low-Income African American and Puerto Rican Fathers' Involvement with Their Children." *Journal of Black Psychology* 24 (3): 351–67.

Jarrett, R. L., K. M. Roy, and L. M. Burton. 2002. "Fathers in the 'Hood': Insights from Qualitative Research on Low-Income African American Men." In *Handbook of Father Involvement: Multidisciplinary Perspectives*, ed. C. Tamis-LeMonda and N. Cabrera, 221–48. New York: Lawrence Erlbaum.

Kreider, R. M. 2008. *Living Arrangements of Children: 2004*. U.S. Department of Commerce, U.S. Census Bureau.

Lamb, M. E. 1986. "The Changing Role of Fathers." In *The Father's Role: Applied Perspectives*, ed. M. E. Lamb, 3–27. New York: Wiley.

McAdoo, J. L. 1988. "The Roles of Black Fathers in the Socialization of Black Children." In *Black Families,* ed. H. P. McAdoo, 257–69. Newbury Park, Calif.: Sage.

McLanahan, S., and G. Sandefur. 1994. *Growing Up with a Single Parent. What Hurts, What Helps.* Cambridge, Mass.: Harvard University Press.

Pitts, L., Jr. 1999. *Becoming Dad: Black Men and the Journey to Fatherhood.* Atlanta: Longstreet.

Wilson, W. J. 1987. *The Truly Disadvantaged: The Inner City, the Underclass, and Public Policy.* Chicago: University of Chicago Press.

DISCUSSION QUESTIONS

1. What factors are important in affecting Black men's parenting roles?

2. What do the authors mean by "social fathers"? Why do they see such roles as important?

3. How does understanding the intersection of race and social class enhance our understanding of how Black men parent?

Section C

Housing and Education
Elizabeth Higginbotham and Margaret L. Andersen

In its 1954 *Brown v. Board of Education of Topeka, Kansas* decision, the Supreme Court declared segregation unconstitutional. The *Brown* decision overturned an 1896 Supreme Court decision, *Plessy v. Ferguson*, that had allowed segregation in public facilities based on the principle of "separate but equal." Of course, public facilities—schools, hospitals, transportation, and so forth—were nowhere near equal. Yet, until the *Brown* decision, segregation was legally sanctioned, that is, **de jure segregation** (meaning segregation by law). Under de jure segregation (known less formally as Jim Crow segregation), the American South, in particular, maintained strictly segregated lives for people based on race. In the North, people were also socially and spatially separated, but rather than the separation being legally enforced, **de facto segregation** (that is, segregation "in fact," if not in law) was the practice. Both systems caused harm through the unequal facilities that people experienced and through the separation in relationships that segregation produced.

Since the *Brown* decision, the nation has slowly recognized that de jure segregation is wrong, but have we dismantled desegregation in our housing, schools, health care systems, and other institutions? Hardly. People of different racial and ethnic backgrounds are still socially and spatially separated from each other to large degrees—and, in some cases, to very high degrees, especially where we live and attend school. The Supreme Court declared that "separate but equal" is unconstitutional, but segregation continues to shape how we live and the educational opportunities available for different groups.

Inequality in housing and education is also fundamental to other life opportunities—whether your home accumulates in value over time (thereby affording you the opportunity to make further investments, such as in the education of your children, the improvement of your property, or the acquisition of more property). Especially in a society so dependent on technological innovation and skilled workers, not providing a quality education to large portions of the citizenry can be a recipe for further problems. More and more, the economy requires skilled, educated workers to keep the nation economically healthy.

Without our own population receiving the education needed, the United States relies more on workers from other nations to fuel the nation's economic health. Meanwhile, millions of our own citizens are left without opportunities for advancement, sometimes leading to the only option they may have: crime.

Segregation in housing and education is thus lethal to the nation's health. Segregation also means that people's daily lives are lived very differently. People's abilities to live in a safe neighborhood, provide for a family, advance educationally, and enjoy the fruits of society can all be thwarted by segregation.

Segregation also shapes the relationships that people have—or do not have. Among other things, as we have seen in the Part III of this volume, segregation can distort people's ideas about each other and make them more susceptible to accepting racial stereotypes. Segregation has consequences for both White people and people of color. With racial segregation, there is less sharing of culture, histories, ideas, and caring across racial lines. Even friendships are affected by segregation. A team of sociologists studied cross-race friendships in several high schools around the nation and found that cross-race friendships increase when school populations are more diverse (Quillian and Campbell 2003). There is a large body of scholarly research indicating that having more diverse working groups— that is, creating work and educational spaces where people are not segregated— means a number of positive outcomes: more learning by everyone (Gurin, Biren, and Lopez 2004), more innovation in ideas (Page 2007), even more profit in companies with a diverse labor force (Herring 2009).

This section explores two areas where segregation remains a huge problem: housing and education. In 1968, the U.S. Congress passed the Fair Housing Act to address discrimination in housing. Yet, **residential segregation** still prevails, isolating people from one another; as well as pushing some people to live in communities with inadequate services and few amenities. Some may even lack nearby grocery stores to provide fresh fruits and vegetables. **Hypersegregation**—that is, the concentration of people of color in poor, urban areas—clusters poverty in certain communities, thereby shielding those in more advantaged areas from understanding the current realities of race and class oppression.

In the articles here, the authors examine different questions about housing and educational segregation. George Lipsitz ("Race, Place, and Power"), expanding on his earlier work on the investment in White privilege, explores how the unfair gains for White people persist and have minimized the effectiveness of civil rights legislation. Lipsitz also builds on the innovative research by Melvin Oliver and Thomas Shapiro (1996) who identified the **wealth gap** between Black and White people, putting the origins of the gap in federal policies and historical practices.

Lipsitz demonstrates that housing options, while rooted in the discriminatory and racist land practices from the past, are still very much alive today. Laws are not enforced and people of color face obstacles that keep them "place-bound," thus vulnerable to other social problems.

Residential segregation sets the stage for segregation in the nation's public school system. One remedy to open up opportunities was court-ordered desegregation following the *Brown v. Board of Education* Supreme Court decision. Now, though, these efforts are being reversed, particularly as the nation's courts have become more conservative and less willing to allow race-specific policies for school desegregation. Many school districts are employing other strategies.

Schools are heavily funded through local property taxes, thus, where one lives matters both in school composition and school resources. When White families with more resources abandon urban neighborhoods, such as in moving to the suburbs, they take tax dollars with them. If well-to-do families that live in urban areas place their children in private schools in their search of a good education, the public schools that remain for everyone else are resource poor. Most parents want to secure a good education for their children, but excellent public schools are less common than in the past and push better-off parents to use their resources on behalf of their own children, inadvertently disadvantaging others (Shapiro 2004).

It is hard to turn the tide, when some processes support racial hierarchies. Attempts at school integration take place within a race and class system, so it is never smooth sailing. Karolyn Tyson ("Desegregation without Integration") documents how North Carolina schools, legally segregated before the Supreme Court decision, are now more racially integrated. Many school districts have introduced academic tracking, often in early grades, so that students are separated by "ability." This process is, however, often racially coded. Tyson finds that students in integrated schools who enroll in Honors and Advanced Placements classes are overwhelmingly White, leaving high-ability students of color in these classes feeling isolated. Meanwhile, the majority of the Black students in these schools are placed in less academically challenging classes. In predominantly Black schools, this is not the case, as Black students are placed in high-ability classes in larger numbers. Tyson shows that a school can look integrated from the outside, but be very segregated on the inside. Not unique to the South, these new patterns of segregation are common around the nation.

Amanda E. Lewis, John B. Diamond, and Tyrone A. Forman ("Conundrums of Integration: Desegregation in the Context of a Racialized Hierarchy") are also concerned with the new forms of segregation that emerge in supposedly integrated schools. In their research, they interviewed students, teachers, parents, administrators, and staff in a suburban high school. They found that the students go to school together, but there is little actual integration, even though everyone seems to be aware of that reality and wants to change it. Class privileges White students whose parents closely monitor their placement while early tracking limits educational opportunities for low-income students whose parents have less ability to navigate the institution. Consequently, racial minority students have less access to Advanced Placement classes, leaving these classrooms mostly White. Talented racial minority students are also often challenged if they look "out of place." Lewis and her colleagues show, however, that teacher awareness and targeted programs can make a difference in students' lives.

The articles in this section suggest that we have to do more to eliminate racial biases in housing and education. New patterns of gentrification, as wealthy people return to urban areas and displace established residents, are not producing integrated communities. We can only dismantle segregated institutions when we understand the scope of the problem and have people willing to recommit to living in a genuinely integrated society.

REFERENCES

Gurin, Patricia, A. Nagda Biren, and Gretchen E. Lopez. 2004. "The Benefits of Diversity in Education for Democratic Citizenship." *Journal of Social Issues* 60 (1): 17–34.

Herring, Cedric. 2009. "Does Diversity Pay? Race, Gender, and the Business Case for Diversity." *American Sociological Review* 74 (2): 208–224.

Oliver, Melvin L., and Thomas M. Shapiro. 1996. *Black Wealth/White Wealth: A New Perspective on Racial Equality*. New York: Routledge.

Page, Scott. 2007. *The Difference: How the Power of Diversity Creates Better Groups, Firms, Schools, and Societies*. Princeton, NJ: Princeton University Press.

Quillian, Lincoln, and Mary E. Campbell. 2003. "Beyond Black and White: The Present and Future of Multiracial Friendship Segregation." *American Sociological Review* 68 (4): 540–566.

Shapiro, Thomas M. 2004. *The Hidden Cost of Being African American: How Wealth Perpetuates Inequality*. New York: Oxford University Press.

FACE THE FACTS: WHO IS SEGREGATED IN SCHOOLS?

Racial Composition of Schools Attended by the Average Student of Each Race, 2005–2006

Percent Race in Each School	White Student	Black Student	Latino Student	Asian Student	American Indian Student
% White	77	30	27	44	44
% Black	9	52	12	12	7
% Latino	9	14	55	21	12
% Asian	4	3	5	23	3
% American Indian	1	1	1	1	35

SOURCE: Orfield, Gary, and Chungmei Lee. 2007. *Historic Reversals, Accelerating Resegregation and the Need for New Integration Strategies.* Los Angeles, CA: UCLA Civil Rights Project.

Think about It: School segregation is increasing in the United States, as shown in one of the readings in this section. But look at which groups are most likely to be segregated, as this table details. What does this suggest to you about how different groups are affected by race relations and the segregation of the schools?

27

Race, Place, and Power

GEORGE LIPSITZ

Even with new legal remedies, the larger picture of racial inequality has not changed. Lipsitz shows how racism stamps the built environment and how racial inequality is expanding with declining investment in public institutions.

How can it be that decades after the adoption of comprehensive civil rights ... laws, racial identity remains the key variable in shaping opportunities and life chances for individuals and groups in the United States? Why does race still matter so much? The most popular answers to these questions lead us in exactly the wrong directions. Since the 1970s, politicians, pundits, and publicists have argued that Black people have shown themselves to be simply unfit for freedom. They argue that in a time when civil rights laws clearly ban discrimination, the persistence of racial inequality demonstrates that Blacks have been unable to take advantage of the opportunities afforded them. Equal opportunity exists, they contend, so unequal outcomes have to be attributed to what they perceive to be the deficient values, beliefs, and behaviors of Black people themselves. At times those who adhere to these positions concede that past generations of Blacks had legitimate grievances about slavery, segregation, vigilante violence, and disenfranchisement, but they argue that the problems that Black people confront today are of their own making. What was once done to them by white racists, this line of argument contends, Blacks are now doing to themselves. Inequality between races today, they claim, exists because Blacks allegedly commit more crimes, have lower rates of marriage, and higher rates of children born out of wedlock. They contend that Black students perform poorly on standardized tests because they and their parents do not value education, and that they are disproportionately poor because their parents either refuse to work or because they foolishly purchase expensive and flashy consumer goods while refusing to save money. Some of these critics even blame these conditions on civil rights laws themselves, arguing that efforts to desegregate schools, to promote fair hiring, and to end housing discrimination have led Blacks to expect special preferences and privileges simply because they are Black. At the same time, these critics complain that society practices reverse racism by punishing hardworking whites and giving unearned rewards to unqualified Blacks.

SOURCE: Lipsitz, George. 2011. *How Racism Takes Place*. Philadelphia, PA: Temple University Press.

In my book *The Possessive Investment in Whiteness*, I showed how focusing on Black disadvantages deflects attention away from the unearned advantages that whites possess. It is not so much that Blacks are disadvantaged, but rather that they are taken advantage of by discrimination in employment, education, and housing, by the ways in which the health care system, the criminal justice system, and the banking system skew opportunities and life chances along racial lines. Moral panics about alleged Black misbehavior, I argued, are designed to obscure the special privileges that whites receive from collective, cumulative, and continuing forms of discrimination.[1]

A large and unrefuted body of research reveals how the economic standing of millions of white families today stems directly from the unfair gains and unjust enrichments made possible by past and present forms of racial discrimination. A wide range of public and private actions protect the assets and advantages that whites have inherited from their ancestors, wealth originally accumulated during eras when direct and overt discrimination in government policies, home sales, mortgage lending, education, and employment systematically channeled assets to whites. For example, at least forty-six million white adults today can trace the origins of their family wealth to the Homestead Act of 1862. This bill gave away valuable acres of land for free to white families, but expressly precluded participation by Blacks.[2] Seventy years later, the 1934 Federal Housing Act distributed federally insured home mortgages to whites in overtly and directly discriminatory fashion, building additional equity in the estates of some thirty-five million white families between 1934 and 1978 while systematically excluding Black families from those opportunities.[3] Moreover, because money is passed down across generations through inheritance, the patterns of the past still shape opportunities in the present. Whites not only inherit the riches that flow across generations because of these policies, but new provisions in the tax code consistently add new forms of favored treatment to inherited wealth while increasing taxes on earned income.

Segregated housing leads to segregated schools that give white people privileged treatment, better facilities and better trained teachers. School and neighborhood networks give them access to insider information which enables them to receive preferential treatment when seeking the 80 to 90 percent of jobs in U.S. society that are never openly advertised to the general public. Over time, these uncompetitive processes shape wealth accumulation. They produce cumulative disadvantages for African Americans, but provide "locked in" advantages for whites.... [W]hites used restrictive covenants, racial zoning, redlining, steering, blockbusting, and mob violence between 1866 and 1948 to monopolize advantages for themselves and their descendants. They acted collectively as a group to gain favored access to homeownership, employment, education, and political power. The Federal Housing Administration and other government agencies translated aspirations for racial power into public policy, channeling home loans to whites while denying them to Blacks. Although many of the practices that secured these gains initially were outlawed by the civil rights laws of the 1960s, the gains whites received for them were already locked in place. Even more important, nearly every significant decision made since then about urban planning, education, employment, transportation, taxes, housing, and health care has served to protect the preferences, privileges, and property that whites first acquired from an expressly and overtly discriminatory market.[4]

Blacks and whites with similar incomes, work histories, and family alignments have very different relationships to wealth. Blacks currently possess merely seven to ten cents for every dollar of net worth that whites possess.[5] Largely because of racialized space, whiteness in this society is not so much a color as a condition. It is a structured advantage that channels unfair gains and unjust enrichments to whites while imposing unearned and unjust obstacles in the way of Blacks. Of course, not all whites benefit equally from the possessive investment in whiteness, but even the poorest of the poor among whites do not face the degree of concentration in impoverished neighborhoods and schools or the levels of exposure to environmental hazards that routinely confront middle-income Blacks.

The wealth that present-day whites acquire from expressly discriminatory and racist land use practices makes a huge difference in their lives. Middle-class whites have between 3 and 5 times as much wealth as equally achieving blacks. Disproportionately large inheritances provide them with transformative assets that enable whites to make down payments on homes, start businesses, and pay for college educations. Inherited wealth is the main reason why whites and Blacks earning exactly the same incomes have widely divergent wealth portfolios.[6] Sociologist Thomas Shapiro shows that between 1990 and 2020, some seven to nine trillion dollars will be inherited by the "baby boom" generation. Almost all of that money is rooted in profits made by whites from overtly discriminatory housing markets before 1968. Adult white wage earners routinely inherit money *from* parents, while adult non-white wage earners routinely send out money *to* their parents to compensate for the low wages and lack of assets they possess because of racial discrimination. Shapiro's research reveals that white inheritance is seven times larger than Black inheritance. One out of three "baby boom" generation whites in 1989 could count on bequests, but only one in twenty Blacks could have similar expectations. In addition, even among those who do inherit wealth, whites are four times as likely as Blacks to receive a sizably significant inheritance. On the average, whites inherit $102,167 more than Blacks. White families are 2.4 times as likely as Blacks to have parents who can provide help with down payments or closing costs. Largely because of assets inherited from the past, Blacks get $2.10 in net worth for every dollar earned, whites get $3.23. Cuts in inheritance and capital gains taxes disproportionately benefit whites and make property income more valuable compared to wage income. The homes that whites do acquire in largely white neighborhoods cost them less than comparable homes purchased by Blacks, but they appreciate in value much more than homes in Black neighborhoods. Only 26 percent of white children grow up in asset-poor households, but 52 percent of blacks and 54 percent of Latinos grow up in these economically fragile households. According to Shapiro, inheritance is more important in determining life chances than college degrees, number of children in the family, marital status, full-time employment, or household composition.[7]

Because these inequalities started with discrimination in the past, one might expect that they would become less important over time, that improvements in race relations would gradually narrow the racial wealth gap. Yet precisely the opposite is the case. Assets that appreciate in value and are transferred across generations *increase* in value over time, especially when their privileged beneficiaries skew public policy to make the fruits and rewards of past discrimination even more valuable in the present. A 2010 study conducted by Shapiro and his colleagues at the Institute on Assets and Social Policy at Brandeis University revealed that the wealth gap between Blacks and whites quadrupled between 1984 and 2007. More than a quarter of African American families have no assets at all. Even *high-income* Blacks average assets of only $18,000 compared to the $74,000 in assets held by *middle-income* whites. These differences are not due to market forces, personal attributes, or family composition, but rather are the consequence of both direct discrimination and the indirect effects of the racial dimensions of state policies designed to provide incentives and subsidies for asset-building activities like homeownership. Seemingly race-neutral changes in public policies have also played an important role in widening the racial wealth gap. Cuts in inheritance and capital gains taxes over the past three decades have augmented the value of past discrimination, increasing the fortunes of the white beneficiaries of past and present housing discrimination. At the same time, deductions allowable for local property taxes produce massive federal subsidies for school taxes in largely white suburbs.[8]...

Privatization of public institutions, cuts in government services, and capital flight to low-wage countries decreases opportunities for upward mobility for most Americans. Under these circumstances, inherited wealth becomes even more important for those positioned to receive it. A 2002 study found that parental income had become a more reliable predictor of children's eventual earnings than it had been in the 1980s.[9] The damaging effects of this racial wealth gap are exacerbated by the massive refusal in our society to desegregate schools or enforce civil rights laws banning discrimination in employment and education. Having civil rights laws on the books is not an effective way of protecting Black rights when white lawlessness is routinely condoned and encouraged by the major institutions in our society....

Relations between races are relations between places.... White identity in the United States is place bound. It exists and persists because segregated neighborhoods and segregated schools are nodes in a network of practices that skew opportunities and life chances along racial lines. Because of practices that racialize space and spatialize race, whiteness is learned and legitimated, perceived as natural, necessary, and inevitable. Racialized space gives whites privileged access to opportunities for social inclusion and upward mobility. At the same time, it imposes unfair and unjust forms of exploitation and exclusion on aggrieved communities of color. Racialized space shapes nearly every aspect of urban life. The racial imagination that relegates people of different races to different spaces produces grossly unequal access to education, employment, transportation, and shelter. It exposes communities of color disproportionately to environmental hazards and social nuisances while offering whites privileged access to economic opportunities, social amenities, and valuable personal networks. The lived experience of race takes place in actual spaces, while the lived experience of place draws its determinate logic from overt and covert understandings of race. Yet as I attempt to demonstrate …, the actual long-term interests of whites are often damaged by spatial relations that purportedly benefit them, while Black negotiations with the constraints and confinements of racialized space often produce ways of envisioning and enacting more decent, dignified, humane, and egalitarian social relations for everyone.

People of different races do not inhabit different places by choice. Housing and lending discrimination, the design of school district boundaries, zoning regulations, policing strategies, the location of highways and transit systems, and a host of tax subsidies do disastrous work by making places synonymous with races. The racial meaning of place makes American whiteness one of the most systematically subsidized identities in the world. It enables whites to own homes that appreciate in value and generate assets passed down to subsequent generations. At the same time, Blacks confront an artificially constricted housing market that often forces them to remain renters unable to take advantage of the subsidies that homeowners receive from the tax code. When they do manage to own homes, Blacks are forced to do so on terms that compel them to pay more for dwellings that are worth less and appreciate in value more slowly than comparable homes inhabited by whites. Housing and school segregation function to channel white children into well equipped classrooms with experienced teachers while crowding Black children into ill-equipped buildings where they are taught by inexperienced teachers and surrounded by impoverished classmates many of whom suffer from lead poisoning, malnutrition, and a variety of undiagnosed and untreated disabilities. The estimated four million violations of federal fair-housing law that take place every year offer whites privileged access to parks, playgrounds, fresh food, and other amenities while relegating Blacks to areas that suffer disproportionate exposure to polluted air, water, food, and land.

Living in segregated inner-city neighborhoods imposes the equivalent of a racial tax on people of color. One important way in which this "tax" is imposed is on the health and well being of Black bodies. The racial wealth gap is also a racial health gap. Michael Marmot, chairman of the World Health Organization's Commission on Social Determinants of Health, offers a vivid illustration of the health consequences of racialized space. "If you catch the metro train in downtown Washington, D.C. to suburbs in Maryland," Marmot observes, "life expectancy is 55 years at the beginning of the journey. At the end of the journey, it is 77 years. This means that there is a 20-year life expectancy gap in the nation's capital, between the poor and predominately African American people who live downtown, and the richer and predominantly non-African American people who live in the suburbs."[10]

Researchers have long established how racial discrimination in housing impacts health as well as wealth. Relegating people of different races to different places artificially skews exposure to toxic hazards. The neighborhoods of people of color become prime sites for the location of garbage and toxic-waste dumps, incinerators, lead-based paint on playground equipment and interior walls, metal-plating shops, and concentrated pollutants from freeways and factories. Segregation-related educational inequality, racialized policing strategies, mismatches between the location of jobs and the residences of communities of color, siting of supermarkets and fast-food outlets, and the constant emergence of new forms of racially targeted exploitation like predatory lending, insurance redlining, foreclosure abandonment, and under-bounding ... combine to undermine the health of ghetto and *barrio* residents....

The cumulative vulnerabilities crafted by centuries of anti-Black racism leave African Americans facing multiple and overlapping economic obstacles. Direct discrimination by insurance agents and mortgage loan officers compounds the already difficult economic situation facing working-class and poor people of color as can be seen from the ways in which segregation into different neighborhoods channels people of different races to different sectors of the banking industry. Banks locate branches disproportionately in suburban neighborhoods, forcing inner-city residents to turn to nonbanking institutions for banking services. Thus they are compelled to pay exorbitant fees for simple needs like cashing checks.[11] Residents of white neighborhoods can expect to do business with mainstream financial service providers. Their neighborhood banks offer them savings and checking accounts, certificates of deposit, prime rate mortgages, individual retirement accounts, and automobile and home improvement loans. People who live in Black neighborhoods, in contrast, find only low-end service providers. They transact business with pay-day lenders, pawn shops, check-cashing establishments, rent-to-own shops, and subprime mortgage lenders who charge them exorbitant fees and rates of interest because they do not have access to the top end of the banking industry....

The home loan industry often attributes Black reliance on subprime lenders to inadequate consumer sophistication rather than admitting to the pervasive nature of discrimination that drives minority consumers to subprime lenders. The current crisis is a direct result, however, of laws that freed the banking industry from regulation, from the 1980 law passed by Congress that removed interest-rate caps on first-lien mortgages to the Banking Reform Act of 1999 and its attendant securitization of the mortgage industry that enabled individuals to make enormous profits by making unsecured loans in a largely unregulated market. Credit-starved Blacks trapped in artificially constrained housing markets proved to be ideal targets for unscrupulous and unregulated lenders....

Of course, racialized space is not simply a matter of Black and white. In many of my previous publications I have described and analyzed the construction of physical places and discursive spaces by Latinos, Asian Americans, and Native Americans.[12] I have written about Chicano poster art and low riders, Asian American music and musicians, and Native American poetry, about interethnic antiracist organizing by the Asian Pacific Environmental Network, Asian Immigrant Women Advocates, the Labor Community Strategy Center, and the Midwest Treaty Network. I have long maintained that race in the United States and around the world is a complex and polylateral phenomenon, that different aggrieved communities have widely varying relations with each other as well as with whites, that the histories they share entail both coalitions and conflicts. The first racial zoning ordinance in the nation was intended to clear Chinese residents of San Francisco out of desirable neighborhoods downtown and confine them to slum neighborhoods adjacent to polluting factories and noxious waste dumps.[13] The restrictive covenants used everywhere to deny housing opportunities to Blacks also blocked Asian Americans, Latinos, and Native Americans from neighborhood choices and homeownership opportunities. Highway construction and attendant urban renewal programs destroyed some sixteen hundred Black communities in the twentieth century, but they devastated many Latino and Asian American neighborhoods as well.... During the first eight years of federally funded urban renewal, more than 75 percent of those displaced were Black or Latino.... The harsh realities of racialized space confront Native Americans in border towns and urban ghettos, while all communities of color suffer from disproportionate proximity to environmental hazards....

I believe that understanding the causes and consequences of racialized space can advance the cause of racial justice. It can help address and redress the injuries that Black people experience from living in a society where not just white property but even white vanity is valued more highly than Black humanity. But the problems produced by racialized space should not be simply the particular and parochial concerns of Blacks. Although the system through which race takes place delivers short-term advantages and benefits to whites, racialized space ultimately hurts everyone. It creates expensive and dangerous concentrations of poverty, pollution, disease, and crime. It misallocates resources by squandering the talents and abilities of deserving Blacks while moving less talented whites into positions they do not deserve. It encourages environmentally unsound patterns of development and transportation, disperses populations inefficiently. It helps produce much of the antisocial behavior that it purports to prevent. It deprives cities, counties, and states of tax revenues by depressing property values artificially. It promotes a suburban culture of contempt and fear that fuels opposition to sensible economies of scale, that encourages each subunit of government to try to win gains against every other subunit. Perhaps most important, it undermines democracy by isolating Black people and the spatial and social imaginaries they have developed over time from potential white allies who would derive great benefit from them—if they could only overcome their allegiances to racial privilege....

NOTES

1. George Lipsitz, *The Possessive Investment in Whiteness: How White People Profit from Identity Politics* (Philadelphia: Temple University Press, 2006).

2. Thomas R. Shapiro, *The Hidden Cost of Being African American* (New York: Oxford, 2004), 190.

3. Kenneth Jackson, *Crabgrass Frontier* (New York: Oxford, 1975), 216.

4. Daria Roithmayr, "Racial Cartels." (University of South California Law and Economics Working Papers Series, Working Paper 66, 2007).

5. Melvin L. Oliver and Thomas M. Shapiro, *Black Wealth/White Wealth: A New Perspective on Racial Equality* (New York: Routledge, 2006), Shapiro, *Hidden Cost*.

6. Oliver and Shapiro, *Black Wealth/White Wealth*.

7. Shapiro, *Hidden Cost*.

8. Richard Rothstein, "How the U.S. Tax Code Worsens the Education Gap." *New York Times*, April 25, 2001, p. A-17.

9. James H. Carr and Nandinee K. Kutty, "Attaining a Just Society." James J. Carr and Nandinee K. Kutty, *Segregation: The Rising Costs for America* (New York: Routledge, 2008), 333.

10. Quoted in Dolores Acevedo-Garcia and Theresa L. Osypuk "Impacts of Housing and Neighborhoods on Health: Pathways, Racial/Ethnic Disparities and Policy Directions," in Carr and Kutty, *Segregation*, 197.

11. John P. Caskey, *Fringe Banking: Check-Cashing Outlets, Pawnshops, and the Poor* (New York: Russell Sage Foundation, 1994), 1, 144.

12. George Lipsitz, *American Studies in a Moment of Danger* (Minneapolis: University of Minnesota Press, 2001), 117–138, 169–184, 213–233. George Lipsitz, *Footsteps in the Dark: The Hidden Histories of Popular Music* (Minneapolis: University of Minnesota Press, 2007), 26–78, 133–153, 211–237. George Lipsitz, "Walleye Warriors and White Identities: Native Americans' Treaty Rights, Composite Identities, and Social Movements," *Ethnic and Racial Studies* 31, no. 1 (January 2008), 101–122.

13. Charles J. McLain, *In Search of Equality: The Chinese Struggle against Discrimination in Nineteenth-Century America* (Berkeley: University of California Press, 1994), 223–233.

DISCUSSION QUESTIONS

1. How is Lipsitz's argument different from the popular explanations for the lack of racial progress?

2. How does attention to wealth change one's understanding of racial inequality?

3. Place or residence is not just important for identity. How does it influence life chances and health?

28

Desegregation without Integration

KAROLYN TYSON

When researchers follow students into schools, they can see how these institutions operate, particularly how the pattern of academic tracking can shape students' vision of their abilities. Tyson asks about the degree to which we have become integrated, especially given the patterns of separation within schools.

In America, nothing matters more for getting ahead than education. The widespread agreement on this point is perhaps why the academic underperformance of racial and ethnic minorities is the focus of so much debate and why public policy, including the No Child Left Behind Act, so often targets these students. The relatively low academic achievement of African American students, in particular, dominates a great deal of the discussion on the achievement gap. Recent figures from the National Assessment of Educational Progress (NAEP) show that, on average, seventeen-year-old black students score 30 points lower than white students on reading and 28 points lower on math.[1] Few issues in the field of education have received as much attention as this one. And everyone—educators, researchers, parents, students, politicians, journalists, and celebrities—seems to have a theory about why black students are not doing better in school. What is most interesting about this chorus of voices is that many loudly repeat some version of the same argument, namely that African American youth are steeped in a culture that ridicules academic achievement because it is equated with acting white.

President Barack Obama's was probably the loudest of these voices, as his remarks were made in a nationally televised speech at the 2004 National Democratic Convention. Commenting on the problems of inner-city neighborhoods, the then-senator emphasized the need to "eradicate the slander that says that a black youth with a book is acting white."[2] Arguably no other explanation of black academic underperformance has become more entrenched. Indeed, it has become part of our commonsense understanding of why black students are underachieving. A former teacher's letter to the editor of the *New York Times* some years ago captures the gist of the public perception about the problem plaguing black students: "Many black children who are serious about their studies are subject to derision by other black students. Acting 'white' is the ultimate put-down. What is so 'white' about being a good student...?"[3]

SOURCE: Tyson, Karolyn. 2011. *Integration Interrupted: Tracking, Black Students, and Acting White after Brown.* New York: Oxford University Press.

Remarks like these initially left me confused. I had spent time conducting research in all-black elementary schools, observing classrooms and talking with students, and these remarks did not match what I had observed. I never heard any student mention acting white, in any context. The students I encountered coveted academic success. There were tears and tantrums when they did not make the honor roll or when they failed to get high marks on an assignment, or when they did not receive a gold star on the board or an invitation to the schoolwide celebration of individual achievement on the state's standardized tests, or when they were not picked for a role in a classroom play. But later, as my research expanded to include adolescents, I began to hear accounts from high-achieving black students about being accused of acting white apparently because of their achievement.

Even now, however, after more than ten years conducting research in schools, what I have learned remains hard to reconcile with claims that black students are "culturally disinclined to do well in school," as linguist John McWhorter claims in his book *Losing the Race*.[4] My empirical analysis shows that explanations such as McWhorter's distort what is really going on. First, many black students do not connect race with achievement. This phenomenon is greatly exaggerated. Second, and perhaps more importantly, the idea that race and achievement are linked is not something that black youth are taught at home or in their communities. This is a connection they learn at school. The popularity of the notion that black students reject achievement as acting white and that this accounts for the black-white achievement gap has helped turn this hypothesis into an accepted fact. But using this argument anecdotally in social discourse is inaccurate and costly. We know surprisingly little about students' use of the acting white slur with respect to achievement, let alone what its empirical connection might be to the achievement gap. When did the association of race with academic achievement emerge among black students? Why have black youth come to equate school success with whiteness?... The students I observed, interviewed, and spent time with during the course of my research generously opened their lives to an outsider's gaze and shared their stories about life at school. They spoke frankly about their successes and disappointments, their trials and tribulations, and their goals, aspirations, and fears.

Consider Sandra, for example. We met in 2002, when she was a junior at Earnshaw School of Excellence. Sandra was one of sixty-five North Carolina public high school students participating in research my colleagues and I were conducting on high-achieving black students. As we walked from one class to another on Earnshaw's sprawling campus, Sandra generally was quiet and reserved. During the interviews, however, she opened up and spoke candidly and at length. Her school, a combined middle and high school, was 49 percent white and 44 percent black. Yet Sandra described her early experiences at Earnshaw as "hell," because she had been "the only black student" in her gifted English and advanced math classes. She explained:

> Well, I—okay, because of my classes that I couldn't take with a lot of
> black students, so I was in, mainly made—I *had* to make friends with a lot

of white students 'cause those are the only people who are in my classes. And those are the people that I tend to sit with at lunch because I had never met anyone else. I mean I had black friends, but because I don't see a lot of them, I made friends with white [students]. And because of that, [the black girls] thought that that meant I was—I didn't want to be with other black people and that I thought that I was better than them, and I was trying to "act white." ... But that's not it, you know.

Some high-achieving black students at other schools described similar encounters with peers over the issue of acting white and particular achievement-related behaviors. For example, Juliana, a senior at Everton High School, recalled that her black friends occasionally joked that she had "turned white on [them]" after she began taking advanced courses, which they called "white people class[es]." Everton's student population was just over one-fifth black, but few advanced classes had more than one or two black students enrolled, and many had none. Another student, Lynden, a senior at City High, remembered being teased in the ninth grade by fellow blacks. They called him "White Pretty Boy" because of his friendships with the "smart" students in his advanced classes, most of whom were white. Although City High's student body was 16 percent black, black students were nearly invisible in the school's higher-level classes. In many of those classes, black students accounted for less than 5 percent of those enrolled.

Stories like these will resonate with readers who are familiar with the contemporary schooling experiences of black youth and the problem of black academic underachievement. What readers may be less familiar with, however, is the unmistakable connection between this peculiar use of the acting white slur and an institutional practice common in secondary schools: tracking. This practice of separating students for instruction, ostensibly based on their ability and prior achievement, often results in segregated classrooms in predominantly white and racially diverse schools like the ones described above. The higher-level classes (gifted, honors, advanced placement) are disproportionately filled with white students, while the lower-level, standard classes are disproportionately filled with black and other minority students. We call this "racialized" tracking, but it is essentially segregation.

With black and white students largely segregated within the schools they attend, racialized tracking has made it possible to have desegregation without integration. It is this school-based pattern of separation that has given rise to students associating achievement with whiteness. This association is found only in schools where racialized tracking is prevalent. In the predominantly black high schools in our study, where advanced classes were majority black, the high-achieving black students we met were not even aware that academic achievement is considered a form of acting white.[5] When we asked Sonya, a senior at Banaker High School (89% black), whether students at the school ever use the acting white slur to refer to high-achieving students in particular, she answered casually, "No, I haven't seen that. When they say you're acting white, they just do if you talk a different way."[6] Sonya and her peers at Banaker had no experience with the type of "hell" Sandra encountered across town at Earnshaw.

The marked difference in the experience of students who attend schools like Sonya's and those who attend schools like Sandra's draws attention to the importance of how schools are organized.... As Sandra's and many other students' experiences suggest, tracking does more than keep black and white students separated during the school day. It also produces and maintains a set of conditions in which academic success is linked with whites: Students equate achievement with whiteness because school structures do.... By focusing on students' equating achievement with acting white as *the* problem, without first trying to understand the cause of this phenomenon, we end up confusing cause with effect. Consequently, we attempt to treat the symptom (equating achievement with whiteness), while the disease (racialized tracking) goes unchecked.... I argue that students' linking achievement with whiteness emerged after desegregation and is a result of racialized tracking, which is part of a historical legacy of strategies used to avoid integration.[7]

In the study of high-achieving black adolescents (*Effective Students*), my research assistants and I shadowed students at school. At racially diverse and predominantly white high schools, the participants were frequently the only black students in their advanced classes. This pattern, which was especially evident in AP courses, did not escape their attention. Without prompting, many mentioned being the "only black" student in their advanced classes. Robin, for example, who attended Shoreline High School (72% white; 15% black), recalled that she did not have another black student in her advanced classes until her junior year. City High (69% white; 16% black) student Courtney called off a list of classes in which he was the only black student: honors physics, AP statistics, AP U.S. history. Jasmine made similar observations about her advanced classes at Lucas Valley High School (65% white; 24% black): "I took advanced English and I was the only black person in [the class both times].... This year I'm the only black person in my physics class." Keisha, who attended Garden Grove High School (54% white; 32% black), complained about being the "only black girl in a sea" of "white people" in her advanced classes....

The participants' reports of their peers' comments indicate that other students at racially diverse and predominantly white schools also were aware of the prevailing pattern of racial segregation across classrooms. Why else would they refer to advanced classes as "white classes" or the "white-people class," or ask peers why they wanted to be in classes or programs in which there were no other black students? These types of comments reflect a general pattern. Reports of racialized ridicule (e.g., being accused of acting white, or being called an "Oreo") for high achievement and for other achievement-related behaviors (e.g., taking advanced classes) always coincide with students' experiences of racial hierarchies in tracking and achievement at the schools they attend. Thus, not all high-achieving black students are taunted for acting white because of their achievement. What I found is that those who are taunted always attend racially diverse or predominantly white schools where racialized tracking makes possible racial isolation and rejection of the kind Sandra experienced....

EXAMINING THE COSTS OF RACIALIZED TRACKING

The pattern of black-white racial segregation in American public schools produced through tracking (and through gifted and magnet programs) is well documented.[8] Indeed, racialized tracking is a common feature of contemporary American secondary schools. Unfortunately, however, it gets relatively little public attention. Americans simply assume that academic placements reflect students' ability and their (and their parents') choices and attitudes toward school. These assumptions are not entirely accurate. There is more to the contemporary high school placement process than meets the eye. Yet in racially mixed schools, what does meet the eye—the image of overwhelmingly black lower-level classes and overwhelmingly white advanced classes—sends powerful messages to students about ability, race, status, and achievement. Linking achievement with whiteness is one consequence of racialized tracking, but there are others that also shape school performance and interracial relations. This book takes a look at how institutional practices such as tracking affect black and other students' schooling experiences. How do students make sense of this pattern of racialized tracking? What does it mean for students' developing sense of self and their decisions and actions at school? How does it affect black students' relationship with same-race and other peers?

In many states, schools' early placement decisions involve some form of ability grouping or the use of academic designations such as "gifted," "gifted and talented," or "advanced." Previous research has shown that these practices contribute to the initial sorting process that sets racial groups on different academic paths in elementary school.[9] Institutional sorting continues more formally and overtly in secondary school. There students are separated for instruction on the basis of a range of criteria, including perceived ability, prior achievement, and/or post-high school occupational plans and aspirations. During adolescence, as students attempt to negotiate the delicate balance between where they fit and where they feel most comfortable, both academically and socially, this sorting reinforces racial patterns and stereotypes. Thus, the schools' early placements and labels have a particularly profound and, for some adolescents, harmful effect....

Racialized tracking and the belief that academic achievement is a "white thing" reflect broader social issues regarding the role of race in education. Each speaks to the continuing significance of race in America. And each also points to obstacles on the path to integration. Since desegregation, we have been witnessing a new form of educational apartheid achieved through tracking.... As part of a historical legacy of strategies to avoid integration, tracking has proven remarkably effective, and, in the wake of desegregation, highly consequential. Black and white adolescents often have very little meaningful contact with one another in school because they are separated for most of their core classes. Not surprisingly, these divisions are often deliberately replicated in other settings. As we observed while shadowing high-achieving black students, in elective classes, at lunch or assemblies, blacks and whites tend to sit apart from one another, occupying different sections of the room.[10] Everyday realities like these help explain why attending a desegregated school does not always provide the benefits we anticipate young people will gain from a racially diverse environment.

TRACKING AND ACTING WHITE AFTER *BROWN*

The Supreme Court's 1954 *Brown v. Board of Education* decision was supposed to eliminate school segregarion.[11] More than five decades after the decision, however, black students and white students throughout much of the country still experience separate and unequal schooling. Not only have American schools been growing more segregated at the school level, they remain overwhelmingly segregated at the classroom level. Numerous scholarly and journalistic accounts detail the startling degree to which black and white students are segregated within the schools they attend.... Yet, within-school segregation has not engendered the same type of urgency, outrage, or national shame that segregation did before and shortly after the *Brown* ruling. There have been no marches or sit-ins, no public outcry, and very little condemnation of school officials or pressure on policy makers. But why should there be protests? Wasn't the fight for equality of educational opportunity won with the *Brown* decision? In some respects it was, if only because the ruling prohibited *de jure* segregation. But *Brown* promised more than desegregation; the decision also promised integration, as it raised hopes that black and white students would come together as equals. Instead, the movement toward integration had barely begun before it was interrupted.

The failure to achieve true integration at school is an enormous loss. No American institution other than schools brings together so many children, from so many diverse backgrounds, for so much time each day, over so many years. Not families, not churches, not neighborhoods. We might expect schools to be where children learn to see past their differences and find commonalities that would allow them to form meaningful relationships with people unlike themselves. Unfortunately, this kind of learning is rare. Instead, schools too often teach young people pecking orders, hierarchies based on status and achievement. The social divisions and animosity that exist between groups in the larger society are reinforced at school. As James Rosenbaum found in his study of tracking at a white working-class school more than three decades ago, "Tracking provides distinct categories that are highly salient."[12] When these categories (e.g., honors, remedial, college prep) mirror the racial, gender, and social class hierarchies in place outside of school, they can lead students to perceive their own and others' assigned placements as accurate and permanent. Indeed, students tend to believe that placements merely reflect racial differences in ability, work ethic, and attitudes toward school.... Thus, for example, some of the high-achieving black students in our study felt that in order to fit in with their peers, they had to project an attitude that downplayed school. Otherwise, they feared, they might be perceived as "acting, like, white or something if you're trying to be smart."

As one student explained, this was because "it's always been [that] ... the smart people are the white people."

The popularity of the belief that a fear of acting white explains black academic underachievement has obscured an important fact. It is only relatively recently that black youth have equated high academic achievement with acting white. The first published accounts of this phenomenon began emerging in the 1980s. Prior to that, there is no indication that acting white included reference to academic achievement and achievement-related behaviors. The earliest published study reporting black students' use of the term documented the students' adjustment to desegregation in a Wisconsin high school in 1969. The black students in the study raised concerns about acting white. They noted that they did not want to lose what was, as the authors put it, "a natural part of Negro behavior as it evolved in this country."[13] The students characterized acting white only as being "more inhibited," "more formal," and "lacking 'soul.'" More recent studies, however, report that among adolescents (blacks and others), acting white currently includes a host of characteristics, including taking honors and advanced placement classes, getting good grades, going to class, and doing schoolwork.[14]

This expansion in meaning of acting white indicates a cultural shift. This is in itself telling, because such shifts typically are associated with other, larger changes in the social world.... In other words, changes in social structure tend to bring about changes in culture. Indeed, patterns like the one Sandra described at Earnshaw suggest that the shift in the meaning of acting white was brought about largely by the change in educational policy that led to school desegregation.[15] Desegregation, and especially attempts to bypass it with tracking and gifted and magnet programs, paved the way for the emergence of distinct racial patterns in student placement and achievement in some schools....

NOTES

1. U.S. Department of Education, National Center for Education Statistics, National Assessment of Educational Progress (NAEP), NAEP 2004 Trends in Academic Progress; unpublished tabulations, NAEP Data Explorer. Rev. March 29, 2010. Available: http://nces.ed.gov/nationsreportcard/nde/.

2. "Transcript: Illinois Senate Candidate Barack Obama," *Washington Post*, July 27, 2004. Rev. October 17, 2005. Available: http://www.washingtonpost.com/wp-dyn/articles/A19751-2004Jul27.html.

3. "Speaking English Properly Is No Cause for Derision." Rev. January 4, 2006. Available: http://query.nytimes.com/gst/fullpage.html?res = 9401EFDE1639F933A25750C0A960958260....

4. McWhorter, *Losing the Race*.

5. Other studies involving black students in predominantly black schools find similar results. See Akom, "Reexamining Resistance as Oppositional Behavior."

6. Prudence Carter also finds that black and Latino youth's most frequent use of the acting white slur is "in reference to speech and language styles." See "Intersecting Identities," 116.

7. See Welner, "Ability Tracking."

8. Important studies of within-school segregation include Clotfelter, *After* Brown; Lucas, *Tracking Inequality*; Lucas and Berends, "Sociodemographic Diversity"; Meier, Stewart, and England, *Race, Class and Education*; Oakes, *Keeping Track*; Orfield and Eaton, *Dismantling Desegregation*; Welner, *Legal Rights, Local Wrongs*.

9. See Brantlinger, *Dividing Classes*; Mickelson and Velasco, "'Bring It On!'"; Oakes, "Two Cities' Tracking"; Oakes and Guiton, "Matchmaking"; Ogbu, *Black American Students in an Affluent Suburb*; Staiger, *Learning Difference*; Tyson, Darity, and Castellino, "It's Not a Black Thing."

10. Reports that black and white students sit apart from one another in the cafeteria and other social spaces at school have persisted for many years. See Tatum, *"Why Are All the Black Kids Sitting Together in the Cafeteria?"*

11. The *Brown v. Board of Education, Topeka, Kansas* (347 U.S. 483) decision over-turned the "separate-but-equal" doctrine legalized in 1896 in *Plessy v. Ferguson* (163 U.S. 537). In 1955, in *Brown II* (349 U.S. 294), the Supreme Court attempted to clarify its position on when and how desegregation should occur with the mandate that schools desegregate "with all deliberate speed."

12. Rosenbaum, *Making Inequality*, 165....

13. McArdle and Young, "Classroom Discussion of Racial Identity."

14. See Bergin and Cooks, "High School Students of Color"; Neal-Barnett, "Being Black"; Peterson-Lewis and Bratton, "Perceptions of 'Acting Black' among African American Teens."

15. Anne Galletta and William Cross make a similar argument about the importance of integration to understanding "oppositionality" among black students. See "Past as Present."

REFERENCES

Akom, A. A. "Reexamining Resistance as Oppositional Behavior: The Nation of Islam and the Creation of a Black Achievement Ideology." *Sociology of Education* 76 (2003): 305–325.

Bergin, David, and Helen Cooks. "High School Students of Color Talk about Accusations of 'Acting White.'" *The Urban Review* 34 (2002): 113–134.

Brantlinger, Ellen. *Dividing Classes: How the Middle Class Negotiates and Rationalizes School Advantage.* New York: RoutledgeFalmer, 2003.

Carter, Prudence. "Intersecting Identities: 'Acting White,' Gender, and Academic Achievement." In *Beyond Acting White: Reframing the Debate on Black Student Achievement*, ed. Erin McNamara Horvat and Carla O'Connor, 111–132. Lanham, MD: Rowman & Littlefield, 2006.

Clotfelter, Charles. *After Brown: The Rise and Retreat of School Desegregation.* Princeton, NJ: Princeton University Press, 2004.

Lucas, Samuel R. *Tracking Inequality: Stratification and Mobility in American High Schools.* New York: Teachers College Press, 1999.

Lucas, Samuel R., and Mark Berends. "Race and Track Location in U.S. Public Schools." *Research in Social Stratification and Mobility* 25 (2007): 169–187.

McArdle, Clare, and Nancy Young. "Classroom Discussion of Racial Identity or How Can We Make It Without 'Acting White.'" *American Journal of Orthopsychiatry* 41 (1970): 135–141.

McWhorter, John. *Losing the Race: Self-Sabotage in Black America.* New York: Free Press, 2000.

Meier, Kenneth J., Joseph Stewart Jr., and Robert E. England. *Race, Class and Education: The Politics of Second-Generation Discrimination.* Madison: University of Wisconsin Press, 1989.

Mickelson, Roslyn, and Anne Velasco. "Bring It On! Diverse Responses to 'Acting White' among Academically Able Black Adolescents." In *Beyond Acting White: Reframing the Debate on Black Student Achievement*, ed. Erin McNamara Horvat and Carla O'Connor, 27–56. Lanham, MD: Rowman & Littlefield, 2006.

Neal-Barnett, Angela M. "Being Black: New Thoughts on the Old Phenomenon of Acting White." In *Forging Links: Clinical Developmental Perspectives*, ed. Angela M. Neal-Barnett, Josefina M. Contreas, and Kathryn A. Kerns, 75–88. Westport, CT: Praeger Publishers, 2001.

Oakes, Jeannie. *Keeping Track: How Schools Structure Inequality.* New Haven, CT: Yale University Press, 1985.

———. "Two Cities' Tracking and Within-School Segregation." *Teachers College Record* 96 (1995): 681–690.

Oakes, Jeannie, and Gretchen Guiton. "Matchmaking: The Dynamics of High School Tracking Decisions." *American Educational Research Journal* 32 (1995): 3–33.

Ogbu, John U. *Black American Students in an Affluent Suburb: A Study of Academic Disengagement.* Berkeley, CA: Lawrence Erlbaum, 2003.

Orfield, Gary, and Susan E. Eaton. *Dismantling Desegregation: The Quiet Reversal of Brown v. Board of Education.* New York: New Press, 1996.

Peterson-Lewis, Sonja, and Lisa Bratton. "Perceptions of 'Acting Black' among African American Teens: Implications of Racial Dramaturgy for Academic and Social Achievement." *The Urban Review* 36 (2004): 81–100.

Rosenbaum, James E. *Making Inequality: The Hidden Curriculum of High School Tracking.* New York: John Wiley and Sons, 1976.

Staiger, Annegret. *Learning Difference: Race and Schooling in the Multiracial Metropolis.* Stanford, CA: Stanford University Press, 2006.

Tatum, Beverly Daniel. *"Why Are All the Black Kids Sitting Together in the Cafeteria?" and Other Conversations about Race.* New York: Basic Books, 1997.

Tyson, Karolyn, William Darity Jr., and Domini Castellino. "It's Not 'a Black Thing': Understanding a Burden of Acting White and Other Dilemmas of High Achievement." *American Sociological Review* 70 (2005): 582–605.

U.S. Department of Education, National Center for Education Statistics, National Assessment of Educational Progress (NAEP), NAEP 2004 Trends in Academic Progress; unpublished tabulations, NAEP Data Explorer. Rev. March 29, 2010. Available: http://nces.ed.gov/nationsreportcard/nde/

Welner, Kevin., *Legal Rights Local Wrongs: When Community Control Collides with Educational Equity*. Albany: State University of New York Press, 2001.

——. "Ability Tracking: What Role for the Courts." *Education Law Reporter* 163 (2002): 565–571.

DISCUSSION QUESTIONS

1. What does Tyson mean by racialized tracking? How does it influence student interactions?

2. Why does Tyson think of these new patterns of segregation in schools as a loss?

29

Conundrums of Integration
Desegregation in the Context of Racialized Hierarchy

AMANDA E. LEWIS, JOHN B. DIAMOND, AND TYRONE A. FORMAN

There is widespread public support for integration, but what does that mean within the context of a society so divided by race and social class? This question motivated Lewis, Diamond, and Forman to interview teachers, staff, students, and parents in a suburban high school that is racially and economically diverse. They discover that patterns within the school still maintain a racial hierarchy that sustains the privileges of White students and create obstacles for the advancement of students of color.

Recent scholarly and public conversations have given renewed attention to integration as a goal, an aspiration, and/or an "imperative." These calls for integration are infused with the conviction that segregation is a linchpin, if not *the* linchpin, of persistent racial hierarchies....

... [M]any suggest that integration is necessary for equality. They do so in part because racial segregation has broad negative consequences (Anderson 2010; Hartman and Squires 2010; Massey and Denton 1993). For instance, blacks and Latinas/os living in segregated communities experience higher crime, lower property values, fewer social services and public facilities, more environmental toxins, and worse health. Similarly, school segregation has consequences for educational quality (Mickelson 2008; Orfield, Frankenberg, and Garces 2008). Majority-minority schools have fewer qualified teachers; less access to technology, instructional materials, and advanced curricula; and lower-quality facilities (Darling-Hammond 2010; Diamond 2013; Lewis and Manno 2011; Mickelson 2003; Oakes 2005). While the costs of persistent segregation remain clear, the call for integration as the unequivocal answer is more contested.

For instance, there remains disagreement about how to achieve integration, or even whether it is the appropriate end goal. The ambivalence about integration stems from many factors, including cynicism about whether one can achieve it. With regard to education, schools have become more segregated in recent decades in part because the courts have largely abandoned desegregation interventions and have become more hostile toward voluntary district efforts to

SOURCE: Lewis, Amanda E., John B. Diamond, and Tyrone A. Forman. 2015. "Conundrums of Integration: Desegregation in the Context of Racialized Hierarchy." *Sociology of Race and Ethnicity*, 1 (January): 37–51.

racially integrate (Orfield et al. 2008; Orfield and Gordon 2001; Wells et al. 2009).

A second set of concerns relates to the assumption that integration is the best means to achieving equality. For example, in 1935, W. E. B. Du Bois (1935:120) argued that integrated schools were the "more natural basis for education of all youth." However, given the racism and ignorance about black culture, historical contributions, and intellectual capacity among the white teachers to be found in integrated schools at the time, he was concerned about the efficacy of integration. He argued, "The Negro needs neither segregated schools nor mixed schools. What he needs is Education" (Du Bois 1935:120)....

While we acknowledge the importance of these concerns, our goal here is not to enter the debate regarding whether integration is the best strategy for achieving equality. We instead ask whether, even where we are "successful" in combating segregation and manage to get diverse bodies into the same social space, the racial hierarchy is disrupted and equity is guaranteed. To do this, we study what many could call an optimal "integrated" space—Riverview High school, a successful suburban high school known for its diversity and commitment to integration. It is a well-funded school in a liberal district in which teachers and parents express a dedication to providing high-quality educational experiences to all. Despite these commitments, what we find at Riverview is that "integration" does not ensure equity. In many ways, the school captures the stark difference between desegregated spaces and truly integrated ones. As Martin Luther King (1986:118) described decades ago in the midst of school desegregation struggles,

> Although the terms desegregation and integration are often used interchangeably, there is a great deal of difference between the two.... The word segregation represents a system that is prohibitive; it denies the Negro equal access to schools, parks, restaurants, libraries and the like. Desegregation is eliminative and negative, for it simply removes these legal and social prohibitions. Integration is creative, and is therefore more profound and far-reaching than desegregation. Integration is the positive acceptance of desegregation and the welcomed participation of Negroes in the total range of human activities. Integration is genuine intergroup, interpersonal doing.

As King noted, while the terms *desegregation* and *integration* are often used interchangeably, they represent quite different realities—one represents the lifting of formal systems of segregation, while the other signals a substantive or "genuine intergroup, interpersonal doing."

The reality we discover at Riverview might well be desegregated but it is certainly not integrated. It is a highly racialized educational terrain in which race shapes student experiences in ways that exacerbate inequality and that seem directly contradictory to the school's (and staff's) intentions. Like many desegregated high schools, it does not live up to its promise (Clotfelter 2004; Tyson 2011). Here we ask why all groups are not experiencing full inclusion in the school's rigorous academic programs (i.e., integration). And even more pointedly, how does a community that expresses an explicit commitment to integration tolerate such patterns? To answer these questions and explore why places like Riverview are not reaching their full potential, we draw on literature from social psychology. Collectively, this work helps us to understand how even those with the best intentions, operating in what many believe to be favorable, if not ideal, conditions, might end up reproducing the racial inequalily they expressly hope to challenge....

SETTING AND DATA

The setting for this research is Riverview, a midsized city located within a large metropolitan area in the United States. Although not as diverse as the large city it abuts, Riverview is, relative to most suburban communities in the metropolitan area, quite diverse. Many flock to Riverview for its diverse population and successful schools. For example, it has very high graduation rates, and nearly 80% attend college (including more than 70% of African American graduates). In 2003–2004, the student body of over 3,000 students was primarily black and white with a significant minority of Latina/o students (just under 10%). Just over 30% of the students come from low-income families. In many ways, the school is a picture of racial integration and high student achievement.

This image, however, belies major differences in school achievement. While almost 90% of white students met or exceeded standards in both reading and math, more than 70% of African American students fell below standards on the same tests. Additionally, blacks make up 40% of the student population in Riverview high school but only 9% of the students taking advanced placement (AP) calculus. In contrast, whites make up 50% of the student body but 82% of the students taking this class by 12th grade.

In terms of resources, Riverview is a largely middle-class community with a median family income of nearly $100,000 (U.S. Census Bureau 2007). However, black and Latina/o families earn less income than white families on average. Black families in Riverview possess 45% of white median family income, and Latina/o families possess 54% of white median income (Table 1). While whites make up 69% of the total population, they make up over 93% of the residents living in the wealthiest census tracts (U.S. Census Bureau 2007). These patterns highlight the extent to which the community is segregated across race and class lines.

Median family incomes for all groups are above national averages and poverty rates are well below national averages, but white families have far more resources than do black and Latino families. These resources are not merely financial: White families tend to have more educational resources, more flexibility in time to spend dealing with their children's education, and more cultural and social resources.

The data for this study come from interviews with over 170 members of the Riverview school community, including students, parents, teachers, administrators, and staff. The interviews lasted between 45 and 150 minutes and were

T A B L E 1   **Key Demographic Characteristics of the Riverview Community**

	Median Family Income, 1999 Dollars	Family Living in Owner-Occupied Housing, %	Individuals below Poverty Line, %
White (non-Hispanic or Latino)	103,145	58.6	7.8
Black (non-Hispanic or Latino)	46,422	44.1	13.9
Hispanic or Latino	55,729	37.8	14.0
Asian/Pacific Islander	63,438	24.1	14.3
Native American	N/A	N/A	N/A

NOTE: N/A, not applicable.

SOURCE: U.S. Census Bureau (2007).

conducted at Riverview High School, private homes, and local coffee shops. They were semistructured interviews conducted by the authors as well as by trained graduate students using a standard protocol....

FINDINGS

So we have great diversity at this school, but it's like two ships traveling in parallel lines—they don't ever really cross. (Ms. Hicks, African American school counselor)

As Ms. Hicks observes, Riverview is very diverse. It has been officially desegregated for decades. However, as she also describes, it is far from a model of integration. Once black, Latino, and white students pass through Riverview's entrance, they mostly pass by each other on the way to different classrooms. As is true of similar high schools across the country, Riverview has an academic hierarchy that is highly racialized; classrooms are internally segregated with high-level courses largely dominated by white students and "regular" classes filled predominantly by black and Latino students (Oakes 2005; O'Connor et al. 2011). In fact, of the many possible measures of "racial achievement gaps" at Riverview, racialized tracking is often the most glaring—it is a physical manifestation of what is otherwise often only represented in abstract statistics. In some ways the racialized academic hierarchy, what other scholars refer to as "second-generation segregation," at Riverview is not new (Mickelson 2001; Oakes 2005; Tyson 2011). Jim, a white school safety officer who grew up in Riverview, described the situation this way: "We instituted integration ... in 1967, okay, when I was bussed to [elementary school] as a white boy. Now we've hardly gone anywhere since 1967 because we just reproduce segregation inside the school." As Jim suggests, such hierarchies have existed here and elsewhere for as long as the school buildings themselves have been desegregated. The question remains, however, how schools like Riverview have what feels like quite "old-fashioned" racial hierarchies embedded within them despite the elimination of formal mechanisms of segregation decades ago.

Such hierarchies are not a manifestation of group-level differences in ability or intelligence.... As Aronson and Steele (2005) argued, academic competence is affected by whether you feel respected, welcomed, and/or treated well. These perceptions not only shape social relations but also influence motivation, performance, and learning. Intelligence is less stable and more fragile than we typically acknowledge, and a whole host of contextual factors shape whether any of us realize our potential. Further, "culture" a typical substitute for "genetics" in discussions of school outcomes, is also dubious as a good explanation of racially stratified academic hierarchies (Darity and Jolla 2009; Lewis 2013; O'Connor, Lewis, and Mueller 2007). Echoing other work, we found little if any evidence that an oppositional culture (one of the more popular "cultural" explanations for achievement gaps) on the part of black students explains these patterns (Diamond, Lewis, and Gordon 2007; Downey 2008; Harris 2011). So what are the racial dynamics generating these patterns?

> "It's racist ... it is inadvertent though." (Ms. McDaniels, African American principal)

The reality of racial dynamics at Riverview was summed up nicely in this statement from one of the school principals. Like most other teachers, administrators, and staff we spoke to at Riverview, Ms. McDaniels both acknowledged that racial issues remain at the school but also sought to make sure we understood that unlike the past, it is usually unintentional or "inadvertent." Longtime white Riverview teacher Don Michaels explained that no one is trying to do anything "purposefully," instead using "institutional racism" to describe what he has seen:

> Sometimes [African American students aren't treated right] and I hate to say it, but it's still true now. And I'm not saying in a sense that people are purposely doing something because of race, but institutional racism is there, I mean it just is.

While these staff referred to "inadvertent" or "institutional" racism, we believe the terms *institutional* and *everyday discrimination* better capture the dynamics these community members describe. *Institutional discrimination* captures such things as highly racialized school practices and structures (e.g., tracking) and the way school practices differently respond to and reward the social and cultural resources (e.g., cultural capital, social networks) students and families bring to school. This kind of discrimination includes "decisions and processes that may not themselves have any explicit racial content but that have the consequence of producing or reinforcing racial disadvantage" (Pager and Shepherd 2008:2). As Pager and Shepherd (2008:20) put it, this frame of institutional discrimination "encourages us to consider how opportunities may be allocated on the basis of race in the absence of direct prejudice or willful bias." *Everyday discrimination* includes all the ways that race-based status beliefs and racial stereotypes shape interactions and expectations—"the subtle, pervasive discriminatory acts experienced by members of stigmatized groups on a daily basis" (Deitch et al. 2003:1299). This everyday discrimination has concrete implications for access to educational opportunities.

So in trying to understand how seemingly "old-fashioned" racial hierarchies persist in putatively "integrated" institutions like Riverview, for the most part we did not find evidence of explicit racism or intentional favoritism. Rather, what we found is a situation in which there is an almost universal espoused commitment to equity but in which almost everything about achievement in the schools is racialized—how school community members understand themselves and each other, how they interact with one another, how decisions are made about which students belong in which classes. Both formal and informal school practices and structures (including tracking, the school discipline system, and many daily exchanges between students, parents, and school personnel) are in multiple and complex ways shaped by racial dynamics. As Mickelson (2003:1076) has noted, "so long as race confers privileges outside schools, it is hard to imagine it does not do the same within schools."

Racial dynamics shape the actual organization of classrooms and spaces. Ms. Hicks, one of the African American school counselors, provided this more elaborate description of the school:

> It's really diverse. Unfortunately, I don't think kids really have the opportunity to take advantage of that diversity, because when you look at the regular level classes, you're going to walk in and 70% of those kids are going to be black. When you look at the honors classes, 70% of those kids are going to be white. So we have great diversity at this school, but it's like two ships traveling in parallel lines—they don't ever really cross. All the black kids play football; the white kids play soccer.... If you look at the cheerleading squad, I think there are two white girls on the cheerleading squad. If you look at the volleyball team, it's almost all white. It's like two high schools. The white kids have found their niche and it's not the same place as the black kids.

Or as Ms. Watson, another long-time African American Riverview teacher, put it,

> As much as it seems like we're together, we're integrated, we're not integrated. We're diverse, but we're not integrated. We just go to school here together.

Both school staff members describe a context of formal integration and informal segregation, "two ships traveling in parallel lines" that rarely cross. Ironically, in trying to capture the extent of separation, Ms. Hicks actually underestimates the extent of the racialization of tracks. White students make up 48% of the student population but nearly 90% of the students taking AP classes and 80% of students taking honors classes....

As we will elaborate, these stark realities are obvious to all, but somehow they become a part of the taken-for-granted everyday reality such that adults and children attend school every day without outrage or surprise.

While we are treating institutional and everyday dynamics as distinct, in practice they overlap and intersect. An example from parental intervention dynamics at the school illustrates this and also reveals a key mechanism in the generation of racial hierarchy—how race and class become conflated and race becomes a key signifier in everyday interpersonal dynamics.

While white families in Riverview generally have more resources than black families, in the daily functioning of the school they do not always have to actively deploy them to gain advantages. Race and class often get conflated such that school personnel tend to assume that most white students come from middle-class or upper-middle-class families and most black and Latino students do not. In other writing, Lewis (2003) has talked about the way this conflation of race and class can pay off for white children because economic capital gets translated into *symbolic capital*. It does not matter whether you specifically are a white student with wealthy parents since your whiteness is the primary signifier being acted upon (Morris 2006; Staiger 2004). Status beliefs are attached to and conflated with what is visible—racial differences. When dealing with white students, Riverview personnel often anticipate parental pressure, intervention, or concern. Thus, while school staff provided multiple examples of the ways white parents deployed resources on behalf of their children, they also described multiple ways that they acted proactively in advance, in anticipation that a parent was likely to be harassing them if they did not. In these ways whiteness became a symbolic resource for white students. As demonstrated in the examples below, none of the actions taken by school personnel are explicitly racial in nature. And even more important for understanding how institutional discrimination works, it is not just that white families had more resources to deploy; rather, the school's interactional practices, rules, and structures made it possible for racial resources to pay off to the advantage of white students.

We first got a sense of these dynamics talking to an African American teacher after a school workshop. She stopped one of us and said she had been thinking throughout the discussion about how she tends to worry more about white students in her class. As she put it, it was not that she cared more about them or wanted them to excel more than her other students but that she "knew" that their parents were likely to be upset if they did not do well. She realized that much of the time parents did not need to come in or say anything—just the possibility that they might was powerful: If a white student was not doing well, she followed up, lest she hear about it from their parents later. Ms. Hicks talked about similar dynamics:

> It's hard, because I would love to say, "Yes, they're all treated the same." But I think the reality is they're not. The squeaky wheel gets the oil. So if you're a parent, and you come, and you say, "My child got a C. You need to help me figure it out." And I know that you're going to call me every day, I'm going to go to that teacher and say, "You know what? What's going on? Tell me what's happening." After I find out what's happening, I'm going to get back to that parent. I'm going to advocate for that kid. Whereas, if you are a kid and your parents are

not going to call me every day or hasn't called me, we have 286 kids on our caseload, it could take me a month to realize that that you have two or three Cs.

As Ms. Hicks describes, her caseload makes it difficult to keep track of hundreds of students' needs. However, when parents make requests, or have made them in the past, or if she recognizes a parent as one who is likely to call every day, then she becomes an advocate. Here she discusses both actual calls and likely calls, both actual and anticipated "squeakiness." White social studies teacher Don Michaels also spoke to this issue of "squeaky" parents:

> It goes back to that issue about which parents are gonna play an active role, and the squeaky wheel and so on. And I know just from conversations with teachers there are some teachers that will simply say you know this kid really didn't deserve the C- but I knew that if I gave him a D+ I was going to get a lot of grief from the parent, and the tradition has tended to be that it's more likely to be a white parent than it's going to be the black parent.

School personnel described different aggregate involvement patterns between white and minority families, but they also described how those experiences generalize to cases where parents do not get involved. Moreover, they note that parental intervention pays off with quite different outcomes for the students. The differential treatment, then, is a matter of school policy—even if is implicitly so. These examples from different school community members capture the many small ways that subtle racial dynamics contribute to different school outcomes (e.g., shaping performance expectations).

RACIALIZED TRACKING—"THE HONORS WHITE SCHOOL AND THE OTHER SCHOOL"

Not only is the distribution of students in tracks at Riverview High School highly racialized, but the tracks themselves have become racialized (O'Connor et al. 2011; Tyson 2011). Thus, the local discourse is not just that some groups tend to be in certain classes but that those have become racialized spaces. Julius, an upper-middle-class, high-achieving black student, described the school this way: "The fact is that Riverview is two schools in one. There is the honors white school, and then there's the other school." Similarly, Richard, a high-achieving white student, speculated, "I mean if you look at the numbers, I'm betting there are more white kids that are in the *honors classes*, and more black kids that are in *minority classes*." As these quotes from Riverview juniors show, the tracks themselves have become identified as *belonging* to different racial groups. They are not just high and low tracks but "honors white" and "minority classes." The fact that white students are more often in high tracks generalizes to those being white tracks, tracks where white students belong and deserve to be. Black and Latina/o students' overrepresentation in regular and remedial classes translates into those being "minority" spaces. These status hierarchies get translated into performance expectations in multiple ways.

When we spoke to African American Vice Principal Mr. Webber, he said, with a shake of his head, "I think that sometimes the expectations [for black and Latino students] are lower." These expectations get communicated to students in lots of subtle ways. A white special education teacher described the recent behavior of one of her peers:

> There was a teacher who had a minority student come into their honors class and you know he was your stereotypical baggy jeans, big shirt, hat turned sideways, you know, and she said to him, 'You know I think you belong in my next period, you're too early' and assumed that he was a general student. And he's like, 'No, no my schedule says I belong here.'

Both teachers and students reported these kinds of interactions between race and performance expectations. For example, in his interview, Julius, a high-achieving African American student, reported his struggle with getting into high track courses:

> My freshman teacher didn't like me. She didn't recommend me for U.S. History AP. My mom had to spend a couple of hours on the phone getting me into the class, even though I'm a kid who takes extremely hard classes, and gets good grades, they just don't let you in. They make it very, very difficult for you to take the classes you want.

And many of the staff of color concurred. As Ms. Tyson, an African American school secretary, said,

> Well, if you are a student of color, could be an African American student or Latino. There are assumptions that you don't care about school, that you ... you don't have the capability of being successful in school. And so those are negative messages that they have to deal with, ah, every day.

It was not only African American students and staff who observed these dynamics. A number of white students reported perceptions among school officials that white students were smarter and better behaved. Gabe, a white junior, reported this:

> I'm white, so I'm expected to be smarter. Usually, when someone sees me, they always think I'm smarter than most people.... I think that usually the perception is ... that black people are dumber than white people and Hispanics are not as smart as everyone else.... So if you have a really smart black person, that's when you see the most, "That's

weird." In one of my classes, there's one black kid in the entire class; there's zero Hispanics. It's all just white people. And that's, it's weird.

As he puts it, he experiences dissonance when exposed to a "really smart black person" because it does not coincide with expectations in a context in which there are large numbers of black students in the school and almost none in his classes. Leah, a white sophomore, put it this way:

> I definitely think that there are stereotypes that go along with all races ... and white not excluded.... So I feel like people see me, I'm like a middle-class, white girl. You know, so ... I feel like people expect ... me to be a certain way. They expect me to be respectful and quiet and intelligent and stuff like that.

For Leah, being white meant that people held high expectations for her in the classroom *because* she was white. People expect a "middle-class white girl" to be "respectful and intelligent." White students like Leah, particularly girls, reported receiving the benefit of the doubt across school contexts, inside and outside of the classroom—a pattern that can lead to enhanced performance through a process called *stereotype lift* (Walton and Cohen 2003).

Students also reported these dynamics in their interactions with other students. Maria, a Latina sophomore, stated:

> Well, there's been times where I've been in classes with white kids, and I tried my best at times. When I do, the white girls, they're always going in their own little clique, and look at the Mexicans as if we were dumb or something. It just makes us feel bad.

Experienced as racial microaggressions, these exchanges are exactly the kinds of stereotype threat "cues" that signal to students that negative racial stereotypes about their group are at play (Lewis, Chesler, and Forman 2000; Steele 2010)....

CONFLATION OF RACE, RIGOR, AND TRACKS

Track placement has consequences not only for students' sense of self but also for the curricula they receive and for their treatment more generally....

... Abundant research over the last 20 years has found that the achievement differences between those in high and low tracks increase over time no matter where students begin in terms of test scores (Oakes 2005). Those placed in low tracks learn less and show fewer gains over time than similarly situated high-track students. High-track students benefit not from the grouping itself but from the enhanced curriculum, special resources, and supports. Tracking exacerbates preexisting inequalities (Condron 2007, 2008). As Sorenson and Hallinan (1986:519) said, "Grouping is not neutral with respect to inequality of educational opportunity." Thus, to the extent that status beliefs shape performance expectations, they also affect actual positional inequalities that result.

Like findings elsewhere, evidence from a number of sources at Riverview indicate that lower-track classes offer a lower-quality educational experience. Ms. Paul, an African American history teacher, put it this way:

> You give that new teacher three classes of the lowest achieving students outside of Special Ed in the school, and you call that setting the students up for success. That's not right. The students that are at the lowest level, at the bottom of the gap, they need the best teachers in school.

In fact, community members widely acknowledged that the upper-level courses provide a stronger educational experience. For instance, white Riverview parent Janet stated, "Well, and everybody told us if she didn't take the honors classes she would be bored silly.... This is what they say is that the teaching quality is not as good amongst the teachers who don't teach honors."

Teachers also reported these patterns. Mike Sellers, a white social studies teacher put it this way:

> I think there's some structural issues just flat out with ... tracking ... whether people admit it or not it is a destructive force in some kids' lives. In particular minority kids.... Some of these classes that are at the lower level aren't taught at the level which would allow kids to do college work in three years or whatever. So there's a huge thing. What are we doing with our [sighs] our so-called "regular ability" students? Are we, you know, expecting enough of those kids to, so that they can meet the challenges that they're going to get [in college] or whatever? ... I think there's too much of a disconnect in expectations.

Here we see a complex set of interactions between meaning and structure such that structural inequalities or institutional discrimination interact with already available racial ideologies to both produce and then justify racial patterns in outcomes.

The question from some will be whether these tracks don't just represent an unfortunate but "real" difference in students' academic potential or commitment. Interestingly, a recent experiment by a relatively new teacher in the school confronted this issue directly. As with all subjects, there are major differences in students' mathematics course taking, which begins early on in their educational careers. During 4th grade, students are tested in mathematics. Based on a combination of these tests and teachers' recommendations, students are placed on two different tracks, one that leads to higher-level mathematics in high school (e.g., calculus by 12th grade) and one that does not. By 5th grade, the vast majority of students placed in the upper-level mathematics sequence are white. By 8th grade, most of these students have taken Algebra I, a critical milestone for students to reach high-level mathematics in high school. Historically, by the time students reached the 12th grade, very few African American and Latino students were in upper-level mathematics courses.

After joining Riverview and being surprised about the racial demographics in AP math, Mr. Bettencourt (a white math teacher) decided to try to address this by starting a new program. Realizing that most African American and Latino students in the high school were never going to have an opportunity to take advanced or AP math courses unless their coursework was accelerated, he invited students doing well in Algebra I (African-American, Latina/o, and white) to spend over four hours a day with him for six weeks in the summer to take geometry so they could catch up to those who had been tracked into the accelerated program years before. Many accepted. Within a few years he had doubled (and eventually tripled) the number of African American and Latina/o students in calculus at the high school. While not systematic in its design, this experiment demonstrates that structural arrangements in the school are serving to narrow options for some students—students who are capable of succeeding in advanced courses into which they have not historically been channeled....

CONSEQUENCES FOR UNDERSTANDINGS OF RACE

The structure of tracking in schools like Riverview is of concern for a number of reasons, both academic and nonacademic. As noted, tracking conveyed complicated messages to those on the bottom and the top about race and ability. Ms. McDaniels, a retiring school principal, spoke about it this way:

> I think kids just don't realize that they can do things. They just kind of accept that role … I don't even know if they really think about it too much anymore … But I think they just kind of accept it.… This is kind of a crass statement [but they must wonder] … "How did everybody white get to be really smart?" And it's another piece … chipping away of … that's the kind of the little insult. "Why am I not with them? We were always together." So. You know … I know that they think about it, but I think another piece of it is we just really don't talk enough about it. You know, it just happens.

What Ms. McDaniels captures is the complexity of institutional discrimination—this kind of chipping away in which there are no clear perpetrators, things just *happen*, but happen in a way that consistently benefits some more than others. Over time students can come to, as she put it, "accept that role," only occasionally wondering what happened to all the white students they used to be in class with. When we spoke to Elaine Peters, a white high-level district administrator, about why there were not more black and Latina/o students in high-level classes, she stated the following:

> Well one it's historical. "I don't belong here." But too it is recruitment. Yeah, I think in a situation, as much as I can imagine what it would be like to be an African American kid in this high school, I wouldn't see— unless I'm being pushed by home or pushed by a teacher and home to do so or a group of friends which is possible, I wouldn't see that as where I needed to end up. But if I'm even an average, strong white student, my kids, my friends are gonna be there. I mean it's gonna kind of sweep me along. So we have to orchestrate the recruitment. You belong here…

CONCLUSION

This research has focused on racial dynamics in a highly resourced, desegregated school within a liberal community to capture some of the conundrums of integration today. One might ask why so many identify Riverview as a desirable educational destination—desirable for both white children and children of color. One answer is that even partial inclusion in a well-resourced context like Riverview provides more opportunity than attending schools in the large majority-minority metropolitan school district nearby. All students benefit somewhat from attending this multiracial school. This is the conundrum of settings like Riverview, and in this way it serves as a cautionary tale. It illustrates the value of attending schools in municipalities that have heavily invested in education at the same time that it demonstrates on multiple levels how, even in these places, race works to advantage white students and disadvantage students of color (Holt 1995; Lewis 2014; O'Connor et al. 2007). The material and symbolic dimensions of racial hierarchy enter into school along with the diverse student body.

... [W]hile the grave costs of segregation are clear, we also have to be attentive to the "perhaps unintended consequences that have been encountered" in trying to realize the potential benefits of integration. Integration or, more accurately, desegregation in schools like Riverview reflects large inequalities in power (along lines of both material resources and cultural belief systems). Therefore, it is not enough to merely include students within the same buildings. For places like Riverview to live up to the hopes for "integration," we need to ensure that students are truly integrated and have full access to all of the educational opportunities offered in these contexts.

While the terms *desegregation* and *integration* are often used interchangeably, they represent quite different realities. As King put it, lifting formal systems of segregation does not automatically signal a substantive or "genuine intergroup, interpersonal doing." Desegregation does not always mean inclusion. While King raised these concerns just over five decades ago in the midst of struggles to implement the famous *Brown v. Board of Education* Supreme Court decision, they remain pertinent today. Despite widespread acceptance of the principle of integration, and a general embrace of the idea of diversity, real commitment to its realization too often remains tenuous at best (Bell and Hartmann 2007).

The reality that integration alone is not an answer does not mean that segregation is not harmful. The fact is that seeking to achieve real integration (whether by race or class, or race *and* class) remains pressing because segregation in our public schools and elsewhere clearly reinforces inequality. If we invested nationally in public schooling, distributed educational resources more equitably, and ensured that all public schools met some minimal quality threshold (what many talk about today as "adequacy") then we might be able to have a substantive debate about whether, all things being equal, diversity itself is beneficial. But we still do not have what Posey-Maddox (2014:145) recently called a "commitment to everyone" that is reflected in a commitment by the state to providing not just an education but a quality education. Thus, separate remains unequal and widespread segregation remains detrimental to justice. However, trying to implement integration within a context of deep inequalities in wealth and resources, entrenched race-based status beliefs, and wide-scale disinvestment in public services presents many challenges.

Thus, while scholars are perhaps correct in identifying segregation as a linch-pin of inequality, if we treat achieving desegregation as an endpoint we will not be able to intervene on the many mechanisms through which racial hierarchies are perpetuated even in desegregated spaces. Only through full attention to all the ways race continues to matter when diverse bodies share space will we make sure it matters less in determining access to opportunities to learn and thrive....

REFERENCES

Anderson, Elizabeth. 2010. *The Imperative of Integration*. Princeton, NJ: Princeton University Press.

Aronson, Joshua and Claude M. Steele. 2005. "Stereotypes and the Fragility of Academic Competence, Motivation, and Self-concept." Pp. 436–56 in *Handbook of Competence and Motivation*, edited by A. Elliott and C. Dweck. New York: Guilford Press.

Bell, Joyce M. and Douglas Hartmann. 2007. "Diversity in Everyday Discourse: The Cultural Ambiguities and Consequences of 'Happy Talk.'" *American Sociological Review* 72(6):895–914.

Clotfelter, Charles. 2004. *After Brown: The Rise and Retreat of School Desegregation*. Princeton, NJ: Princeton University Press.

Condron, Dennis J. 2007. "Stratification and Educational Sorting: Explaining Ascriptive Inequalities in Early Childhood Reading Group Placement." *Social Problems* 54(2):139–60.

Condron, Dennis J. 2008. "An Early Start: Skill Grouping and Unequal Reading Gains in the Elementary Years." *The Sociological Quarterly* 49(1):363–94.

Darity, William A. and Alicia Jolla. 2009. "Desegregated Schools with Segregated Education." Pp. 99–117 in *Integration Debate: Competing Futures for America's Cit*ies, edited by C. Hartman & G. Squires. Florence, KY: Routledge.

Darling-Hammond, Linda. 2010. *The Flat World and Education: How America's Commitment to Equity Will Determine Our Futu*re. New York: Teachers College Press.

Deitch, Elizabeth A., Adam Barsky, Rebecca M. Butz, Suzanne Chan, Arthur P. Brief, and Jill Bradley. 2003. "Subtle Yet Significant: The Existence and Impact of Everyday Racial Discrimination in the Workplace." *Human Relations* 56(11): 1299–324.

Diamond, John B. 2013. The Resource and Opportunity Gap: The Continued Significance of Race for African American Student Outcomes. Pp. 97–111 in *Contesting the Myth of a Post-racial Era: The Continued Significance of Race in Education*, edited by D. J. C. Andrews and F. Tuitt. New York: Peter Lang Publishers.

Diamond, John B., Amanda E. Lewis, and Lamont Gordon. 2007. "Rethinking Oppositionality: Positive and Negative Peer Pressure among African American Students." *International Journal of Qualitative Studies in Education* 20(6):655–79.

Downey, Douglas B. 2008. "A Funny Thing Happened on the Way to Confirming Oppositional Culture Theory." Pp. 298–311 in *Minority Status, Oppositional Culture, & Schooling*, edited by John Ogbu. New York, New York: Routledge.

DuBois, W.E.B. 1935. *Black Reconstruction: An Essay toward a History of the Part Which Black Folk Played in the Attempt to Reconstruct Democracy in America, 1860–1880.* New York: Harcourt Brace.

Harris, Angel. 2011. *Kids Don't Want to Fail: Oppositional Culture and the Black-white Achievement Gap.* Cambridge, MA: Harvard University Press.

Hartman, Chester and Gregory D. Squires. 2010. "Integration Exhaustion, Race Fatigue, and the American Dream." Pp. 1–8 in *The Integration Debate: Competing Futures for American Cities*, edited by C. Hartman and G. Squires. New York: Routledge.

Holt, Thomas C. 1995. Marking: Race, Race-making, and the Writing of History. *American Historical Review* 100(February):1–20.

King, Martin Luther, Jr. 1986. "The Ethical Demands for Integration." Pp. 117–125 in *A Testament of Hope: The Essential Writings and Speeches of Martin Luther King, Jr.*, edited by J. M. Washington. San Francisco: Harper & Row.

Lewis, Amanda E. 2003. *Race in the Schoolyard: Negotiating the Color Line in Classrooms and Communities.* New Brunswick, NJ: Rutgers University Press.

Lewis, Amanda E. 2013. "The Nine Lives of 'Oppositional Culture': A Review of Angel Harris' *Kids Don't Want to Fail* and Karolyn Tyson's *Integration Interrupted*." *Du Bois Review* 10(1):279–289.

Lewis, Amanda E., Mark Chesler, and Tyrone Forman. 2000. "The Impact of Color-blind Ideologies on Students of Color: Intergroup Relations at a Predominantly White University." *Journal of Negro Education* 69(1/2):74–91.

Lewis, Amanda E. and Michelle J. Manno. 2011. "The Best Education for Some: Race and Schooling in the United States Today." Pp. 93–109 in *State of White Supremacy*, edited by M-K. Jung, J. C. Vargas, and E. Bonilla-Silva. Palo Alto, CA: Stanford University Press.

Lewis, L'Heureux. 2014. *Inequality in the Promised Land.* Stanford, CA: Stanford University Press.

Massey, Douglas S. and Nancy A. Denton. 1993. *American Apartheid: Segregation and the Making of the Underclass.* Cambridge, MA: Harvard University Press.

Mickelson, Roslyn. 2001. "Subverting Swann: First and Second-generation Segregation in the Charlotte-Mecklenburg Schools." *American Educational Research Journal* 38(2):215–52.

Mickelson, Roslyn. 2003. "When Are Racial Disparities in Education the Result of Racial Discrimination? A Social Science Perspective." *Teachers College Record* 105(6):1052–86.

Mickelson, Roslyn. 2008. "Twenty-first Century Social Science on School Racial Diversity and Educational Outcomes." *Ohio State Law Journal* 69:1173.

Morris, Edward. 2006. *An Unexpected Minority.* New Brunswick, NJ: Rutgers University Press.

Oakes, Jeannie. 2005. *Keeping Track: How Schools Structure Inequality.* 2nd ed. New Haven, CT: Yale University Press.

O'Connor, Carla, Amanda E. Lewis, and J. Mueller. 2007. "Researching African-American's Educational Experiences: Theoretical and Practical Considerations." *Educational Researcher* 36(9):541–52.

O'Connor, Carla, Jennifer Mueller, R. L'Heureux Lewis, and Seneca Rosenberg. 2011. "'Being' Black and Strategizing for Excellence in a Racially Stratified Academic Hierarchy." *American Educational Research Journal* 48(6):1232–57.

Orfield, Gary, Erica Frankenberg, and Liliana M. Garces. 2008. "Statement of American Social Scientists of Research on School Desegregation to the U.S. Supreme Court in *Parents v. Seattle School District* and *Meredith v. Jefferson County.*" *The Urban Review* 40(1):96–136.

Orfield, Gary and Nora Gordon. 2001. *Schools More Separate: Consequences of a Decade of Resegregation.* Cambridge, MA: Civil Rights Project, Harvard University.

Pager, Devah and Hana Shepherd. 2008. "The Sociology of Discrimination: Racial Discrimination in Employment, Housing, Credit, and Consumer Markets." *Annual Review of Sociology* 34:181–209.

Posey-Maddox, Linn. 2014. *When Middle-class Parents Choose Urban Schools: Class, Race, and the Challenge of Equity in Public Education.* Chicago: University of Chicago Press.

Sorenson, Aage and Maureen Hallinan. 1986. "Effects of Ability Grouping on Growth in Achievement." *American Educational Research Journal* 23(4):519–42.

Staiger, Annegret. 2004. "Whiteness as Giftedness: Racial Formation at an Urban High School." *Social Problems* 51(2):161–81.

Steele, Claude. 2010. *Whistling Vivaldi: And Other Clues to How Stereotypes Affect Us.* New York: W.W. Norton.

Tyson, Karolyn. 2011. *Integration Interrupted: Tracking, Black Students, & Acting White After Brown.* New York: Oxford University Press.

U.S. Census Bureau. 2007. *American Community Survey.* Generated by John Diamond using American FactFinder.

Walton, Gregory M. and Geoffrey Cohen. 2003. "Stereotype Lift." *Journal of Experimental Social Psychology* 39(5):456–67.

Wells, Amy Stuart, Jennifer Jellison Holme, Anita Tijerina Revilla, and Awo Korantemaa Atanda. 2009. *Both Sides Now: The Story of School Desegregation's Graduates.* Berkeley: University of California Press.

DISCUSSION QUESTIONS

1. What are some of the unintended consequences that have arisen while integrating the school studied here?

2. How does this investigation show you the link between individual social decisions and larger social processes?

3. Given what you have learned from the research on Riverview High School, what two recommendations would you make to promote greater integration in the high school?

Section D

Health Care

Elizabeth Higginbotham and Margaret L. Andersen

R acial inequality, along with class and gender inequality, is a huge part of the inequality that we see in the health of American people. How long you live, the nutrition in your diet, what kind of care you receive, even how you die are all things that are related to your race. On average, non-Hispanic White men live five years longer than non-Hispanic Black men, and non-Hispanic White women live three years longer than non-Hispanic Black women (Miniño 2013). Scholars recognize these differences as **health disparities**, differences that are rooted in differential access to care. Health insurance is a critical issue. Even after the passage of the Affordable Care Act, which provides health care insurance to millions of adults, 25 percent of Hispanics still lack health insurance, compared to 14 percent in the U.S. overall (Krogstad and Lopez 2014).

Segregation and racial bias shape the options and opportunities for people based on race. The limited access to jobs and segregation in housing and education, as well as the challenges faced by families of color, are all foundations for inequality in other institutions. Health care is directly related to other structural inequalities. Scholars have documented many health problems today that are rooted in a legacy of racial disparities in access to health care, including age-specific mortality rates and treatment outcomes (Smedley, Smith, and Nelson 2003).

How care is delivered, who is served, and how much people pay and understand about their health reveals a lot about what the society values. In some nations, people would find it unimaginable to have to pay for health care; in others, not only do people pay, but they pay a lot. In some societies, health care is provided through a national health care program; in others, health care is privatized and managed through large, profitable corporations. In some societies, sick people have access to care regardless of their ability to pay; in others, the quality of care depends on one's economic and social resources. The system in the United States reflects our cultural values of individualism and personal responsibility, but it also has built into it the structural inequalities of racism.

People with privileges often take health care for granted. Think about your own health history—were you born in a hospital? Did you have regular checkups and get vaccinations with a pediatrician? If you have an acute illness, how was it handled? If you had a chronic illness, were your parents able to secure treatment for you? You have been interacting with health care institutions over your life-time. We often take getting health care for granted, but race and social class have much to do with how you interact with the system. Privileges are often invisible to those who have them, while people who lack privileges can see the glaring inequities in who gets health care and how it is delivered. We all need an accurate picture of the state of our health care system.

Experts think infant mortality is a good indicator of a nation's health. On this measure, the United States ranks low relative to other nations—in fact, the number of infant deaths is higher in the United States than in any other indus-trialized nation. Furthermore, our status in this regard has dropped significantly since 1960, when we were twelfth among a group of other nations (number one having the lowest infant mortality); now we are number thirty (Save the Chil-dren 2013). A major problem for this country is the high number of babies born prematurely, putting them at risk for death during their first year of life, a prob-lem that is directly related to access to prenatal care.

Prior to the **Civil Rights Act** in 1964, access to health care institutions was directly related to race. Not only were people of color more likely to be in occupations that did not offer them health care insurance, but they also faced challenges when seeking medical assistance. Even in the 1960s, many southern health care facilities were segregated; hospitals had Black and White wards and separate entrances. Many hospitals around the nation did not grant privileges to Black and Mexican American physicians, thus prohibiting them from overseeing the treatment of their patients in hospitals. Patients would be treated by White staff members who often held negative stereotypes about them, reflected in the care they received. Rebecca Skloot's (2009) amazing story of Henrietta Lacks who died unnecessarily from cancer, documents the racism a Black woman faced securing health care in segregated Baltimore in the 1950s. Black Americans and other vulnerable groups have been used as research subjects in medical experiments (Washington 2006), often without consent or knowledge. The leg-acy of discrimination in health care results in distrust, leaving many African Americans less eager to seek medical help. It is well documented that negative experiences with the health care system can influence health-seeking behaviors (Smedley, Smith, and Nelson 2003).

Implemented in 1965, Medicaid and Medicare enhanced access to health care for poor people and many people of color, but biases against people of color con-tinued. Describing this era, Alondra Nelson (2011) has found that although poor and low-income people of color finally have access to decent health care, they are more likely to be seen in clinics and emergency rooms, facing long waits to receive rushed and inconsistent care. Hospital care is available in university-affiliated training hospitals where professionals are more interested in getting experience with the disease than providing care to people of color. Such circumstances have prompted many to seek alternative health care, which was popular as early as the 1960s (Nelson 2011). It is not just access to care, but the quality of that care is important to ending health disparities. The increase in the numbers of people of color providing treatment has changed as we have come to understand that medical education has to include "cultural competency" so that health care providers will be more sensitive to the needs of diverse populations.

Living in a racist society is also a source of illness. Coping with segregation and coping with integration are challenges in a racist society. Living in segregated space means increased vulnerability to health hazards and obstacles to health. Such environments are less safe, more likely to be food deserts, and are less likely to have good health facilities. For example, rates of childhood asthma are clustered in low-income minority communities, especially where there is hypersegregation. High blood pressure stems from dealing with chronic poverty, low-wage employment, and other sources of stress. In short, inequality can make you sick.

Even securing an education and a decent professional job with access to health insurance does not necessarily mean you will have a healthy life. As long as negative stereotypes persist, even middle-class people of color will face microaggressions in schools and workplaces, as well as facing racial profiling in public places. Such daily assaults are sources of illness and health vulnerabilities. Racial disparities in health do not just disappear just because people of color become middle class.

The authors in this section explore some of the major issues involved with race-based health disparities. David R. Williams and Selina A. Mohammed ("Racism and Health: Pathways and Scientific Evidence") document the persistence of institutional and cultural racism, charting the many pathways that influence health for different groups. They show how experiences of racial discrimination influence health status.

Earlier in this book, we learned that there is no "race gene" (see Graves in Part I), but there are frequent claims in the media about specific diseases being linked to genetic differences between racial groups. Connections claimed between race, genes, and illness can reinforce historical notions of race and biology. Jamie D. Brooks and Meredith King Ledford ("Geneticizing Disease: Implications for Racial Health Disparities") address this issue as they look at how simplistic thinking about genetics misreads the complex understanding of the relationship between society and genetic causation. Brooks and Ledford remind us that nature and nurture interact; we should not ignore the social determinants of health care even in the context of new genetic research.

How we measure and collect data on "race" has to consider the various social locations of individuals. A Latina college professor will have a different relationship to the health care system than a Black male factory worker. We have to think about what it means to live in a race *and* gendered body where you encounter obstacles and supports that directly affect your health. Nancy López ("Contextualizing Lived Race-Gender and the Racialized-Gendered Social Determinants of Health") argues for such an intersectional analysis. She uses her autobiography as an Afro-Latina who encounters race-gender profiling while trying to secure health care for herself and her family. That even an educated professional has to cope with such inequities shows how racism is very much alive in the health care system.

REFERENCES

Krogstad, Jens Manual, and Mark Hugo Lopez. 2014. "Hispanic Immigrants More Likely to Lack Health Insurance than U.S. Born." Washington, DC: Pew Research Center, September 26. www.pewresearch.org.

Nelson, Alondra. 2011. *Body and Soul: The Black Panther Party and the Fight Against Medical Discrimination*. Minneapolis: University of Minnesota Press.

Miniño, Arialdi M. 2013. *Death in the United States, 2011*. Data Brief No. 115. Hyattsville, MD: National Center for Health Statistics.

Save the Children. 2013. *State of the World's Mothers*. Westport, CT: Save The Children.

Smedley, Brian D., Adrienne Y. Smith, Alan R. Nelson. 2003. *Unequal Treatment: Confronting Racial and Health Disparities in Health Care*. Washington, D.C.: National Academy Press.

Skloot, Rebecca. 2009. *The Immortal Life of Henrietta Lacks*. New York: Crown Publishers.

Washington, Harriet A. 2006. *Medical Apartheid*. New York: Doubleday.

FACE THE FACTS: INFANT MORTALITY BY RACE AND ETHNICITY, 2012

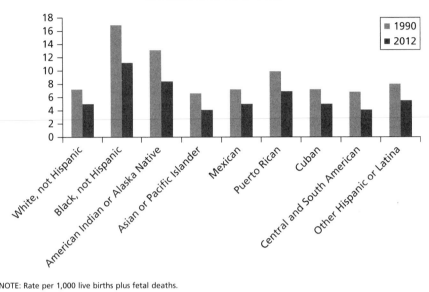

NOTE: Rate per 1,000 live births plus fetal deaths.

SOURCE: U.S. Department of Health and Human Services. 2015. *Health United States, 2014.* Hyattsville, MD: National Center for Health Statistics. www.cdc.gov

Think about It: As you can see, much progress has been made since 1990 in reducing the rate of infant death at birth. Still, differences remain between different racial and ethnic groups. What factors explain the difference? What can be done to further reduce the rate for every group?

30

Racism and Health

Pathways and Scientific Evidence

DAVID R. WILLIAMS AND SELINA A. MOHAMMED

Racism in many forms remains embedded in the society and has negative outcomes for health of people of color in various ways. Both institutional and cultural racism play a role in creating a negative environment for the delivery of health care. Living in a racist society can also be a source of health problems in its own right.

In the United States, as in other racialized countries in the world, racially stigmatized and disenfranchised populations have worse health than their more advantaged counterparts (D. Williams, 2012). The poorer health of these racial minority populations is evident in higher rates of mortality, earlier onset of disease, greater severity and progression of disease, and higher levels of comorbidity and impairment. In addition, disadvantaged racial populations tend to have both lower levels of access to medical care and to receive care that is poorer in quality. In U.S. data, these patterns tend to be evident for African Americans (or Blacks), American Indians (or Native Americans), Native Hawaiians and other Pacific Islanders, and economically disadvantaged Hispanic (or Latino) and Asian immigrants with long-term residence in the United States (D. Williams, 2012). These striking disparities are persistent over time and, although reduced, are evident at every level of income and education (Braveman, Cubbin, Egerter, Williams, & Pamuk, 2010; D. Williams, 2012). In recent years, increased attention has been given to the role of racism as a determinant of these patterns of racial inequality in health....

OVERVIEW OF THE NATURE OF RACISM
AND ITS PERSISTENCE

Racism is an organized system premised on the categorization and ranking of social groups into races and devalues, disempowers, and differentially allocates

SOURCE: Williams, David R., and Selina A. Mohammed. 2013. "Racism and Health: Pathways and Scientific Evidence." *American Behavioral Scientist* 57 (8): 1152–1173.

desirable societal opportunities and resources to racial groups regarded as inferior (Bonilla-Silva, 1996; D. Williams, 2004). Racism often leads to the development of negative attitudes (prejudice) and beliefs (stereotypes) toward nondominant, stigmatized racial groups and differential treatment (discrimination) of these groups by both individuals and social institutions. These multiple dimensions of racism do not always co-occur. For example, it is possible for racism to exist in institutional structures and policies without the presence of racial prejudice or negative racial stereotypes at the individual level.

Despite progress in the reduction of explicit public support of racism in the United States, there is also strong evidence of its persistence. National data on the racial attitudes of Whites reveal positive changes over time in support of the principle of racial equality (Schuman, Steeh, Bobo, & Krysan, 1997)....

Documenting the persistence of racism is a challenge because the nature of racism in contemporary society has also changed in ways that make it not readily recognizable to most adults. Scientific evidence indicates that in addition to conscious, deliberate cognitive processes, humans also engage in implicit (unconscious), effortless, automatic, evaluative processes in which they respond to a stimulus based on images stored in their memory (Dovidio & Gaertner, 2004).... *Aversive racism* is one of the terms used to characterize contemporary racism (Dovidio & Gaertner, 2004). An aversive racist lacks explicit racial prejudice (that is, has sympathy for those who were victimized by injustice in the past and is committed to principles of racial equality) but has implicit biases that favor Whites over Blacks. Research suggests that almost 70% of Americans have implicit biases that favor Whites over Blacks (Nosek et al., 2007). The pattern is most pronounced among Whites but is also evident for Asians, Hispanics, and American Indians. These high levels of implicit bias suggests that discrimination is likely to be commonplace in American society, with much of it occurring through behaviors that the perpetrator does not experience as intentional (Dovidio & Gaertner, 2004).

Discrimination

Second, racial discrimination persists in contemporary society, with Whites continuing to self-report that they discriminate against minorities (Pager & Shepherd, 2008). In addition, there is considerable high-quality scientific evidence documenting the persistence of racial discrimination. A recent review of audit studies—those in which researchers carefully select, match, and train individuals to be equally qualified in every respect but to differ only in race—provide striking examples of contemporary racial discrimination (Pager & Shepherd, 2008). For example, audit studies in employment document that a White job applicant with a criminal record is more likely to be offered a job than a Black applicant with an otherwise identical resume whose record was clean. Similarly, job applicants with distinctively Black names (e.g., Aisha, Darnell) are less likely to get callbacks for job interviews than applicants with identical resumes who have distinctively White names (e.g., Alison, Brad). Other audit studies reveal racial discrimination in renting apartments, purchasing homes and cars, obtaining mortgages and medical care, applying for insurance, and hailing taxis. Research has also found that even the price of a fast food meal increases with the percentage Black of a zip code (Pager & Shepherd, 2008)....

Institutional Racism

Third, racial discrimination also persists in institutional mechanisms and processes. Residential segregation by race is a prime example (Massey & Denton, 1993). *Segregation* refers to the physical separation of the races in racially distinctive neighborhoods and communities that was developed to ensure that Whites were safeguarded from residential closeness to Blacks (Cell, 1982). This enforced residence in separate areas developed in both northern and southern urban areas in the late 19th and early 20th centuries and has remained strikingly stable since then but with small declines in recent years (Glaeser & Vigdor, 2001; Lieberson, 1980; Massey & Denton, 1993). Although segregation has been illegal since the Fair Housing Act of 1968, it is perpetuated today through an interlocking set of individual actions, institutional practices, and governmental policies....

Cultural Racism

The persistence of institutional and interpersonal discrimination is driven by the racism that remains deeply ingrained in American culture. Ideas of Black inferiority and White superiority have historically been embedded in multiple aspects of American culture, and many images and ideas in contemporary popular culture continue to devalue, marginalize, and subordinate non-White racial populations (Dirks & Mueller, 2007). Moreover, anti-Black ideology and representation is distinctive because it is typically the benchmark to which other groups are compared....

MECHANISMS BY WHICH RACISM CAN AFFECT HEALTH AND EVIDENCE OF HEALTH EFFECTS

Figure 1 outlines the multiple pathways by which racism can affect health. It indicates that racism is one of several fundamental or basic determinants of health, and it gives emphasis to institutional and cultural racism (D. Williams, 1997). The model emphasizes the importance of distinguishing basic causes from surface or intervening causes (proximal pathways). Whereas changes in fundamental causes lead to changes in outcomes, interventions in the intermediate or proximal pathways, without corresponding changes in fundamental causes, are unlikely to produce long-term improvements in population health. The model argues that race and other social status categories, such as SES, gender, age, and marital status, are created by the larger macro forces in society and are linked to health through several intervening mechanisms. Racism and other fundamental causes operate through multiple mechanisms to affect health, and the pathways through which

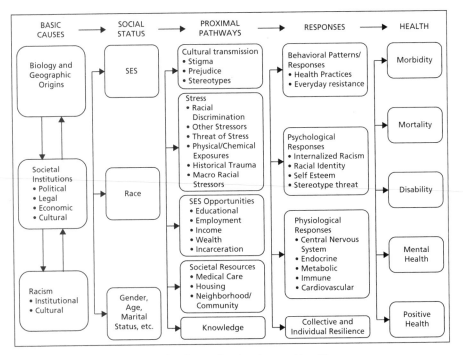

FIGURE 1 A framework for the study of racism and health.

distal causes affect health can change over time. Institutional and cultural racism can adversely affect health through stigma, stereotypes, prejudice, and racial discrimination. These aspects of racism can lead to differential access to SES and to a broad range of societal resources and opportunities. Racism is not the only determinant of intervening mechanisms, but its presence as a fundamental cause in a society can alter and transform other social factors and can exacerbate the negative effects of other risk factors for health. For example, stress is posited as one of the intervening pathways. Racism creates some types of stressors, such as discrimination and historical trauma, but it can also affect the levels, clustering, and impact of stressors, such as unemployment, neighborhood violence, or physical and chemical exposures in residential and occupational environments.

The model acknowledges that social inequalities in knowledge and communication play an insufficiently recognized role in contributing to and exacerbating social inequalities in health (Viswanath, 2006). Much of the contemporary disease burden is linked to behaviors that are potentially modifiable with the appropriate opportunities and access to preventive care and health information. Communication factors that shape health knowledge, attitudes, and behavior, such as access to and the use of various media sources, attention to health information, trust in the sources of information, and the processing of information, all vary by race-ethnicity and SES. Moreover, members of stigmatized racial groups are less able to act on and benefit from relevant health knowledge because they often lack the necessary resources to do so.

Much research on the determinants of health focuses on the responses (behavioral, psychological, physiological) to the proximal pathways. Figure 1 reminds us that these responses can be optimally understood and contextualized in the light of the upstream factors that initiate and sustain the conditions that population groups are responsive to. It also indicates that attention should be given to both individual and collective resistance and resilience. For example, some recent research suggests that some unhealthy behaviors of minority populations may reflect everyday resistance—an effort to express opposition to the larger society, assert independence, and reject the dominant society's norms (Factor, Kawachi, & Williams, 2013).

Institutional Racism and Health

Residential segregation is a potent institutional legacy of racism that is a driver of the persistence of racial economic inequality and thus racial inequities in health (D. Williams & Collins, 2001). Segregation was one of the most successful domestic policies of the 20th century in the United States (Cell, 1982), and it can affect health through multiple pathways (D. Williams & Collins, 2001). First, it restricts socioeconomic mobility by limiting access to quality elementary and high school education, preparation for higher education, and employment opportunities....

Segregation is also associated with residence in poorer-quality housing and in neighborhood environments that are deficient in a broad range of resources that enhance health and well-being, including medical care. The concentration of poverty in segregated environments can lead to exposure to elevated levels of chronic and acute stressors....

Cultural Racism and Health

... Cultural racism is likely to be a major contributor to negative racial stereotypes and the absence of positive emotion for stigmatized racial groups that can shape the policy preferences of the larger society and contribute to lack of political will to address racial inequalities in society, including those in health....

Cultural racism can trigger unconscious bias that can lead to unequal access to health-enhancing economic opportunities and resources. Many Whites have automatic, rapid, and unconscious emotional and neural reactions to Blacks, noticing an individual's race and whether he or she is trustworthy in less than 100 ms (Fiske, Bergsieker, Russell, & Williams, 2009). Research indicates that when one holds a negative stereotype about a group and meets someone who fits the stereotype, he or she will discriminate against that individual (van Ryn et al., 2011)....

Unconscious (as well as conscious) bias can also lead to unequal access to high-quality medical care. A 2003 report from the Institute of Medicine concluded that across virtually every therapeutic intervention, ranging from high-technology procedures to the most basic forms of diagnostic and treatment interventions, Blacks and other minorities receive fewer procedures and poorer-quality medical care than Whites (Smedley, Stith, & Nelson, 2003). Strikingly, these differences persist even after statistical adjustment for variations in health insurance, SES, stage and severity of disease, co-occurring illness, and the type of health care facility are taken into account. Analyses of data from a large, volunteer, and non representative sample of persons who took the Implicit Association Test (IAT) reveal that physicians have an implicit preference for Whites over Blacks, similar to the pattern observed for other professionals (lawyers and others with PhDs) and the general population (Sabin, Nosek, Greenwald, & Rivara, 2009). Research reveals that higher implicit bias scores among physicians is associated with biased treatment recommendations in the care of Black patients (van Ryn et al., 2011), although the pattern is not uniform (Haider et al., 2011)....

Experiences of Discrimination

Individuals are aware of at least some of the experiences of discrimination created by institutional and cultural racism. Research reveals that these subjective experiences of discrimination are psychosocial stressors that adversely affect a very broad range of health outcomes and health risk behaviors (Pascoe & Richman, 2009; D. Williams & Mohammed, 2009). For example, Tené Lewis and colleagues have shown that chronic everyday discrimination is positively associated with coronary artery calcification (Lewis et al., 2006), C-reactive protein (Lewis, Aiello, Leurgans, Kelly, & Barnes, 2010), blood pressure (Lewis et al., 2009), giving birth to lower-birth-weight infants (Earnshaw et al., 2013), cognitive impairment (Barnes et al., 2012), subjective and objective indicators of poor sleep (Lewis et al., 2012), visceral fat (Lewis, Kravitz, Janssen, & Powell, 2011), and mortality (Barnes et al., 2008).

Research on discrimination has also shed light on some puzzles in the literature. For example, prior research reveals that African Americans are more likely than Whites to manifest no blood pressure decline or a blunted blood pressure decline during sleep, a pattern that has been associated with increased risk for mortality and cardiovascular outcomes (Profant & Dimsdale, 1999). Recent studies reveal that exposure to discrimination contributes to the elevated levels of nocturnal blood pressure among Blacks (Brondolo et al., 2008; Tomfohr, Cooper, Mills, Nelesen, & Dimsdale, 2010). Decreases in blood pressure dipping during sleep have also been associated with low SES and other psychosocial stressors (Tomfohr et al., 2010). Prior research has also found lower levels of health care seeking and adherence behaviors among racial minorities, and research on discrimination now documents that racial bias is a contributor to these patterns. Moreover, research in the United States, South Africa, Australia, and New Zealand reveals that discrimination makes an incremental contribution over SES in accounting for racial disparities in health (D. Williams et al., 2008; D. Williams & Mohammed, 2009)....

The backdrop of cultural racism can also racialize presumably nonracial societal events so that they have negative health consequences. A study of birth outcomes in California found that infants born to Arab American women 6 months after September 11, 2001 (a period of increased discrimination of Arab Americans), had an increased risk of low birth weight and preterm birth compared to those born in the 6 months before (Lauderdale, 2006). Women of other racial and ethnic groups in California had no change in birth outcome risk, pre- and post-September 11. Personal experiences of abuse and discrimination linked to September 11 were also positively associated with psychological distress and poor health status and inversely associated with happiness among Arab Americans (Padela & Heisler, 2010). Some evidence also suggests that immigration policies hostile to immigrant groups can adversely affect their quality of life (Garcia & Keyes, 2012). Analyses of data collected in California in 2001 (a time of multiple anti-immigrant legislative proposals) found that inconsistent with prior research, Latino and some other immigrant groups reported higher psychological distress than native-born respondents (D. Williams & Mohammed, 2008). Research is needed to systematically assess the health consequences, if any, of the negative climate and hostile policies toward immigrants....

CONCLUSION

A major research challenge is the need for the conceptual and analytic models used to study racism and other determinants of disparities in health to reflect the actual clustering of diseases and their determinants. The co-occurrence of multiple diseases is commonplace, increases with age, and is evident at younger ages among low-SES and minority populations (Barnett et al., 2012). Moreover, advantage and disadvantage, resources and risks, tend to co-occur with each other and to cumulate within the same individuals and social spaces over time. Failure to model this accumulation of adversity may fail to capture, and effectively address, the full burden of social exposure. Unhealthy behaviors, such as poor nutrition, physical inactivity, cigarette smoking, and excessive alcohol intake, are clustered in individuals with and without chronic diseases (Héroux et al., 2012). Similarly, discrimination and other psychosocial stressors are clustered with each other and co-occur more frequently in disadvantaged racial populations (Sternthal et al., 2011). Inadequate research attention has been given to the ways in which multiple aspects of racism relate to each other and combine, additively and interactively, with other psychosocial risks and resources to affect health. New analytic models that reflect the complexity of the determinants of health and the clustering and accumulation of risk factors and health outcomes are urgently needed (Adler, Bush, & Pantell, 2012). The model in Figure 1 suggests that there are likely to be multiple causal pathways by which a given distal upstream factor, such as racism, can affect health status. Thus, the configuration of intervening mechanisms may vary over time, in different contexts, and for different outcomes.

Historically, racial variations in health were often viewed as genetic or biological, and some current observers view them as intractable and deeply embedded in cultural values and behaviors. The research reviewed here indicates that racism in its institutional and cultural forms have been and continue to be major contributors to initiating and sustaining racial inequalities in a broad range of societal outcomes that combine to create inequalities in health. We need a deeper understanding of how cultural norms and institutional policies and pro-cedures with regard to race shape interpersonal relations and the quality of living conditions in ways that affect health. It follows that we are unlikely to make significant progress in reducing the well-documented large racial disparities in health without intensive, comprehensive, and sustained initiatives to eliminate racial inequalities in a broad range of social, political, and economic indicators. We therefore need more concerted efforts to develop the science base that would enable us to effectively intervene to reduce and ultimately eliminate the pathogenic effects of racism and health. We consider the evidence and research opportunities for effective intervention in a companion article (D. Williams & Mohammed, 2013)....

Funding

The author(s) disclosed receipt of the following financial support for the research, authorship, and/or publication of this article: Preparation of this article was supported by grants P50 CA 148596 from the National Cancer Institute, P01 AG020166 from the National Institute of Aging, and R01 AG 038492 from the National Institute of Aging.

REFERENCES

Adler, N., Bush, N. R., & Pantell, M. S. (2012). Rigor, vigor, and the study of health disparities. *Proceedings of the National Academy of Sciences, 109* (Suppl. 2), 17154–17159.

Barnes, L. L., de Leon, C. F. M., Lewis, T. T., Bienias, J. L., Wilson, R. S., & Evans, D. A. (2008). Perceived discrimination and mortality in a population-based study of older adults. *American Journal of Public Health, 98*(7), 1241–1247.

Barnes, L. L., Lewis, T. T., Begeny, C. T., Yu, L., Bennett, D. A., & Wilson, R. S. (2012). Perceived discrimination and cognition in older African Americans. *Journal of the International Neuropsychological Society, 18*(5), 856–865.

Barnett, K., Mercer, S. W., Norbury, M., Watt, G., Wyke, S., & Guthrie, B. (2012). Epidemiology of multimorbidity and implications for health care, research, and medical education: A cross-sectional study. *The Lancet, 380*(9836), 37–43.

Bonilla-Silva, E. (1996). Rethinking racism: Toward a structural interpretation. *American Sociological Review, 62*(3), 465–480.

Braveman, P. A., Cubbin, C., Egerter, S., Williams, D. R., & Pamuk, E. (2010). Socioeconomic disparities in health in the United States: What the patterns tell us. *American Journal of Public Health, 100*, S186–196.

Brondolo, E., Libby, D. J., Denton, E.-G., Thompson, S., Beatty, D. L., Schwartz, J., & Gerin, W. (2008). Racism and ambulatory blood pressure in a community sample. *Psychosomatic Medicine, 70*(1), 49–56.

Cell, J. W. (1982). *The highest stage of White supremacy: The origin of segregation in South Africa and the American South.* New York, NY: Cambridge University Press.

Dirks, D., & Mueller, J. C. (2007). Racism and popular culture. In J. Feagin & H. Vera (Eds.), *Handbook of racial and ethnic relations* (pp. 115–129). New York, NY: Springer.

Dovidio, J. F., & Gaertner, S. L. (2004). Aversive racism. In M. Zanna (Ed.), *Advances in experimental social psychology* (Vol. 36, pp. 1–51). San Diego, CA: Academic Press.

Dumont, D. M., Brockmann, B., Dickman, S., Alexander, N., & Rich, J. D. (2012). Public health and the epidemic of incarceration. *Annual Review of Public Health, 33*(1), 325–339.

Earnshaw, V., Rosenthal, L., Lewis, J., Stasko, E., Tobin, J., Lewis, T., & Ickovics, J. (2013). Maternal experiences with everyday discrimination and infant birth weight: A test of mediators and moderators among young, urban women of color. *Annals of Behavioral Medicine, 45*(1), 13–23.

Factor, R., Kawachi, I., & Williams, D. R. (2013). The social resistance framework for understanding high-risk behavior among non-dominant minorities: Preliminary evidence. *American Journal of Public Health*, e1–e7.

Fiske, S. T., Bergsieker, H. B., Russell, A. M., & Williams, L. (2009). Images of Black Americans. *Du Bois Review: Social Science Research on Race, 6*(1), 83–101.

Garcia, A. S., & Keyes, D. G. (2012). *Life as an undocumented immigrant: How restrictive local immigration policies affect daily life.* Washington, DC: Center for American Progress. Retrieved from http://www.americanprogress.org/issues/2012/03/pdf/life_as_undocumented.pdf

Glaeser, E. L., & Vigdor, J. (2012). The end of the segregated century: Racial separation in America's neighborhoods, 1890–2010. *Civic Report, 66.* Retrieved from http://www.manhattan-institute.org/html/cr_66.htm

Haider, A. H., Janel, S. N. S., Cooper, L. A., Efron, D. T., Swoboda, S., & Cornwell, E. E., III. (2011). Association of unconscious race and social class bias with vignette-based clinical assessments by medical students. *JAMA: The Journal of the American Medical Association, 306*(9), 942–951.

Héroux, M., Janssen, I., Lee, D.-C., Sui, X., Hebert, J. R., & Blair, S. N. (2012). Clustering of unhealthy behaviors in the Aerobics Center Longitudinal Study. *Prevention Science, 13*(2), 183–195.

Lauderdale, D. S. (2006). Birth outcomes for Arabic-named women in California before and after September 11. *Demography, 43*(1), 185–201.

Lewis, T. T., Aiello, A. E., Leurgans, S., Kelly, J., & Barnes, L. L. (2010). Self-reported experiences of everyday discrimination are associated with elevated C-reactive protein levels in older African-American adults. *Brain, Behavior, and Immunity, 24*(3), 438–443.

Lewis, T. T., Barnes, L. L., Bienias, J. L., Lackland, D. T., Evans, D. A., & Mendes de Leon, C. F. (2009). Perceived discrimination and blood pressure in older African American and White adults. *Journals of Gerontology Series A: Biological Sciences and Medical Sciences, 64A*(9), 1002–1008.

Lewis, T. T., Everson-Rose, S., Powell, L. H., Matthews, K. A., Brown, C., Karavolos, K., & Wesley, D. (2006). Chronic exposure to everyday discrimination and coronary artery calcification in African-American women: The SWAN Heart Study. *Psychosomatic Medicine, 68*, 362–368.

Lewis, T. T., Kravitz, H. M., Janssen, I., & Powell, L. H. (2011). Self-reported experiences of discrimination and visceral fat in middle-aged African-American and Caucasian women. *American Journal of Epidemiology, 173*(11), 1223–1231.

Lewis, T. T., Troxel, W. M., Kravitz, H. M., Bromberger, J. T., Matthews, K. A., & Hall, M. H. (2012). Chronic exposure to everyday discrimination and sleep in a multiethnic sample of middle-aged women. *Health Psychology*. Advance online publication.

Lieberson, S. (1980). *A piece of the pie: Black and White immigrants since 1880.* Berkeley: University of California Press.

Massey, D. S., & Denton, N. A. (1993). *American apartheid: Segregation and the making of the underclass.* Cambridge, MA: Harvard University Press.

Nosek, B. A., Smyth, F. L., Hansen, J. J., Devos, T., Lindner, N. M., Ranganath, K. A., & Banaji, M. R. (2007). Pervasiveness and correlates of implicit attitudes and stereotypes. *European Review of Social Psychology, 18*, 36–88.

Padela, A. I., & Heisler, M. (2010). The association of perceived abuse and discrimination after September 11, 2001, with psychological distress, level of happiness, and health status among Arab Americans. *American Journal of Public Health, 100*(2), 284–291.

Pager, D., & Shepherd, H. (2008). The sociology of discrimination: Racial discrimination in employment, housing, credit, and consumer markets. *Annual Review of Sociology, 34*, 181–209.

Pascoe, E. A., & Richman, L. S. (2009). Perceived discrimination and health: A meta-analytic review. *Psychological Bulletin, 135*(4), 531–554.

Profant, J., & Dimsdale, J. E. (1999). Race and diurnal blood pressure patterns: A review and meta-analysis. *Hypertension, 33*(5), 1099–1104.

Sabin, J. A., Nosek, B. A., Greenwald, A. G., & Rivara, F. P. (2009). Physicians' implicit and explicit attitudes about race by MD race, ethnicity, and gender. *Journal of Health Care for the Poor and Underserved, 20*(3), 896–913.

Schuman, H., Steeh, C., Bobo, L., & Krysan, M. (1997). *Racial attitudes in America: Trends and interpretations* (Rev. ed.). Cambridge, MA: Harvard University Press.

Smedley, B. D., Stith, A. Y., & Nelson, A. R. (2003). *Unequal treatment: Confronting racial and ethnic disparities in health care.* Washington, DC: National Academy Press

Sternthal, M. J., Slopen, N., & Williams, D. R. (2011). Racial disparities in health: How much does stress really matter? *Du Bois Review, 8*(1), 95–113.

Tomfohr, L., Cooper, D. C., Mills, P. J., Nelesen, R. A., & Dimsdale, J. E. (2010). Everyday discrimination and nocturnal blood pressure dipping in Black and White Americans. *Psychosomatic Medicine, 72*(3), 266–272.

van Ryn, M., Burgess, D. J., Dovidio, J. F., Phelan, S.M., Saha, S., Malat, J., & Perry, S. (2011). The impact of racism on clinician cognition, behavior, and clinical decision making. *Du Bois Review, 8*(1), 199–218.

Viswanath, K. (2006). Public communications and its role in reducing and eliminating health disparities. In G. E. Thomson, F. Mitchell, & M. B. Williams (Eds.), *Examining the health disparities research plan of the National Institutes of Health: Unfinished business* (pp. 215–253). Washington, DC: Institute of Medicine.

Williams, D. R. (1997). Race and health: Basic questions, emerging directions. *Annals of Epidemiology, 7*(5), 322–333.

Williams, D. R. (2004). Racism and health. In K. E. Whitfield (Ed.), *Closing the gap: Improving the health of minority elders in the new millennium* (pp. 69–80). Washington, DC: Gerontological Society of America.

Williams, D. R. (2012). Miles to go before we sleep: Racial inequities in health. *Journal of Health and Social Behavior, 53*(3), 279–295.

Williams, D. R., & Collins, C. (2001). Racial residential segregation: A fundamental cause of racial disparities in health. *Public Health Reports, 116*(5), 404–416.

Williams, D. R., Gonzalez, H. M., Williams, S., Mohammed, S. A., Moomal, H., & Stein, D. J. (2008). Perceived discrimination, race and health in South Africa: Findings from the South Africa Stress and Health Study. *Social Science and Medicine, 67*(3), 441–452.

Williams, D. R., & Mohammed, S. A. (2008). Poverty, migration, and health. In A. C. Lin & D. R. Harris (Eds.), *The colors of poverty* (pp. 135–169). New York, NY: Russell Sage Foundation.

Williams, D. R., & Mohammed, S. A. (2009). Discrimination and racial disparities in health: Evidence and needed research. *Journal of Behavioral Medicine, 32*(1), 20–47.

Williams, D. R., & Mohammed, S. A. (2013). Racism and health II: A needed research agenda for effective interventions. *American Behavioral Scientist, 57*(8): 1200–1266.

DISCUSSION QUESTIONS

1. What different pathways to causes of health care disparities do Williams and Mohammed identify? Which do you think are the most compelling?

2. What role does the experience of racial discrimination play in a person's health?

31

Geneticizing Disease

Implications for Racial Health Disparities

JAMIE D. BROOKS AND MEREDITH KING LEDFORD

The public has many questions about how genes are related to diseases, but there is not a simple causal relationship between genetics and health disparities. Integrating genetics into research on racial disparities requires caution until we have a clear understanding of what race means.

Today it is almost impossible to pick up a newspaper or open a Web browser without finding an article that links a specific gene to a certain medical condition. In fact, a simple Google search of "gene linked" in November last year pulled up hits with genes linked to depression risk, restless leg syndrome, autism, breast cancer, childhood asthma, and type 1 diabetes in children. This is only on the first page of results from a total of 30,600,000 hits.

Increasingly, genes are being linked in the mainstream press, on the Web and also in prestigious medical journals not only to medical conditions but also to behavioral conditions such as narcissism, aggressiveness, and in some instances to voting behavior. Linking disease to specific genes is becoming progressively more common among the American public, too. The increasing perception is that an individual's genes are the main cause of disease.

The "geneticizing" of disease is used most appropriately in those instances where we know that genes or gene variants alone can cause disease—such as Tay-Sachs disease, which is prevalent among the descendants of Eastern European Jews but not just this one ethnic group, or sickle cell anemia, which is common among Africans and African Americans but also in other ethnic groups that have faced the scourge of malaria over countless generations. Yet that is a real stretch in other instances when genes are linked to health conditions that become labeled as race specific, since this has the potential to distort the discussion on racial health disparities.

The implication in the press is that race is the determining factor in these and other possibly "race-based" diseases. Health professionals and the public must be wary of oversimplifying the idea that "x" gene equals "y" medical condition

SOURCE: Jamie D. Brooks, and Meredith King Ledford. 2008. *Geneticizing Disease*. Washington, D.C.: Center for American Progress.

since millions of genetic variations may exist and identifying them all, and how genes interact with one another, has yet to be determined.

Indeed, researchers within the medical industry are wary of the oversimplification of geneticizing disease. Consider the growing concern among a consortium of scientists that genes are operating in a much more complex way than previously believed. Findings from the National Human Genome Research Institute, for example, suggest that it may be inaccurate to say that a gene can be linked to a single function like a predisposition to heart disease. This is critical information since the portrayal of genetic research and disease within the mass media often presents this information as mostly based on simple genetic predispositions.

If one examines the research on genes, race, and disease more closely, most research points only to a correlation of genes to disease, which is significantly different from a gene-based disease. Genes may predispose a person to certain health ailments, but health conditions are a combination of environment, lifestyle impositions, personal decisions, and access to affordable, quality health care. As geneticist Francis Collins observes, "associations often made between race and disease only occasionally have anything to do with DNA [and] most diseases are not single-locus genetic diseases and often are quite complex, involving many genetic loci as well as environmental factors."

In short, it has been well documented that disease is a combination of nature and nurture. Health care policymakers must ensure that a correlation between a gene pattern and a medical condition does not become a proxy for the causation of that medical condition as some in the medical and pharmaceutical industries move toward geneticizing and racializing disease.

Perhaps the issue of most concern in this shift to geneticize disease is the inclusion of race into the research and development of medications in an attempt to combat health disparities. The inclusion of race into medical research is not novel, nor is the controversy surrounding it. In fact, opposing sides of the debate use the same argument—those in favor of eliminating racial categories and those in favor of using racial categories in medical research argue that such a move is problematic. Yet both sides of the debate express legitimate concerns on whether to include race in medical research.

The problem with including race in gene-based medical research is that recent scientific developments undermine the notion that race as a biological fact is still in question. While a lively debate about the biological underpinnings of race ensues under projects such as the HapMap project, a partnership of scientists and funding agencies from around the world to help researchers find genes associated with human disease that respond to pharmaceuticals, elsewhere world renowned geneticists such as Craig Venter and Francis Collins declare that race is not biologic. Craig Venter, who along with Collins helped map the human genome, states that "skin colour as a surrogate for race is a social concept not a scientific one."

Still, some scientists rely on biological theories that oversimplify genetic variation between groups of people and confuse this with socially defined races. For example, if you ask three different scientists to define race, you will most likely get three different answers.

This misconception about race has taken hold due to the tendency to racially categorize people based upon physical appearance, in most instances skin color and hair-type (the "I know it when I see it," or phenotypical, reasoning). To date, however, the variations known as races are best explained by genetic drift, or the subtle changes within culturally breeding subpopulations over geologic time, and gene flow, or asymmetrical exchanges that are the byproducts of conquest between human subpopulations. Due to genetic drift and gene flow the existence of races is today a "social reality," one which public policies ... reinforce.

This social reality is scientifically misleading, yet these socially constructed racial categories can be used to measure health disparities between different racial groups to determine the health status of different populations. It is well documented, for example, that people of color suffer from health disparities such as shorter life spans, higher infant mortality rates, and higher prevalence rates of many chronic conditions compared with their white counterparts. But using race to measure health disparities is very different from using unproven genetic differences to account for these same health disparities.

Problems arise when race as a social reality and race as a scientific "fact" are conflated in medical research. These efforts to geneticize or racialize disease have several dangerous implications:

- They may skew research by placing individuals on a short list of socially constructed, government-defined racial categories, thereby increasing the risk of perpetuating health disparities.

- They may compromise the health of people of color by eliminating from medical consideration the social determinants of health problems.

- They may contribute to the reemergence of scientific racism through an emphasis on linking genes to disease and race.

The dangers implicit in all three of these efforts to racialize disease require health professionals to come to grips with the reasons for health disparities due to race as a social reality and the misconceptions about health disparities due to race as a scientific myth....

DISCUSSION QUESTIONS

1. Sociologists recognize that race is a social construction, but how are medical professionals likely to think about race?

2. Why do the authors ask for caution in how we include race in medical research that seeks remedies for specific illnesses?

32

Contextualizing Lived Race-Gender and the Racialized-Gendered Social Determinants of Health

NANCY LÓPEZ

López would like to see greater attention to the social construction of race and gender and less attention to physical appearance when treating patients. She talks about how she is race-gendered profiled in encounters with health professionals and how they are resistant to more nuanced understanding of social location that also address social class and power.

As an Afro-Latina and a sociologist of racial and gender stratification, I am ... viscerally aware of the importance of collecting data and analysis of data on "race" and ethnicity. ... [O]ne way of pursuing high-quality research on race and inequality in a variety of domains including health, education, and beyond is to take the social construction of race seriously.... While it is tempting to equate ethnicity with racial status, the conceptual and analytical distinction between race and ethnicity is of particular importance, as studies have found qualitatively different treatment and health outcomes for Latinos who self-identify or are socially defined as Black as opposed to White, or "some other race" (LaVeist-Ramos et al. 2011; Jones et al. 2008; Gravlee and Dressler 2005). For example, I was born and raised in a New York City public housing project and Spanish is my first language. Although I share the same ethnic background of my immigrant Dominican parents, my father, who is light-skinned, and not of discernable so-called African phenotypes, occupies a very different racial status than my mother and me. In most social circumstances in the U.S. my mother and I are classified as Black (... Rodriguez 2000; Vidal-Ortiz 2004).[1] The distinction between ethnicity and "race" is not trivial. As argued by Griffith (2012, 110), "In the context of men's [and women's] health, distinguishing between race and ethnicity can help researchers disentangle health outcomes that may be due to environmental constraints and contexts that vary by race from the cultural traditions, beliefs and habits and practices that vary by ethnicity."

SOURCE: López, Nancy. 2013. "Contextualizing Lived Race-Gender and the Racialized-Gendered Social Determinants of Health." Pp. 179–211 in *Mapping "Race": Critical Approaches to Health Disparities Research*, edited by Laura E. Gómes and Nancy López. New Brunswick, NJ: Rutgers University Press.

In an effort to explore the separate effects of ethnicity from "race" in health disparities research, LaVeist-Ramos et al. (2011) used the National Health Interview Survey to disentangle whether Black Hispanics are more similar to their co-ethnics or to Black non-Hispanics. They found that co-ethnics regardless of race shared similar health outcomes; however, for health services outcomes, Black Hispanics occupy the same stigmatized racial status as U.S.-born Blacks. This means that Black Hispanics did not receive the same type of treatment as their White Hispanic counterparts when they access health care: "The common cultures among black and white Hispanics people may motivate similar values, beliefs, attitudes, behaviours. On the other hand, that race exerts greater influence on both health status and health services of black Hispanics may reflect the impact of societal forces. Black Hispanics visual similarity with non-Hispanic blacks may lead to similar social status and subject them to similar levels of discrimination" (LaVeist-Ramos 2011, 5).… Here LaVeist-Ramos et al. underscore the value added to health disparities research by disentangling ethnicity (culture, values, behaviors, and so on), from "race" as a social status that is analytically distinct from ethnicity or cultural background.…

… [I]n order to understand the historic and ongoing health disparities among racially stigmatized groups, we must anchor our analysis in an examination of what I term "lived-race gender" and the "racialized-gendered social determinants of health." The racialized-gendered social determinants of health is a framework that interrogates intersecting systems of stratification at multiple levels, including the micro/individual level or what I call lived race-gender, the meso/institutional level, for example neighborhoods, schools or other local social contexts, and the macro/structural level of society, including state and federal policies and political economic structures at the national and global levels.…

I propose the concept of "racialized-gendered social determinants of health" as a key concept for understanding and ameliorating health disparities.… The racialized-gendered social determinant of health framework consists of two major concepts: "lived-race gender" and "racialized-gendered pathways of embodiment." Lived-race gender refers to the everyday experiences related to one's intersecting ascribed racial and gender social status in society. Examining the unearned privileges or disadvantages related to one's intersecting race-gender social status in a given context can capture lived race-gender. The racialized-gendered pathways of embodiment refer to the cumulative and life course effect of everyday microaggressions as well as the impact of racialized-gendered contexts in shaping health status and health outcomes.[2]…

… Given that everyone is simultaneously racialized and gendered in a given society, whether these racialized and gendered inequalities translate into cumulative unearned social advantages or cumulative disadvantages in either a particular social setting or over the life course requires empirical scrutiny and meticulous contextualization of pathways of embodiment (Chapman and Berggren 2005). And now we turn to my own experiences with lived race-gender as examples of the social forces of intersecting race and gender hierarchies in a given context to shape access to health care and health outcomes.

AUTOBIOGRAPHICAL ENCOUNTERS WITH
LIVED RACE-GENDER

Over the course of the last few decades, I have often engaged in conversations about the meaning of race with the physicians and medical practitioners I have encountered. I inquired about medical practitioners' personal views as well as what was relayed to them about the "race" concept in their formal medical training. To my chagrin, I often encountered troubling essentialist and biodeterministic conceptualizations. For example, in the spring of 1988, I completed my first year at Columbia University in New York City and, a causality of overindulgence in dorm food, I embodied the infamous "freshman fifteen." By the beginning of the fall semester of my sophomore year, I had successfully lost the weight, but I began to experience acute abdominal pain in my upper-right abdomen. After several months of reporting this chronic discomfort, I finally ended up in an emergency room where a sonogram revealed that I had many small gallstones. When I returned to my primary care physician, a middle-aged White man, to discuss my treatment options, he joked that he had not suggested a sonogram for gallstones because I was not a typical textbook case—in his words, "I was not fat, fair, and forty." Because I was a dark-skinned nineteen year-old who was not overweight, my physician could not "see" the clear-cut, classic symptoms of gallstones I was manifesting. Had the physician inquired about other triggers for gallstone formation, such as rapid weight loss, I might have received better treatment: "race-gender profiling" impeded my timely, access to quality health care (see Epstein 2007)....

One day in my early thirties and in the third trimester of my first pregnancy, I reached the door of the apartment complex where I lived at the same time as an older White man. Since I had multiple grocery bags in tow, I thanked him for opening the door and proceeded to enter the building. The man berated me for not ringing the doorbell so I could be buzzed in. Smiling, I said, "What makes you think that I don't live here?" The man was speechless and apologetic. Again, it may be the case that in the apartment building where I lived in Albuquerque, there were few if any people of African descent or dark-skinned Latinos, and this man was simply hoping to "protect" his home from potential intruders. However, this example also brings into sharp relief the reality that middle-class privilege does not protect people of color from race-gender microaggressions in their own homes (Feagin and Sikes 1994; López 2003; see also Bonilla-Silva 2003 for an example of a Black Latino professor being mistaken for a construction worker while gardening in his own home or the national coverage on Professor Henry Louis Gates's experience of being mistaken for a thief as he entered his home in Massachusetts). This reality has particular relevance for women of color of all class backgrounds who are pregnant, and have to deal with daily microaggressions that may be related to being "pregnant while Black" or "pregnant while Brown."

Another story that serves as a window to lived race-gender is an experience I had when I delivered my first daughter in 2002 at one of the hospitals in

Albuquerque. Several hours after my prolonged early morning delivery, the nurses all of whom were White, instructed me to go to the basement of the hospital to fill out paternity papers. I assumed that this was standard protocol, so my husband, a Chicano who is frequently assumed to be a recent immigrant from Mexico despite the fact that his family has lived in New Mexico for several centuries, and I dutifully made our way to the basement.[3] When we got there, the clerk a woman of color, asked us if we were married. When we said yes, she informed us that we didn't have to fill out the papers after all. Apparently, the nursing staff had assumed that I was an unwed mother. The controlling image of women of color, and Black women in particular, as unwed mothers has become so pervasive that no one bothered to ask (Collins 2009).

These autobiographic vignettes bring into sharp relief one of the key dimensions of contextualized "lived race-gender" and the "racialized-gendered social determinants of health." What is relevant here is that lived race-gender whether in the form of privilege or stigma depends on context and adds up across time as life course embodiment (Walters et al. 2011). The added stresses faced by women of color, and Black women in particular, may be part of the puzzle of why even middle-class Black women give birth to children with lower birth weights than their White counterparts, as pregnancy is a site of racialization (Bridges 2011; Geronimus et al. 2006)....

I also encountered race-gender profiling in medicine when I queried my pediatrician about my daughter's umbilical hernia. The pediatrician, whom I believe was Asian American, said that umbilical hernias were more common among people of African ancestry than among people of European ancestry. But how did my pediatrician reach this conclusion? She relied on my daughter's and my physical appearance, which was enough for her to summarily deduce that there was concordance between our phenotype and our genetic profiles. Most importantly, even if she had evidence of my so-called African "ancestry informative markers (AIMs)," these AIMs would have only included select segments of populations in Africa.... Even if my daughter's pediatrician had the AIMs that traced my daughter's genetic history to Africa, correlation does not indicate causation (Zuberi and Bonilla-Silva 2008). It would still be unlikely that my daughter's genes alone "caused" her to have an umbilical hernia. It would be equally important to explore the environment and social forces shaping my daughter's embodied health status, including during my gestation and labor. Could umbilical hernias be more common among people of African ancestry in the U.S. context because of concrete differences in race-gender experiences of stigma and privilege and pathways and mechanisms of embodied health, such as induced labor, prolonged delivery, cumulative race-gender microaggressions and discrimination during pregnancy and delivery? We don't know the answers to these questions because physicians and researchers have not asked them: they revert to thinking of race in essentialist, biological terms or what Montoya (2011, 184) terms "bioethnic conscription." In short, genetic reductionist explanations generally fail to account for the interaction of genes with other social forces, such as race-gender profiling, microaggressions and social context in shaping illness and disease (Duster 2006; ... LaVeist 1994; Roberts 1997).

Another increasingly common yet equally troubling conceptualization of race that circulates in the medical industry is the antiessentialist, color-blind understanding of race. While seeking treatment for my daughter during an asthma attack, I mentioned to the resident doctor in the emergency room that I was organizing a multidisciplinary National Institutes of Health (NIH) Workshop on Best Practices for Conceptualizing and Operationalizing "Race" in Health Policy Research at the University of New Mexico. This physician, a young woman who I believe was Asian American, seemed perplexed about the premise of the workshop. She informed me that her medical training precluded her from inquiring about any patient's race; she reminded me that under the Hippocratic oath she was sworn to treat "prisoners and millionaires" (read: people of color and Whites) the same. In other words, her training advocated a so-called color-blind approach to medical practice (Bonilla-Silva 2003; Frankenberg 1993). Since the Human Genome Project discovered that "race" is genetically meaningless, this approach maintains that it makes no sense to invoke race as an analytical concept at all. In a similar vein, the American Anthropological Association (1997) asserts that we should eliminate the use of the term "race" from data collection and instead use the term "ethnicity." The pitfalls of this seemingly antiessentialist, color-blind approach to race is that, like the genetic reductionist understanding of race, it disregards the reality of historic and ongoing racialization processes that stigmatize entire groups of people over long periods of time that cannot be deduced to ethnicity. A more productive conceptualization of race would begin from the social constructionist approach (Morning 2011).

And finally, perhaps the most common conceptualization of race I have come across when interacting with physicians and other medical personnel is that idea that race is an uncomplicated proxy for ethnicity and or social class. What these two hegemonic discourses share is that belief that racial disparities in health are really just due to "cultural" and "behavioral differences" or variations that vary according socioeconomic status (income, educational attainment, occupational status, net worth, and so on) or ethnicity, national origin, and culture. The problem with these conceptualizations is that they reduce race to an epiphenomenon of something else, either ethnicity or class (Omi and Winant 1994). The conceptualization of race as simply a proxy for ethnicity also ignores the differences in power and lived experiences among entire categories of people by race that cannot be reduced to cultural practices or national origin or social class and wealth....

... How can we do a better job of collecting and analyzing data on racial disparities in health and beyond that can help us eradicate health disparities? Among the most important changes that can occur in the collection of "race" data for mapping health disparities is a paradigm shift that contextualizes the racialized-gendered social determinants of health. This will mean that all health providers will have to be trained in the social determinants of health as well as the social construction of race and gender as intersecting systems of inequality such as social class. While I recognize that there will never be a perfect system for collecting "race" data for health disparities, at the very least we should clarify what dimension or level of the social construction of race we are researching....

NOTES

1. In New Mexico, a state with a Black population of less than 3 percent, people of so-called African phenotype are seen as a novelty. For example, at a new faculty reception in 2001, a white male administrator pointed at me and smiled from across the room. Upon meeting, he remarked how much I resembled Secretary of State Condoleezza Rice. While I can imagine that this man was trying to create conversation and connect with me by comparing me to a highly accomplished African American woman, this episode reveals how race and racial status in particular is not equivalent to ethnicity. It also points to the reality that faculty of color have qualitatively different experiences than their White counterparts who may not be subjected to jokes about their racial status.

2. The concept of "lived race-gender" builds on my previous "race-gender experience framework" and the concepts of "race-gender experiences" and "race-gender outlooks," which more directly relate to the realm of education (López 2003, 6). The concepts of "lived race-gender" and "racialized-gendered social determinants of health" more directly capture the impact of the intersection of race-gender processes in the field of health (see also Brown et al. 2006b for a discussion of lived race and health status among Latinos).

3. Although my husband's family can trace its genealogy over several centuries in New Mexico before it became a U.S. state in 1912, in many social circumstances in the United States, he is assumed to be an immigrant. This is in large part due to racist media projects depicting dark-skinned Mexicans as criminal "illegal aliens." My husband recalls stopped and frisked by police officers on a number of occasions because he fit the profile of suspected criminals.

REFERENCES

American Anthropological Association (AAA). 1997. "American Anthropological Association Response to OMB Directive 15: Race and Ethnic Standards for Federal Statistics and Administrative Reporting." Washington, DC: AAA. http//www.aaanet.org/gvt/ombdraft.html, retrieved July 19, 2008.

Bonilla-Silva, E. 2003. *Racism without Racists: Color-Blind Racism and the Persistence of Racial Inequality in the United States*. 2nd ed. Lanham, MD: Rowan and Littlefield.

Bridges, K. 2011. *Reproducing Race: An Ethnography of Pregnancy as a Site of Racialization*. Berkeley: University of California.

Brown, S. J., S. Hitlin, G. H. Elder Jr. 2006b. "The Greater Complexity of Lived Race: An Extension of Harris and Sims." *Social Science Quarterly* 87 (2): 411–13.

Chapman, R., and J. Berggren. 2005. "Radical Contextualization: Contributions to an Anthropology of Racial/Ethnic Health Disparities." *Health: An Interdisciplinary Journal for the Social Study of Health, Illness, and Medicine* 9 (2): 145–67.

Collins. P. H. 2009. *Black Feminist Thought: Knowledge, Consciousness, and the Politics of Empowerment*. 3rd ed. New York: Routledge.

Duster, T. 2006. "Comparative Perspectives and Competing Explanations: Taking on the Newly Configured Reductionist Challenge to Sociology." *American Sociological Review* 17: 1–5.

Epstein, Steven. 2007. *Inclusion: The Politics of Difference in Medical Research.* Chicago: University of Chicago Press.

Feagin, J., and M. Sikes. 1994. *Living with Racism: The Black Middle-Class Experience.* New York: Routledge.

Frankenberg, R. 1993. *White Women, Race Matters.* Minneapolis: University of Minnesota.

Geronimus, A., M. Hicken, D. Keene, and J. Bound. 2006. "Weathering and Age-Patterns of Allostatic Load Scores among Blacks and Whites in the United States." *American Journal of Public Health* 96 (5): 826–33.

Gravlee, C., and W. Dressler. 2005. "Skin Pigmentation, Self-Perceived Color, and Arterial Blood Pressure in Puerto Rico." *American Journal of Human Biology* 17: 195–206.

Griffith, D. 2012. "An Intersectional Approach to Men's Health." *Journal of Men's Health* 9 (2): 106–12.

Jones, C., B. Truman, L. Elam-Evans, C. Y. Jones, R. Jiles, S. Rumisha, and G. Perry. 2008. "Using 'Socially Assigned Race' to Probe White Advantages in Health Status." *Ethnicity and Disease* 18: 496–504.

LaVeist, T. 1994. "Beyond Dummy Variables and Sample Selection: What Health Services Researchers Ought to Know about Race as a Variable." *Health Services Research* 29: 1.

LaVeist-Ramos, T., J. Galarraga, R. Thorpe Jr., C. Bell, and C. Austin. 2011. "Are Black Hispanics Black or Hispanic? Exploring Disparities at the Intersection of Race and Ethnicity." *Journal of Epidemiological Community Health* 10: 1136.1–1136.5.

López, N. 2003. *Hopeful Girls, Troubled Boys: "Race" and Gender Disparity in Urban Education.* New York: Routledge.

Montoya, M. 2011. *The Making of the Mexican Diabetic.* Berkeley: University of California.

Morning, A. 2011. *The Nature of Race: How Scientists Think and Teach About Human Difference.* Berkeley: University of California.

Omi, M., and H. Winant. 1994. *Racial Formation in the United States: From the 1960s to the 1990s.* New York: Routledge.

Roberts, D. 1997. *Killing the Black Body: Race, Reproduction, and the Meaning of Liberty.* New York: Pantheon.

Rodriguez, C. 2000. *Changing Race: Latinos, the Census, and the History of Ethnicity in the United States.* New York: New York University Press.

Vidal-Ortiz, S. 2004. "On Being a White Person of Color: Using Autoethnography to Understand Puerto Ricans' Racialization." *Qualitative Sociology* 27 (2): 179–203.

Walters. K., S. Mohammed, T. Evans-Campbell, R. Beltrán, D. Chae, and B. Duran. 2011. "Bodies Don't Just Tell Stories, They Tell Histories: Embodiment of Historical Trauma among American Indians and Alaska Natives." *Du Bois Review: Social Science Research on Race* 8 (1): 179–89.

Zuberi. T., and E. Bonilla-Silva, eds. 2008. *White Logic, White Methods: Racism and Methodology.* Boulder, CO: Rowman and Littlefield.

DISCUSSION QUESTIONS

1. What does Lopez mean by race-gender profiling? Can you think about instances where health care providers looked at you and made assumptions that interfered with your diagnosis and treatment?

2. What does she mean by the concepts of race-gender and living in a racialized-gendered context?

3. How do these factors influence the possible health issues for people and their outcomes when seeking treatment?

Section E

Crime, Citizenship, and the Courts

Elizabeth Higginbotham and Margaret L. Andersen

In a democratic society, all individuals should be treated fairly by the law, law enforcement officials, and the courts. Beginning with the Fourteenth Amendment to the U.S. Constitution (adopted in 1868), no state shall "deny to any person within its jurisdiction the equal protection of the laws." Furthermore, the nation has laws in place, such as the **Civil Rights Act of 1964**, that prohibit discrimination based on race, creed (that is, religion), national origin, and sex. Yet we know that, even after decades of people mobilizing for equal rights, race still plays a role in the treatment of individuals in all institutions—perhaps nowhere more starkly than in the courts and the criminal justice system. There is racial profiling on the highways, on city streets, in airports, and in other public places. Police and other law enforcement officials can be a hostile presence in minority communities. African Americans, Native Americans, and Latinos are more likely than White people to be brought into the criminal justice system and, once in "the system," their treatment reflects the racism in the society.

In this final section on social institutions, we ask, how just is the justice system? Is it mostly operating to protect people's rights or is it mostly operating as a system of social control? Reams of research document that African Americans have a unique relationship with law enforcement, as has been vividly shown in national coverage of the large number of police shootings of Black men. Police shootings in places like Ferguson, Missouri, Baltimore, Maryland, New York City, and other locations shows that surveillance of minority communities and differential treatment before the law is a fact for people of color (Alexander 2010). Now—at a time when the nation is fearful of "foreigners" and sees itself as under threat—minority men in particular have been harshly subjected to stereotypes as potentially violent criminals.

The society is granting more power to the criminal justice system, even though this is a poor substitute for social reforms such as education, mental health care, and social services. National immigration policy focuses more on border control and the criminalization of immigrants than on services that would help immigrants gain citizenship. Middle Eastern citizens are subjected to perceptions

293

that they are potential terrorists (see Susan M. Akram and Kevin R. Johnson, Part III). Many low-income, inner-city neighborhoods are routinely portrayed in evening newscasts as violent, dangerous spaces—not places where people are trying to make a living and keep their families well and safe. Our nation spends far more on prisons than on education. Altogether, these realities mean that for many, criminal justice means injustice; for others, the law has become a source of intimidation and harassment, not a rational system of dispute resolution or protection from harm.

In her groundbreaking book, *The New Jim Crow: Mass Incarceration in the Age of Colorblindness*, Michelle Alexander (2010) shows that, although the crime rate in the United States is dropping, people are still being imprisoned at high rates, especially people of color. As she notes, our "incarceration rate is six to ten times greater than that of any other industrialized nation" (pp. 7–8). The inmates are overwhelmingly people of color who, after their release, are denied basic rights of citizenship. What are the consequences of imprisoning so many of our citizens?

Bruce Western ("Incarceration, Inequality, and Imagining Alternatives") questions the cycle of incarceration and identifies the problems for the wider society. Western examines what incarceration means for individuals and their families, and how it promotes inequality and separates people in society, particularly along racial lines. Increasingly, many people do not see the criminal justice system as legitimate. Others do not question it, particularly when the media reinforce the idea that we arrest and convict people who are guilty. Western suggests that current practices of incarceration mark people even after their release, making their economic survival tenuous. He concludes that our nation's incarceration practices pose a threat to civic life and to citizenship itself.

Although native-born people of color are easy targets of the criminal justice system, immigrants are also at risk for incarceration, mostly because of stereotypes in the media. Latino immigrants have been particularly marked as "illegal" and thus criminal (Chavez 2008). Jamie Longazel ("Subordinating Myth: Latino/a Immigration, Crime, and Exclusion") debunks the social myth that immigrants are prone to criminal behavior. He examines the research literature on Latino/a immigrants and shows that immigrants, including those who are undocumented, are less likely to engage in criminal behavior than American-born members of their ethnic groups. Immigrants work to curtail crime rather than to engage in it. In fact, these immigrants are often the victims of crimes because of their vulnerability to labor exploitation. Media presentations of immigrants as criminals support punitive legislation that excludes them from material, cultural, and political resources.

The nature of racism in the criminal justice system means that the police arrest and the courts convict and sentence many people of color to prison. Often the public is reassured about the effectiveness of our institutions, only to find out years later that the system has convicted the wrong person. **Exoneration**, the process of clearing people from blame, has become important in challenging the growing power of the criminal justice system. Earl Smith and Angela J. Hattery ("Race, Wrongful Conviction, and Exoneration") document how many people of color, particularly African American men, have been imprisoned only to be found innocent with further investigation of these cases. Looking at the cases of 250 people who have been exonerated through the Innocence Project, most often with DNA evidence, they find that those released after spending years in prison are disproportionately African American men. This finding is striking evidence of the biases against them in our system. The fact that most of these cases involve a White female victim demonstrates how stereotypical images of African American men as criminals works against them in the criminal justice system. Unchecked power influences who is likely to be prosecuted, found guilty, and sentenced.

Genuine justice can only follow the dismantling of segregation in major institutions. This will also require unraveling the ideologies and stereotypes that legitimate the targeting of people of color. As long as our institutions do not operate fairly for all our citizens and residence, inequalities persist and we are not a land of opportunity for all.

REFERENCES

Alexander, Michelle (2010) *The New Jim Crow: Mass Incarceration in the Age of Colorblindness*. New York: The New Press.

Chavez, Leo (2008) *The Latino Threat: Constructing Immigrants, Citizens, and the Nation*. Stanford, CA: Stanford University Press.

FACE THE FACTS: IMPRISONMENT RATE, STATE AND FEDERAL PRISONERS, 2013

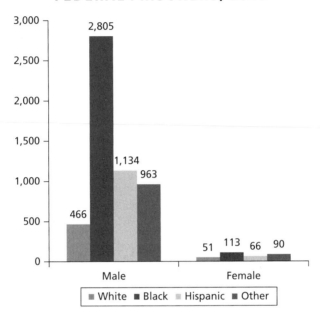

NOTE: Rate is per 100,000 residents in the U.S. population. White category excludes those of Hispanic or Latino origin; Other includes American Indians, Alaska Natives, Asians, Native Hawaiians, Pacific Islanders, persons of two or more races, and additional racial categories not covered by other categories.

SOURCE: Carson, E. Ann. 2014. *Prisoners in 2013*. Washington, DC: U.S. Department of Justice. www.bjs.gov

Think about It: What factors explain the different races of imprisonment that you see here? What impact do these races of imprisonment have on different racial-ethnic communities, including such things as family stability, political participation, community health, and the potential for violence?

33

Incarceration, Inequality, and Imagining Alternatives

BRUCE WESTERN

Incarceration has emerged as a major policy and response to complex social issues. Incarceration unfortunately targets members of racial minority groups and jeopardizes their social standing, even when they are released. Incarceration introduces and sustains more racial inequality into the society.

Citizenship is a public declaration of equality. Regardless of the inequalities that internally divide a society, citizens enjoy a common set of rights—perhaps to protest or vote or run for office. The content of citizenship rights has varied greatly across time and place. In the stylized history of T. H. Marshall (1950/ 1992), rights to speech and access to the courts were among the earliest pillars of citizenship, followed by rights to the franchise, and finally by rights to social welfare with the emergence of modem social policy.

Marshall's account of the historical development of citizenship describes a virtuous circle in which the pool of citizens grows as the rights of citizenship become more extensive. Civil rights empower citizens to press for voting rights. Once the male working class was enfranchised in Europe, unions and labor parties set about expanding social rights embodied in the welfare state. In Marshall's account, universal education was the key breakthrough, but safety net programs, national health care, and public pensions could also be added to the list.

… [M]ass incarceration in the United States has reconfigured civic life. The virtuous circle of citizenship—never fully developed in the United States to begin with—has been interrupted. The punitive turn in criminal justice policy amounts to a transformation of the quality of citizenship, in which the state plays an active role in deepening, not reducing, inequality. The virtuous circle turned vicious.

The vicious circle of mass incarceration has three main elements. First, incarceration deepens inequality because its negative social and economic effects are concentrated in the poorest communities. Deep race and class inequalities in incarceration are well documented. Recent cohorts of African American men are more likely to go to prison than to graduate from college with a four-year degree.

SOURCE: Western, Bruce. 2014. "Incarceration, Inequality, and Imagining Alternatives." *The Annals of the American Academy of Political and Social Science* 651 (January): 302–306.

In 2010, young black male dropouts were more incarcerated (37 percent) than employed (26 percent) (Western and Pettit 2009). It is among these men that we observe the reduced wages, impaired health, and family instability following time in prison.

... [P]eople who have been incarcerated experience heightened depression, anxiety, and fear. The prevalence of psychiatric symptoms among the formerly incarcerated is about twice as high as in the general population. ... [Others investigate] the acute material hardship faced by the children of incarcerated parents. ... [C]hildren's risk of homelessness roughly doubles when their fathers are incarcerated. These effects are particularly clear for African American children. Even if confident causal conclusions are elusive, the strong intercorrelation of severe economic disadvantage, poor mental health, and childhood hardship with incarceration is unmistakable.

Second, much of the social and economic inequality associated with incarceration is invisible. Becky Pettit provides a detailed treatment of this in her book, *Invisible Men* (2012).... Our data systems are unsuited to measure populations that are significantly institutionalized. They skew our assessments of social inequality. Household surveys reporting on employment, wages and educational attainment make the well-being of African Americans look artificially good because the poorest are uncounted. The invisibility of inequality associated with incarceration is also sustained in a related but different way. The experience of incarceration is so deeply stratified, its reality is confined to a small fraction of the population for whom the prison has become ubiquitous. The lived experience of urban poverty has been transformed by the growth of the penal system, but this is largely unknown to the mainstream of American society whose relationship to criminal justice institutions looks much as it did before the prison boom.

Third, mass incarceration and its social and economic consequences contribute to a legitimacy gap between the poor and the rest of American society. In the mainstream perspective, incarceration and its accompanying socioeconomic disadvantage stems more from the criminal conduct of offenders than the policy choices of government officials. The institutions themselves are not questioned. This is the sense in which Marshall (1950/1992, 7) observed that "citizenship is the architect of legitimate social inequality." The inequalities arising under a citizenship regime are traced to the defects of individuals and not the institutions themselves. In poor, high-incarceration communities, criminal justice and other state institutions have become de-legitimated....

These elements of the vicious circle—deeper and invisible inequalities combined with a widening legitimacy gap between the poor and the mainstream—make the social exclusion of mass incarceration intractable. Exclusion through incarceration further impoverishes poor communities and breeds cynicism about the civic institutions that might bring about change.

How can the vicious circle of mass incarceration be broken? Much of the political significance of incarceration lies in its intimate connection with extreme poverty. Because of its social and economic effects and because of the sense of injustice it fuels, incarceration adds mightily to the social distance between the poor and the affluent. Against this great social distance, the extension of citizenship is an act of imagination. Outsiders must be reconceived as insiders. For this to happen, the insiders must recognize something of themselves in the citizens to be.

In April 2012, a research team at Harvard University began a series of interviews with 135 men and women who were leaving prison and returning to neighborhoods in the Boston area. One of my first interviews was at a minimum security prison with Peter, an African American man in his 40s who had spent most of the last 15 years in prison. Peter had a long history of violence that was episodically associated with drinking and drug use. Throughout his life, he had dealt drugs, stolen cars, and gotten in fights. The first time we spoke, in prison, he was reluctant to share his social security number (which we use for record linkage to unemployment insurance) or many other personal details.

Over the following year, after his release, we saw in detail many of the challenges confronting those who leave prison. In the first week out, Peter applied to a community program for work clothes and a mass transit card. At first he divided his time between a friend's place and his older sister's crowded home that also housed her fiancé, daughter, and granddaughter. Over the months, Peter began to develop a relationship with two of his three children and their two mothers. Closer to one mother and her child than to the other, he seemed to focus more attention on his 10-year-old son, though he did what he could to buy them both clothes and school supplies. Peter was enrolled in food stamps program and MassHealth, the Massachusetts Medicaid program. Each month he gave his $200 in food stamps to his sister, though his eligibility for that program has now expired and he is hoping to be reinstated. Through MassHealth, he attended mental health and substance abuse treatment. He also had a physical and visited the dentist.

Peter could not find work but did enroll in employment programs doing maintenance and operating machines. The programs paid less than minimum wage, but he thought it was important to get into a daily routine. For six months, he would rise at 5:00 a.m., start work at 6:30 a.m., and work through until 3:00 p.m. Every two weeks he would report to his probation officer and on weekends he would do community service in lieu of a $65-a-month probation fee. Peter gets from place to place by mass transit and walking, concerned that riding in a car with friends might get him arrested. (If the car is stopped by police, his probation status might come up and this might trigger a violation if others in the car have felony convictions or if drugs or guns are discovered.) He is yet to find a job, unable to move forward in the hiring process once his criminal record is disclosed.

With more interviews and phone calls, we learned a little of Peter's family history. He grew up in the Lennox Street projects, a public housing complex in Boston's South End. He was a runaway. Starting around the age of 12, he would leave the house for weeks at a time, living with friends, often drinking, using drugs, and getting in trouble. One of his earliest memories of serious violence was of a brawl at midnight between blacks and whites outside an Irish bar in the neighborhood of Mattapan. Peter was 12, and he and his uncle were involved in the fight. In that melee, Peter saw a white man stabbed to death, just outside the diner where we talked some 30 years later. At 13, Peter stole a car and drove it to New York and stayed there for a while. Shortly after, his father took him to live in Los Angeles, but they returned after about a year and Peter began to live with his mother.

Mass incarceration looks a lot like this. It is not just a burgeoning prison population but how American poverty has come to be lived. The poor do not just live below an income threshold. Low income accompanies the tightly correlated adversity of violence, addiction and mental illness, childhood trauma, school failure, labor market discrimination, housing instability, and family complexity. On top of all this, we have overlaid lengthy periods of penal confinement.

The future of citizenship for the poor depends a lot on how much the insiders of American society see of themselves in Peter. The totality of his life seems light years away. He is struggling now for sobriety, housing security, economic independence, and reconciliation with his children. Mass incarceration yields few degrees of freedom. Arrest and reincarceration are often close at hand. Can we imagine a citizenry that includes Peter as much as it includes the rest of society?

If Peter's life circumstances are unfamiliar, he shares in a history that is thoroughly recognizable. The Lennox Street projects are one of the many large public housing complexes that formed much of the context for U.S. race relations over the last half century. His daily struggles to make ends meet and provide for his children are basic human challenges increasingly swamping the American middle class. Of course, the likelihood and consequences of failure for Peter are far greater, but the struggle is familiar.

So perhaps his place in a shared history and the effort he makes to find a righteous place in the world can close the social distance enough to help start a different conversation about prison policy, crime, and American poverty. Building such a conversation requires a detailed analysis of poverty as it is lived. Here we should acknowledge all the agency, effort, and creativity of those engulfed in the penal system, just as we emphasize the weighty constraint of social forces. This would seem to be a basic part of the project of humanization that might disrupt the vicious circle.

Peter has sold drugs, been on both sides of serious violence and has been locked up for most of his adult life. For all that, he gets up at 5:00 a.m., looks for work or goes to programs, and meets his son after school. Street life, he says, is well behind him. Still, the future is hard to predict for poor men with long criminal histories. Could he once again wave a gun in someone's face, abandon his children, or relapse to addiction? Sure. A good part of Peter's humanity rests in that possibility, too.

REFERENCES

Marshall, T. H. 1950/1992. *Citizenship and social class.* London: Pluto.

Pettit, Becky. 2012. *Invisible men: Mass incarceration and the myth of black progress.* New York, NY: Russell Sage Foundation.

Western, Bruce, and Becky Pettit. 2009. Technical report on revised population estimates and NLSY79 analysis tables for the Pew Public Safety and Mobility Project. Unpublished manuscript, Harvard University, Cambridge, MA.

DISCUSSION QUESTIONS

1. What does Western suggest are the major social and personal outcomes of incarceration?
2. What is the connection between imprisonment and poverty?
3. How does imprisonment exclude felons from mainstream social life?

34

Subordinating Myth

Latino/a Immigration, Crime, and Exclusion

JAMIE LONGAZEL

Social science research challenges the media images of Latino/a immigrants as criminals. These stereotypical images target them for abuses by the criminal justice system and complicate their struggle for rights.

INTRODUCTION

Political and popular rhetoric in the contemporary United States has virtually synonymized the terms "immigrant" and "criminal" thus contributing to the passage and implementation of laws at various levels of government that punish immigrants, primarily those from Mexico and other parts of Latin America. Based in large part on assumptions of entrenched immigrant criminality, the U.S.-Mexico border is now staffed with substantially more agents, reinforced with walls and fences, and patrolled with cameras, flood-lights, motion sensors, and drones (e.g. Inda 2006; Nevins 2010). Immigration law has become so punitive that it has converged with criminal law to form a hybrid doctrine legal scholars have termed "crimmigration law" (Stumpf 2006; see also Welch 2002; Newton 2008). And many states and local municipalities have gone so far as to implement laws of their own, many of which rely on an "attrition through enforcement" strategy designed to make life so difficult for supposedly criminal immigrants that they ultimately choose to "self-deport."…

All the while, a "scholarly consensus" (Lee and Martinez 2009) has been emerging that runs contrary to this conventional wisdom: *Immigrants do not increase crime; they may actually help reduce it.* Scholars conducting research on various levels, amongst diverse immigrant groups, and using multiple methodologies have found negative relationships between immigration and crime in the United States (e.g. Hagan and Palloni 1999; Nielsen et al. 2005; Reid et al. 2005; Sampson et al. 2005; Rumbaut and Ewing 2007; Sampson 2008; Butcher and Morrison Piehl 2008; Lee and Martinez 2009; Ousey and Kubrin 2009). This also holds true for Latina/o immigrants and undocumented immigrants, groups who have been

SOURCE: Longazel, Jamie. 2013. "Subordinating Myth: Latino/a Immigration, Crime, and Exclusion." *Sociology Compass* 7 (February): 87–96.

subjected to vilification in recent years (Sampson et al. 2005). In fact, research finds that first generation immigrants are less crime-prone than their second- and third-generation counterparts, suggesting that, if anything, it is *increased "American-ness"* rather than a failure to assimilate that would lead immigrants to crime (e.g. Sampson 2006; Rumbaut and Ewing 2007). "Cities of concentrated immigration," in short, "are some of the safest places around" (Sampson 2008: 30).

To explain what emerges as a glaring contradiction—low rates of immigrant criminality on one hand, hyper-enforcement driven primarily by the assumption that immigrants are crime-prone on the other—this article will discuss the criminalization of Latina/o immigrants as a *subordinating myth*, a social construction based on a false premise that "forcefully contributes to preserving racial [and ethnic] stratification through exclusion ... [by misallocating] political, cultural, and material goods" (Haney López 2010: 1045). The exploitation of immigrant laborers is indeed widely acknowledged as a means by which primarily white elites reap the material benefits produced by immigrants of color (e.g. Estrada et al. 1981; Calavita 1992, 1994; Massey 2007), but what is often overlooked is the extent to which criminalization serves similar purposes. More specifically, by reviewing recent research on Latino/a immigration, crime, and social control in the context of racial/ethnic stratification, this article will reveal how the false premise of entrenched immigrant criminality creates profits, engenders political benefits for elites, degrades minority groups, and controls exploitable populations.

Despite rapidly growing rates of immigrants and refugees arriving in the United States from other global regions (e.g., Semple 2012), my primary focus is on immigrants from Mexico and other Latin American countries, whom I'll refer to throughout as simply "Latino/a immigrants." This group has borne the brunt of immigrant criminalization in recent years and the American public perceives them more negatively in comparison to other groups (Timberlake and Williams 2012)....

RACIAL STRATIFICATION AND IMMIGRATION

Understanding race/ethnicity from the perspective of racial stratification ... entails seeing race/ethnicity "not simply as an emotional eruption, efficient shortcut, cognitive habit, or institutional tic, but as a central means of ordering and rationalizing the distribution of resources, broadly conceived" (Haney López 2010: 104). To that end, racial/ethnic stratification is achieved through *exploitation*— "when people in one social group expropriate a resource produced by members of another social group" (Massey 2007: 6)—and *opportunity hoarding*—where "beneficiaries do not enlist the efforts of outsiders but instead exclude them from access to relevant resources" (Tilly 1998: 91).

Racial stratification theory serves as a useful framework for helping us to understand the contemporary subordination of immigrants. It is widely acknowledged that the history of immigration into the United States, particularly Latino/a immigration, is in many respects a history of exploitation, as this group has historically "formed a reserve labor pool that could be called up as the situation dictated" (Estrada et al. 1981: 110 …). And although the exploitive intentions of modern capitalists may be less explicit than those of the past, there is no denying that such practices are widespread today (e.g., Massey 2007). My focus here, however, is on the relevance of hoarding; I seek to explicate how criminalization has served to exclude Latina/o immigrants from access to material, cultural, and political resources.

A useful jumping off point then is Ian Haney López's (2010) recent trenchant analysis, *Post-Racial Racism*, which links the war on crime and attendant mass incarceration to racial hoarding. He notes that punitive crime policies contribute to racial exclusion in principally four ways: through *profit* (i.e., the reemergence of prison labor and the privatization of prisons, both of which have produced massive profits for corporate elites at the expense of communities of color), *politics* (i.e., felon disenfranchisement laws and the political tendency to play to the racial anxieties of whites), *degradation* (i.e., the forging of symbolic linkages between people of color and crime which alter the "cultural meanings tied to racial categories" [1048]), and *population control* (i.e., exerting direct control over minority populations by transforming the prison into a "surrogate ghetto" [Wacquant 2009: 195])….

CRIMINALIZING FOR PROFIT

Accounting for the increased punitiveness in immigration law is the emergence of an "immigration industrial complex" which, much like the pervasive "prison industrial complex," is characterized by laws and policies that serve the interests of government and corporate elites (Golash-Boza 2009). Uncovering the workings of privately operated immigration detention facilities puts the immigration industrial complex clearly on display. Currently, just under 400,000 immigrants are detained per year in the United States (Department of Homeland Security 2011), an all-time high, and almost half of these detainees—most of whom are Latina/o—are housed in private facilities (Kirkham 2012). In fact, the private prison industry was pulled out of financial peril in the late 1990s by a surge in immigrant detention (Kirkham 2012). Today, thanks to increased border enforcement and contracts with the federal government, the industry now thrives on immigrant detention….

Immigrants have thus quite literally become the "raw materials" of a now booming for-profit prison industry (see Christie 2000: 87). Under these arrangements, private prisons and their stockholders desire increased incarceration *regardless* of whether it is empirically misguided (Selman and Leighton 2010). This often means lobbying intensely for punitive laws that will assure prisons remain "well-stocked" and thus profitable….

THE POLITICS OF CRIMINALIZATION

Racial threat theorists have long acknowledged that rising minority populations coupled with economic decline tend to provoke anxiety amongst the white majority (Blumer 1958; Quillian 1995). As immigrant populations increase and white workers confront painful economic shifts such as demanufacturing, the stage is thus set for harsh anti-immigrant backlash.... Thus in addition to disenfranchisement, which obviously precludes undocumented immigrants from political participation, potential class-based political allies become formidable foes as white workers are encouraged to point their fingers at working class immigrants as the source of their declining economic and social position....

CRIMINALIZATION AS DEGRADATION

Considering how criminalization contributes to degradation and enables population control unveils how punitive immigration policies play out "on the ground." Frighteningly, scholars have empirically documented how anti-immigrant ordinances stoke resentment and lead to increases in the actual discrimination experienced by Latino/as....

In discussing group degradation as a mechanism of exclusion, Haney López (2010) invokes the concept of microaggression—"stunning, automatic acts of disregard that stem from unconscious attitudes of white superiority and constitute a verification of black inferiority" (Davis 1989: 1576). This concept brings the devastating effects of discrimination more clearly into view: degrading day-to-day encounters breed humiliation and painfully reaffirm one's subordinate status....

CONTROLLING CRIMINALIZED IMMIGRANT POPULATIONS

The criminalization of immigrants also engenders population control to the extent that immigrants find themselves, as Susan Coutin (2007: 9) so eloquently put it in her ethnography of Salvadoran migrants, "physically present but legally absent, existing in a space that is not actually 'elsewhere' or beyond borders but that is rather a hidden dimension of social reality." Latina/o immigrants, that is to say, experience "entrapment," as Guillermina Núñez and Josiah Heyman (2009: 354) similarly point out in their study of how migrants (documented and undocumented) "are not just enclosed inside the country as a whole, but are also impeded from moving around locally to access vital resources and join with loved ones."

Much of this restriction stems from precarious legal status and hyper-enforcement—what Nicholas De Genova (2002) terms *deportability*—which puts migrants and their family members in a constant state of fear. A recent Center for American Progress report reveals the astounding portion of immigrants who are understandably anxious about engaging in seemingly safe, mundane tasks such as driving a car (67%) and walking in public (64%). Indeed, most respondents report avoiding public places at all costs, and, when that is not possible, going to great lengths to blend in (García and Keyes 2012)....

CONCLUSION

Beginning with the criminalization of Latino/a immigrants on one hand and scholarly research that refutes connections between immigration and crime on the other, I have discussed in this article how what appears as a glaring contradiction can be understood as a *subordinating myth*. Using racial stratification theory as a guide, I reviewed recent research on Latino/a immigration, crime, and social control as a way to illuminate how criminalization has contributed to exclusion in the areas of profit, politics, degradation, and population control. Specifically, I discussed how private detention facilities and other governmental and corporate elites have capitalized financially on the increased incarceration of predominately Latina/o immigrant populations; I described the political benefits afforded to those who scapegoat Latina/o immigrants as criminals; I drew attention to the degrading features of immigration enforcement; and I highlighted how harsh immigration laws enable population control....

Exposing the myth of immigrant criminality is thus only a first step. As scholars, we must continue to unpack the relationship between criminalization and exclusion. This entails envisioning crime not just as an act an individual chooses to commit, but rather a political tool that serves racial purposes. Such an approach has the potential of building a powerful counter-narrative around issues of immigration and crime that can make the simplistic, empirically inaccurate immigrant-as-criminal narrative appear derisory in comparison. That is to say, if we are to *subordinate the myth* of immigrant criminality, we must place the criminalization of immigrants within a broader historical, racial, and sociological context that illuminates how criminalization contributes to exclusion....

REFERENCES

Blumer, Herbert. 1958. 'Race Prejudice as a Sense of Group Position.' *Pacific Sociological Review* 1: 3–7.

Butcher, Kristin F. and Anne Morrison Piehl. 2008. 'Crime, Corrections, and California: What Does Immigration Have to Do with It?' *California Counts* 9(3): 1–23.

Calavita, Kitty. 1992. *Inside the State: The Bracero Program, Immigration, and the I.N.S.* New York, NY: Routledge.

Calavita, Kitty. 1994. 'U.S. Immigration and Policy Responses: The Limits of Legislation.' Pp. 55–82 in *Controlling Immigration: A Global Perspective*, edited by

Wayne A. Cornelius, Phillip L. Martin and James Frank Hollifield. Stanford, CA: Stanford University Press.

Calavita, Kitty. 2005. *Immigrants at the Margins: Law Race, and Exclusion in Southern Europe.* United Kingdom: Cambridge University Press.

Christie, Nils. 2000. *Crime Control as Industry: Towards Gulags, Western Style* (3rd Edition). New York, NY: Routledge.

Coutin, Susan Bibler. 2007. *Nation of Emigrants: Shifting Boundaries of Citizenship in El Salvador and The United States.* Ithaca, NY: Cornell University Press.

Davis, Peggy C. 1989. 'Law as Microaggression.' *The Yale Law Journal* 98: 1559–77.

De Genova, Nicholas P. 2002. 'Migrant "Illegality" and Deportability in Everyday Life.' *Annual Review of Anthropology* 31: 419–47.

Department of Homeland Security. 2011. 'Immigration Enforcement Actions: 2010.' *Annual Report.* [Online]. Retrieved on 10 July 2012 from: http://www.dhs.gov/ xlibrary/assets/statistics/publications/enforcement-ar-2010.pdf

Estrada, Leobardo F., Chris Garcia, Reynaldo Flores Macis and Lionel Maldonado. 1981. 'Chicanos in the United States.' *Daedalus* 110: 103–13.

García, Angela S. and David G. Keyes. 2012. 'Life as an Undocumented Immigrant: How Restrictive Local Immigration Policies Affect Daily Life.' *Center for American Progress.* [Online]. Retrieved on 10 July 2012 from: http://www.americanprogress .org/issues/2012/03/life_as_undocumented.html/

Golash-Boza, Tanya. 2009. 'A Confluence of Interests in Immigration Enforcement: How Politicians, the Media, and Corporations Profit from Immigration Policies Destined to Fail.' *Sociology Compass* 3(2): 283–94.

Hagan, John and Alberto Palloni. 1999. 'Sociological Criminology and the Mythology of Hispanic Immigrant Crime.' *Social Problems* 46: 617–32.

Haney López, Ian. 2010. 'Post-Racial Racism: Racial Stratification and Mass Incarceration in the Age of Obama.' *California Law Review* 98: 1023–74.

Inda, Jonathan Xavier. 2006. *Targeting Immigrants: Government, Technology, and Ethics.* Malden, MA: Blackwell Publishing.

Kirkham, Chris. 2012. 'Private Prisons Profit From Immigration Crackdown, Federal and Local Law Enforcement Partnerships.' *Huffington Post.* [Online]. Retrieved on 10 July 2012 from: http://www.huffingtonpost.com/2012/06/07/private-prisons-immigration-federal-law-enforcement_n_1569219.html

Lee, Matthew and Ramiro Martinez. 2009. 'Immigration Reduces Crime: An Emerging Scholarly Consensus.' Pp. 3–16 in *Sociology of Crime, Law, and Deviance*, edited by William McDonald. New York, NY: Elsevier.

Massey, Douglas S. 2007. *Categorically Unequal: The American Stratification System.* New York, NY: Sage.

Nevins, Joseph. 2010. *Operation Gatekeeper and Beyond: The War on "Illegals" and the Remaking of the U.S.-Mexico Boundary.* New York, NY: Routledge.

Newton, Lina. 2008. *Illegal, Alien, or Immigrant: The Politics of Immigration Reform.* New York, NY: New York University Press.

Nielsen, Amy L., Matthew T. Lee and Ramiro Martinez Jr. 2005. 'Integrating Race, Place, and Motive in Social Disorganization Theory: Lessons from a Comparison of Black and Latino Homicide Types in Two Immigrant Destination Cities.' *Criminology* 43: 837–72.

Núñez, Guillermina Gina and Josiah McC. Heyman. 2007. 'Entrapment Processes and Immigrant Communities in a Time of Heightened Border Vigilance.' *Human Organization* 66(4): 354–65.

Ousey, Graham C. and Charis E. Kubrin. 2009. 'Exploring the Connection Between Immigration and Violent Crime Rates in U.S. Cities, 1980–2000.' *Social Problems* 56(3): 447–73.

Quillian, Lincoln. 1995. 'Prejudice as a Response to Perceived Group Threat: Population Composition and Anti-Immigrant and Racial Prejudice in Europe.' *American Sociological Review* 60(4): 586–11.

Reid, Lesley Williams, Harold E. Weiss, Robert M. Adelman and Charles Jaret. 2005. 'The Immigration-Crime Relationship: Evidence Across US Metropolitan Areas.' *Social Science Research* 34: 757–80.

Rumbaut, Rubén G. and Walter A. Ewing. 2007. 'The Myth of Immigrant Criminality and the Paradox of Assimilation: Incarceration Rates among Native and Foreign-Born Men.' *Immigration Policy Center*. [Online]. Retrieved on 10 July 2012 from: http://www.ime.gob.mx/2007/immigrant_criminality.pdf

Sampson, Robert J. 2006. 'Open Doors Don't Invite Criminals: Is Increased Immigration Behind the Drop in Crime?' *New York Times*. [Online]. Retrieved on 10 July 2012 from: http://www.nytimes.com/2006/03/11/opinion/11sampson.html

Sampson, Robert J. 2008. 'Rethinking Crime and Immigration.' *Contexts* 7(1): 28–33.

Sampson, Robert J., Jeffrey D. Morenoff and Stephen Raudenbush. 2005. 'Social Autonomy of Racial and Ethnic Disparities in Violence.' *American Journal of Public Health* 95: 224–32.

Selman, Donna and Paul Leighton. 2010. *Punishment for Sale: Private Prisons, Big Business, and the Incarceration Binge*. Lanham, MD: Rowman and Littlefield.

Semple, Kirk. 2012. 'In a Shift Biggest Wave of Migrants is now Asian.' *New York Times*. [Online]. Retrieved on 10 July 2012 from: http://www.nytimes.com/2012/06/19/us/asians-surpass-hispanics-as-biggest-immigrant-wave.html?_r=1

Stumpf, Juliet P. 2006. 'The Crimmigration Crisis: Immigrants, Crime, and Sovereign Power.' *bepress Legal Series*. [Online]. Retrieved on 10 July 2012 from: http://law.bepress.com/cgi/viewcontent.cgi?article=7625&context=expresso

Tilly, Charles. 1998. *Durable Inequality*. Berkeley, CA: University of California Press.

Timberlake, Jeffery M. and Rhys H. Williams. 2012. 'Stereotypes of U.S. Immigrants from Four Global Regions.' *Social Science Quarterly*. [Online]. Retrieved on 10 July 2012 from: http://onlinelibrary.wiley.com/doi/10.1111/j.1540-6237.2012.00860.x/pdf

Wacquant, Loïc. 2009. *Punishing the Poor: The Neoliberal Government of Social Insecurity*. Durham, NC: Duke University Press.

Welch, Michael. 2002. *Detained: Immigration Laws and the Expanding I.N.S. Jail Complex*. Philadelphia, PA: Temple University Press.

DISCUSSION QUESTIONS

1. How does Longazel's account of the scholarship on Latino/a immigrants differ from what you might have learned about immigrants, even undocumented immigrants in the media?

2. How does criminalizing immigrants connect with the prison industrial complex?

3. How might racial profiling impact the lives of all Latinos/as regardless of immigration status and citizenship?

35

Race, Wrongful Conviction, and Exoneration

EARL SMITH AND ANGELA J. HATTERY

*The majority of people exonerated are African American men, meaning they
are very likely to be arrested and convicted of crimes they did not do.
Smith and Hattery explore how that happens and the role of racism.
Understanding the systemic problems can help us prevent abuses of power
by the system.*

Imagine that when you are 18 or 19 years old, or even 20, and you have
... your whole life ahead of you, possibly you have a passion that you think
you can translate into a paycheck or even a career. Maybe you already have your
eye on a potential person you think you could settle down with and call your-
selves a family, maybe you are simply enjoying the freedom that so many of us
enjoy in those short years between adolescence and adulthood, imagine that in
the blink of an eye all of your dreams and hopes come crashing down on top of
you. Imagine that your worst nightmare has come true. Imagine that you are not
only arrested but convicted of a crime you didn't commit. Imagine that you sit
before a jury being called the most filthy and vile names: rapist, murderer,
because the crime you are accused of is heinous. Imagine that you believed,
you were taught, that the laws that give rights to defendants are there to prevent
what is happening before your eyes. Imagine that you stand before a judge who
sentences you to spend the rest of your life in prison, or worse to stand in line to
be executed. If you are one of the more than 250 men and women who have
been exonerated at the time of the writing of this paper you don't have to imagine. This is what happened to you. This paper seeks to examine the relationship
between race and exoneration; in essence we ask the question is race a "risk factor" for wrongful conviction?...

When we first began to talk about this research we got one of two reactions:
(1) since most of those who are incarcerated are African American men, doesn't
it make sense that this population is also the most likely to be wrongfully convicted and exonerated and (2) won't the DNA techniques developed and

SOURCE: Smith, Earl, and Angela J. Hattery. 2011. "Race, Wrongful Conviction, and
Exoneration." *Journal of African American Studies* 15 (March): 74–94.

strengthened in the last decade prevent this from happening in the future?... It is precisely these two misconceptions that we address in this [article].

AN OVERVIEW OF THE POPULATION
OF EXONOREES

As of January 1, 2010 there have been approximately 250 exonerations in the United States. Most, or over 90%, of these post-conviction exonerates are men. Approximately 75% are members of minority groups (Mauer 2009) and on average have spent 13 years in prison for crimes they did not commit....

Among the most recent phenomenon in the areas of crime and the law is the use of scientific forensic evidence—primarily DNA—to exonerate individuals who were wrongly convicted and incarcerated (Gross 2008; National Academy of Sciences 2009). At the time of the writing of this ..., there were 250 exonerees. This is a number that is constantly in a state of flux as more and more individuals are granted the tools and the opportunity to gain their freedom....

Scientifically speaking we don't know how many people there are sitting in our jails and prisons who are factually innocent. There is no systematic way of gaining an exoneration. As stated by Gross exonerations are accidental (Gross and O'Brien 2007). But, typically exonerations result from the dedicated work of attorneys ... who believe their client is innocent, ... investigative journalists who pay attention to serious inconsistencies in the evidence—as in the case of both Darryl Hunt and Roy Brown—and often the inmate himself, like Ronald Cotton, who do their own detective and legal work trying to prove their innocence. Many, but not all, of these cases finally catch the attention of the Innocence Project whose mission is to find and free wrongly convicted innocent people who rot for decades in American prisons (Irwin 2009).

Because the cases are handled on an individual basis, it is hard to estimate, but some experts suggest as much as 6% of our incarcerated population is actually innocent (Gross 2008). If that statistic is accurate, of the 2.2 million people who are currently incarcerated as many as 140,000 may be factually innocent....

Exoneration: The Role of DNA in Shaping Exonerations

We cannot underestimate the impact of the role that the science of DNA has played in exonerations; in fact *all* of the 250 exonerations to date have been gained at least in part through DNA analysis. Because of the particulars of DNA it profoundly shapes exonerations. DNA evidence is present, collected, and analyzed primarily in murder and rape cases. And, though these are perhaps the two most serious personal crimes, this limitation significantly shapes exoneration. Specifically, because DNA is *not* routinely collected and analyzed when other crimes occur—assault, robbery, or non-violent property or drug crimes—when innocent people are incarcerated for these crimes they seldom have any avenue for seeking exoneration....

FINDINGS

... African American men are disproportionately represented among the population of exonerees, in fact of the 150 cases in which we have reliable race data for the offender, 105 or 70%, are African American.

... [T]hough African Americans are disproportionately among the incarcerated population—they comprise about 40-50%—African Americans men account for 70% of the exonerees....

... [W]e examined the rates of exoneration for African Americans taking in to consideration their disproportionate likelihood of being incarcerated. ... [T]he rate of exoneration for African Americans is clearly and statistically significantly greater than the overall rate of incarceration for this same population. Clearly, when we incarcerate 2.3 million people, mistakes will be made. But, if the mistakes are *random* they will be distributed in a pattern that is similar to the phenomenon itself. In other words, patterns in exoneration would mimic patterns in arrest, conviction, and incarceration. Whenever a phenomenon exhibits patterns that are different than those that exist in the population ... then we have reason to suspect that something systematic and non-random is at work.

... [T]he picture that emerges is disturbing. African American and White men have reverse experiences; African American men make up far more of the exonerated population than they do the incarcerated population. For White men the trend is the opposite, they make up significantly fewer of the exonerations than their overall representation among those incarcerated. This suggests that African American men are disproportionately among the wrongly convicted. The question is why? The relationship between race and exoneration becomes even more profound when we focus our attention on the particular crimes, rape and murder, that produce the vast majority of the exonerations.

... First, ... about equal numbers of African American and White men commit homicide. Second, homicide is an overwhelmingly *intraracial* crime: people murder and are murdered by others in the same race/ethnic group. Though African Americans are slightly more likely than Whites to be the perpetrators in interracial homicides, only 11% of all homicides are interracial, whereas 89% are intraracial.

... [W]e note that contrary to popular myths about African American men (see Angela Davis' discussion of "The Myth of the Black Rapist" in her book *Women, Race and Class*), White men make up 50% of all men incarcerated for the crime of rape or sexual assault. When we examine the patterns inside the data we see that in cases where the victim is a White woman, 50% of the time White men are the perpetrators and in only 16% of the cases are African American men the perpetrators. (Those whose race could not be identified by the victim make up the remainder.) When African American women are the victims, nearly half of the time (43%) African American men are the perpetrators. Thus, as with homicide, rape is also predominately an *intraracial* crime. Given these trends, we were curious about the potential interaction among race of the perpetrator, race of the victim, crime committed and the likelihood of exoneration.

... [T]he overwhelming majority, 84% of the 87 cases on which we have race data on both the exonoree and the victim, involve an African American man being exonerated for the rape and/or murder of a White woman.... Exonerations follow a pattern that is *exactly the opposite* of the pattern of actual crimes that are committed. African American men commit only 16% of the rapes against White women, yet this crime accounts for 68 of the 87 (78%) of all exonerations.

It's also interesting to note that when we segregate out crimes in which the victim is only raped (and not murdered), 58 of the 87 (65%) exonerations involve a crime that occurs only 16% of the time! In other words *African American men are four times more likely to be exonerated for raping White women compared to the number of times they actually commit this crime.*

WRONGFUL CONVICTION: THE HUMAN COSTS

... Wrongful conviction has in many regards claimed the lives of the 250 men and women who have been wrongly incarcerated. The average exonoree served 13 years in prison. The Innocence Project estimates that nearly 3,000 years have been collectively served by these 250 men and women, including the 17 exonorees who served time on death row.

These years are, by all accounts, the best years of one's life. The average age at which the exonorees were incarcerated is 26 years old, but many were sent to prison for life while they were still in their late teens or early 20s. These are the years in which most Americans build their adult lives; they finish their education, they start working in their professions or occupations, they find life partners, they begin childbearing if they so choose, those with resources buy their first home and so on. By and large the 250 men and women who have been exonerated spent most or all of these critical years in prison. Most of the exonorees had not married or otherwise entered committed relations, most had not started families, most had not bought their own homes. Some had started their professional lives, but many had not. Regardless of the total number of years lost, these individuals were systematically denied the freedom to do the things that most Americans take for granted. Not because they gave up that right, but because the system failed them.

In addition to their own lives, collectively, families and communities have been denied fathers and husbands and sons (Arditti 2009; Charles and Luoh 2010). And, regardless of the actual innocence of the wrongly incarcerated father or mother, we can assume that for those who did leave children behind, these children suffer from the same risks that all children of incarcerated parents face, including increased likelihood for being incarcerated themselves (Foster and Hagan 2009).

As a community the wrongful conviction of just these 250 individuals amounts to 7 million hours of lost work, $42 million dollars in lost wages, and the $87 million dollars used to incarcerate these individual who were factually innocent. Finally, and very important we also see the delay of true justice for the victims of the crimes for which these men and women were wrongly incarcerated. For example, in the case of Darryl Hunt, he was tried twice and sought a third trial during his nearly 20 years of being incarcerated. This is typical in exoneration cases because they are often riddled with errors and suspicions that result in a judge granting a new trial. Whereas this is helpful to the exoneree as he or she may ultimately be able to prove their innocence in this manner, it is devastating for the victims' families who have to relive the traumatic events of the crime, and if they are still alive for the victims who are often required to testify in multiple trials across multiple decades.... Whether we are sympathetic to the person wrongfully convicted or not, the costs to our society are great. And, perhaps the greatest is the threat to public safety that we all live with when the real perpetrator is free to roam our communities raping our mothers and sisters, molesting our daughters and murdering our loved ones.

Exoneration: Causes

In short, wrongful convictions—and the exonerations they produce—are a microcosm of the social world; with systems such as capitalism and racial domination playing a substantial role in shaping these patterns....

... We offer several explanations, that when taken together, help us to understand this disturbing phenomenon.

1. *The fallibility of eyewitness testimony*: As the research of others has documented, in 70% of the 250 wrongful convictions, the conviction hinged on eye-witness testimony that was later documented to be faulty. Eyewitness misidentification can happen for many reasons. As psychologists have demonstrated using experimental designs even in low-stress situations with relatively long exposures to the "target," eyewitness accounts are ridiculously unreliable. In addition these cases force us to examine the issue of cross-race identifications. As noted, the vast majority of the exoneration cases involve a White victim who mis-identifies an African American man....

2. *The misuse of forensic evidence*: The Innocence Project has carefully catalogued and documented that in 52% of the exoneration cases there was improper use of forensic evidence. There are a variety of mistakes that can be made, including failure by the police department to collect and preserve evidence properly, mistakes at the laboratories the conduct the tests (National Academy of Sciences 2009), as well as mistakes in interpretation of the data. The conclusion then is that even when individuals have access to DNA testing as part of the criminal investigation and trial, there are no assurances that it is

properly done. Not only are wrongful convictions a result of this flaw, but lawyers who are aware of this may counsel their clients *not* to have the DNA testing done because they fear that flawed analysis which point to their client will be impossible for the defense to overcome. Because DNA is believed to reveal the ultimate truth, especially by jurors, a DNA conviction would be nearly impossible to overturn and as a result the risk of having DNA tested when the processes for analysis and interpretation have been demonstrated to be faulty is great....

3.   Another problem that occurs over and over again is the lack of access to DNA analysis for most of the incarcerated individuals. In fact we are shocked at the frequency of this. An analysis of the cases of those who were exonerated reveals that the assumption that DNA analysis is available to those facing serious charges or those fighting a wrongful conviction is flat out wrong....

4.   *False confessions*: Though it is very difficult for most people who have not seen the under-belly of the criminal justice system to imagine, in nearly a quarter (23%) of the exoneration cases catalogued and analyzed by the Innocence Project, the wrongly convicted person gave a false confession. How on earth does this happen? There are two different "causes" of false confession: individual causes and policy causes. In a fair number of cases the false confessions were obtained from individuals who were unable to fully understand the police interrogation. For example, false confessions have been obtained from juveniles who lacked the intellectual development to distinguish between the hypothetical and the actual as well as adults with diminished capacity or mental impairment, who like the juveniles cannot distinguish a story a police officer is painting from the truth.... The second class of false confessions involves people who are not of diminished capacity but who are convinced by legal interrogation tactics that it is in their best interest to confess.... These cases are especially difficult for exonerees because for a period of time they actually convince themselves they have committed some horrible crime: remember all the exonerees were convicted of rape and/or homicide.... All of these practices of extracting a false confession, taken together scream for serious revisions to legal interrogation practices....

5.   *Snitches*: In more than 15% of exonerations an informant or "snitch" was used to testify in such a manner that contributed significantly to a wrongful conviction. The most common scenario involves snitches being paid to testify or testifying in exchange for being released from prison....

6.   *The racial history of the United States*: As many scholars have noted, and we have discussed in our own research, the boundary between White and Black sexuality has been the most heavily patrolled and controlled throughout U.S. history (Smith and Hattery 2009). And, accusations of the rape of White women by African American men have been a cornerstone of race relations and the justice system for centuries. Angela Davis (1983) documents

the fact that the mere accusation of rape of a White woman would send vigilante mobs in search of an African American man to lynch. And, more than 10,000 African American men were lynched between 1880 and 1930. Davis documents that only a handful involved an actual rape and a handful more involved consensual relations between White women and African American men. Thus, we argue, that the long-standing myth of the black rapist and the lynching of 10,000 African American men, mostly without cause, provides the historical context, the back drop if you will, for the way in which the police, the criminal justice system, and even the public deal with African American men accused of rape....

In short, wrongful convictions—and the exonerations they produce—are a microcosm of the social world; with systems such as capitalism and racial domination playing a substantial role in shaping these patterns. Thus, a careful, systematic, sociological analysis of wrongful convictions and exonerations lays bear one of the most extreme and horrific outcomes of systems of oppression at work in the United States since slavery....

REFERENCES

Arditti, J. (2009). Families and incarceration: an ecological approach. *The Journal of Contemporary Social Services, 86*, 251–260.

Charles, K. K., & Luoh, M. C. (2010). Male incarceration, the marriage market and female outcomes. *Review of Economics and Statistics*. Forthcoming.

Davis, A. (1983). *Women, race, and class*. New York: Vintage Books.

Foster, H., & Hagan, J. (2009). The mass incarceration of parents in America: issues of race/ethnicity, collateral damage to children, and prisoner reentry. *ANNALS, 623*, 179–194.

Gross, S. (2008). Convicting the innocent. *Annual Review of Law and Social Science, 4*, 173–179.

Gross, S., & B. O'Brien. (2007). Frequency and predictors of false conviction: Why we know so little, and new data on capital cases. *Michigan State University College of Law, Legal Studies Research Paper No. 05-14.*

Irwin, J. (2009). *Lifers: Seeking redemption in prison*. New York: Routledge.

Mauer, M. (2009). Two tiered justice: Race, class and crime policy. In C. Hartman & G. Squires (Eds.), *The integration debate: Competing futures for American cities* (pp. 169–183). New York: Routledge.

National Academy of Sciences. (2009). *Strengthening Forensic Science in the United States: A path forward*. Washington, DC: National Academies Press.

Smith, E., & Hattery, A. J. (2009). *Interracial intimacies: An examination of powerful men and their relationships across the color line*. Durham: Carolina Academic Press.

DISCUSSION QUESTIONS

1. What accounts for the disproportionate number of African American men wrongly convicted and imprisoned only to be exonerated with DNA and other evidence?

2. Identify three problems in the justice system that account for false testimony that can lead to a wrongful conviction.

3. How can we calculate the loss for individuals and communities of men spending 13 to 20 years in prison for a crime they did not commit because of the failure of the criminal justice system?

Student Exercises

1. Think about your first experience of finding employment. Did you start in the informal economy, perhaps babysitting or doing yard work, or in the formal economy, perhaps at a fast food restaurant or a retail job? At what age did you start to work? How much money did you earn, and what did you do with it? What role did your community or your family network play in your securing employment? Once you have thought about your history and these questions, share your experiences with the whole class, or in groups of eight or ten, so that there will be some diversity of experiences heard. How do you think your own work experience is related to your race, gender, social class, and residential location? What lessons can you learn from this exercise about how race, gender, social class, residential location, and other social factors are related to a young person's job search?

2. Ideas about relationships outside of one's own race and ethnicity are changing. We are also seeing a growing number of people who are multiracial or multiethnic. What is happening in your own family? What are your attitudes about being friends with someone from another race and ethnicity group? What about dating or perhaps marrying someone from another race and ethnicity group? How do people in your friendship circle think? What about your parents' expectations and possible acceptance? Families have many different reasons for wanting their children to marry within their racial or ethnic group, while others are more readily supportive of their children having relationships outside of their groups. What are the dynamics in your own family? Can you relate your family's position to historical or contemporary trends, such as patterns of segregation and integration, as well as to your family's history?

PART VII

Moving Forward

Building a Just Society

ELIZABETH HIGGINBOTHAM
AND MARGARET L. ANDERSEN

Racial hierarchies, like all social hierarchies, change over time, but how do they change? They can change because the political economy needs different groups of workers with particular skills, which can result in dominant groups changing their views of others. For example, African Americans were barred from industrial work as it developed in the late 19th century, but were then recruited for those same jobs when European immigration was curtailed by legislation in the early 20th century. Access to industrial work not only changed their opportunities but also brought them into the cities in great numbers and gave them the resources to create new representations of themselves, as in the flowering of the Harlem Renaissance—a period spanning the 1920s and into the 1930s when there was a burst of African American culture centered in urban areas, such as Harlem (Marks 1989).

More often, however, change occurs because of actions that oppressed people take. Dominant groups use their economic and political power to develop ideologies that make the current racial order appear natural. Disadvantaged groups in turn mobilize not only to challenge those ideologies but also perhaps to change laws and create a more inclusive and just society. Such actions show us that, although systems of inequality may be established and protected by those with the most power, they are not just quietly accepted. Rather, oppressed groups and their allies contest racial injustice. For example, notions of racial and cultural inferiority justified slavery, the mistreatment of American Indians, and the denial of rights to Chinese immigrants. People in these groups interpreted their situations differently. Many slaves rejected the system, escaping to form maroon colonies in the South; some engaged in armed rebellions; others resisted through whatever means possible, including traveling to the northern United States and Canada for freedom. American Indians knew that European settlers had guns, while they had bows and arrows; they also knew that the new laws of the land gave advantages to White people, who had economic and political power. Still, Native Americans contested and challenged these views and continue to do so now. Chinese Americans, historically denied the opportunity to become citizens, used the state and federal court systems to fight the erosion of their rights as human beings (Takaki 1993). All oppressed racial and ethnic groups have a long history of active resistance, even as the elites used the power of government to control others.

Change can also occur because of changing demographic conditions in society, such as a high rate of immigration, a slowing of the birthrate, and other factors. Currently in the United States, change in race and ethnic relations is coming from the increased diversity in the population—the result of immigration, intermarriage, and the growing number of people who identify as multiracial. Multiracialism changes the traditional "color line" that has divided American society into mostly a Black-White dichotomy. Racial lines are now more fluid— and will likely continue to be, a point made in the opening article of this section by Jennifer Lee and Frank D. Bean ("Reinventing the Color Line: Immigration and America's New Racial/Ethnic Divide").

New questions arise from the change in the so-called color line in U.S. society. Will the nation become more accepting of people of color? Will some people of color become "whitened" while others remain "other" or "blackened" by their ethnicity? Some argue that there will be a "tri-racial divide," as Lee and Bean point out (Bonilla-Silva 2004), while others think that the category "Black" will be expanded to include some Latinos/as, Native American, and other immigrant groups who fall outside the bounds of perceived whiteness.

These questions raise new possibilities for movements for social justice, even while challenging how we think about race and how we work to eliminate its deleterious effects. Belief in a color-blind society is not enough … not as long as racial inequalities persist. What kind of mobilization will transform society?

Reducing hate is one answer, as articulated in the article here from the Southern Poverty Law Center ("Ten Ways to Fight Hate"). This Center, long focused on monitoring and eliminating hate groups in the United States, offers some actions for individuals to take. They offer ten actions that anyone can take to have an immediate impact on those around them. The privileges of many people are often invisible. People saddled with disadvantages can usually see the privilege and advantage of members of the dominant group who frequently do not recognize their own advantages. Changing the perspectives of all parties is important in working for social change.

Change, then, must happen at the individual level, but that is not enough. Sometimes people mobilize for collective action, as was certainly true throughout the long history of the civil rights movement and in other movements where people have fought for freedom and inclusion. Change efforts continue now through the work of many organizations, communities, and individual efforts. One example is provided here in the article by David Naguib Pellow and Robert J. Brulle ("Poisoning the Planet: The Struggle for Environmental Justice"). Pellow and Brulle document that people of color are more likely than other groups to live near hazardous waste sites. Pellow and Brulle also show, however, how people of color have mobilized through an **environmental justice movement**, the term used to refer to the broad coalition of local groups that take action against the disposal of toxic waste in their communities. Pellow and Brulle also show how people who are denied participation in the mainstream society build their own communities. Racial segregation can foster community building when people in segregated communities question the controlling images imposed by the dominant group to justify inequality. Within minority communities there have long been alternative explanations for their lack of resources and opportunities. Such messages contradict explanations from the ruling system of power that typically defines people of color as somehow less able. Alternative explanations identify the structural sources of inequality and the power differentials; such alternatives to dominant ways of thinking can motivate oppressed people and their allies to demand change. Disadvantaged groups also often build communities to survive and raise the next generation. Through such actions, people challenge dominant images and beliefs that better represent themselves and foster new visions of possibilities (Takaki 1993).

Change through legal reform has been another avenue for moving toward racial justice. Major legal changes in such laws as the Civil Rights Act of 1964 and the Voting Rights Act of 1965 have been critical in dismantling some of the most egregious forms of racial oppression. Vigilance is needed though as some of the most important changes in law could be eroded by further resistance to change. Fighting discrimination is important, but as Kimberlé Williams Crenshaw argues in the part of her law review article reprinted here ("Race, Reform, and Retrenchment: Transformation and Legitimation in Antidiscrimination Law"), antidiscrimination law may not go far enough. She points to two different views within antidiscrimination law: the restrictive view and the expansive view. Her argument asks whether law in and of itself is enough to guarantee the rights of citizens.

Finally, as we see in this book's final article, by E. Earl Parson and Monique McLaughlin ("Citizenship in Name Only: The Coloring of Democracy While Redefining Rights, Liberties, and Self Determination for the 21st Century"), some of the progressive changes of the past—for example, the Voting Rights Act of 1965—can be eroded by practices that are ongoing and that suppress the votes of people of color. For a democracy to work, all citizens must participate in the democratic process. When powerful interest groups impede the political participation of some, as documented in Parson and McLaughlin's article, democracy fails. Especially as the population is becoming more diverse with people of color and their allies committed to a more fully inclusive society, powerful people in the dominant group will often fight to preserve their interests. Voting is one way that people can make change, but as the articles here collectively show, change has to happen on many fronts to be effective. All of us have our own strengths to draw upon in promoting change. Our hope is that education through a book like this can be one part of that complex transition toward racial justice.

REFERENCES

Bonilla-Silva, Eduardo. 2004. "From Bi-racial to Tri-racial." *Ethnic and Racial Studies* 27 (6): 931–950.

Marks, Carole. 1989. *Farewell—We're Good and Gone: The Great Black Migration.* Bloomington: Indiana University Press.

Takaki, Ronald. 1993. *A Different Mirror: A History of Multicultural America.* Boston: Little, Brown and Company.

FACE THE FACTS: RACIAL PERCEPTIONS OF WHAT NEEDS TO BE DONE

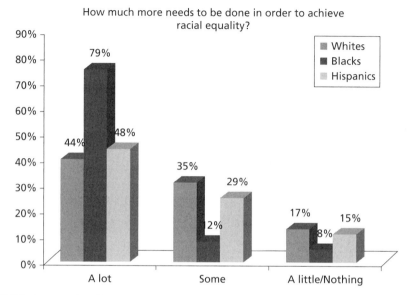

How much more needs to be done in order to achieve racial equality?

SOURCE: Pew Research Social & Demographic Trends. 2014. "King's Dream Remains an Elusive Goal: Many Americans See Racial Disparities" (August 22, 2013). Washington, DC: Pew Research Center. www.pewsocialtrends.org

Think about It: What patterns do you see in the national survey data presented in this graph? Why do you think there are such sharp differences in what people perceive as the need for change?

36

Reinventing the Color Line

Immigration and America's New Racial/Ethnic Divide

JENNIFER LEE* AND FRANK D. BEAN

The authors here discuss some of the implications of the rapidly changing and increasingly multiracial U.S. population. They query how this transformation will change the traditional Black–White "color line" that has characterized American society.

In 1903, the prominent African American social theorist W.E.B. Du Bois prophesied that the "problem of the twentieth-century is the problem of the color line," by which he meant the relatively impermeable bicategorical black-white fault line that had historically divided the country ([1903] 1997:45). Owing to the practice of slavery, the persistence of white prejudice and discrimination resulting from slavery, and the legacy of black social and economic disadvantage, the central organizing principle of race/ethnic relations in the United States has revolved around the axis of the black-white color line (Bobo 1997; Brown et al. 2000; Drake and Cayton [1945] 1993; Myrdal 1944). This delineation consigns blacks and whites to different positions in the social order and attaches a different set of rights and privileges to each group. The unique deprivations imposed on blacks and the tensions spawned by the uneasy history of black-white relations provided stark reminders of the strength of the divide throughout much of the 20th century.

During the latter third of the 20th century, however, the United States moved far beyond black and white, partly as a result of the 1965 Hart-Celler Act—legislation that eliminated national origin quotas and opened the nation's doors to increased flows of nonwhite immigrants. Today, immigrants and their children number almost 66 million, or about 23 percent of the U.S. population (... Lee and Bean 2004; U.S. Bureau of Census 2002). Unlike the immigrants who arrived at the turn of the 20th century, today's newcomers are mainly non-European. The shift in national origins—from Europe to Latin America,

We wish to thank the Russell Sage Foundation for generously supporting the research on which this paper is based (#88-01-11)....

SOURCE: Lee, Jennifer, and Frank D. Bean. 2007. "Reinventing the Color Line: Immigration and America's New Racial-Ethnic Divide." *Social Forces* 86 (December): 561–586.

Asia and the Caribbean—is the single most distinctive aspect of "new immigration" in the United States (Bean and Bell-Rose 1999; Waldinger and Lee 2001). Today's new arrivals have left an indelible imprint on the nation's racial/ethnic scene, transforming it from a largely black-white society at the end of World War II to one now consisting of multiple racial and new nonwhite ethnic groups (Alba and Nee 2003; Bean and Stevens 2003; Sears et al. 2003). In 1970, Latinos and Asians comprised only 5 and 1 percent of the nation's population respectively, but in 2005, these percentages rose to 13 and 4 percent. Moreover, America's Latino and Asian populations are continuing to expand. According to National Research Council projections, by the year 2050, they are likely to constitute about 25 and 8 percent of the U.S. population respectively (Smith and Edmonston 1997).

While today's immigration dramatizes the analytical inadequacy of the black-white color line, other social trends are also augmenting the racial/ethnic diversity of the United States, most notably the rise in intermarriage and the growth of the multiracial population. Intermarriage soared more than 20-fold over a 40-year period, from 150,000 such marriages in 1960 to 3.1 million in 2000 (Jacoby 2001 ...). Today, about 13 percent of American married couples involve someone whose partner is of a different race, a significant increase from earlier lower levels that cannot be attributed to changing racial composition alone (Bean and Stevens 2003). In turn, the upswing in interracial marriage is responsible in large part for a growing multiracial population, which became highly visible when the 2000 U.S. Census allowed Americans to identify themselves as belonging to more than one race. Currently, 1 in 40 Americans identifies himself or herself as multiracial, and by the year 2050, this ratio could soar to one in five (Farley 2002; Smith and Edmonston 1997).

Each of these phenomena—increasing nonwhite racial/ethnic diversity occurring through immigration, rising intermarriage, and the growing multiracial population—suggests that the traditional black-white color line may be losing salience. Given that today's immigrant newcomers from Latin America and Asia may not identify themselves as either black or white, the traditional black-white model of race relations may not adequately depict the character of race/ethnic relations for Asians and Latinos nor accurately portray the structure of today's color line. Consequently, a pressing question in the current sociology of race/ethnic relations is: are the incorporation experiences of America's newest nonwhite immigrant groups tracking those of their European predecessors, or are these groups becoming racialized minorities who see their experiences as more akin to those of African Americans? In other words, do Asians and Latinos more closely resemble whites or blacks in the United States at this point in time? Answers to such questions will help to reveal whether the black-white color line of the past is morphing into a white-nonwhite divide, a black-nonblack divide, or a new tri-racial hierarchy. If the problem of the 20th century was the color line, the question of the 21st century could be one of multiple color *lines*....

THEORY AND PREVIOUS RESEARCH

A White-Nonwhite Divide

One possible emergent color line might be a white-nonwhite divide. Such a divide has been legally enforced throughout the history of the United States, well into the 20th century. In 1924, for example, the state of Virginia passed a Racial Integrity Law that created two distinct racial categories: "pure" white and all others. The statue defined a "white" person as one with "no trace whatsoever of blood other than Caucasian," and emerged to legally ban intermarriage between whites and other races. While blacks were clearly nonwhite under the legislation, Asians and Latinos were also consigned to the nonwhite side of the strict binary divide. The statute reflected the Supreme Court rulings of *Takao Ozawa v. United States* (1922) and *United States v. Bhagat Singh Thind* (1923), in which persons of Asian origin were not only classified as nonwhite but also considered ineligible for U.S. citizenship. In the first case, Takao Ozawa (a Japanese citizen of the United States) filed for U.S. citizenship under the Naturalization Act of June 29, 1906, which allowed whites and persons of African descent or African nativity to naturalize. Rather than challenging the constitutionality of the racial restrictions to U.S. citizenship, Ozawa argued that his skin color made him a "white person" and that Japanese persons should be classified as "white." The Supreme Court ruled that only Caucasians were white, and because the Japanese were not of the Caucasian race, they were not white but rather members of an "unassimilable race."…

Solidifying the placement of nonwhites with blacks were administrative policies adopted in the latter half of the 1960s following the Civil Rights movement. Most prominently, Civil Rights administrators extended affirmative action policies to minority groups they perceived as "analogous to blacks" with respect to physical distinctiveness and to having "suffered enough" to be similarly categorized (Skrentny 2002). According to these criteria, Latinos, Native Americans and Asians became eligible for affirmative action programs while disadvantaged white ethnics did not. Perhaps one unintended consequence of such policies was that Latinos and Asians were identified as racialized minorities more akin to blacks than to whites. In essence, these federal policies placed Asians and Latinos on the nonwhite side of the divide, helping to foster a delineation between whites and nonwhites.

Further cementing the divide was the introduction of the label "people of color," which gained momentum and popularity in the late 1980s (Hollinger 2005). This umbrella term combines all nonwhite groups on the basis of their racialized minority status and connotes that they share a similar subordinate status vis-à-vis whites. By homogenizing the experiences of all nonwhite groups, the "people of color" rubric indicates that the boundaries among nonwhite groups are less distinct and salient than the boundary separating whites and nonwhites. Accordingly, in a white-nonwhite model of racial/ethnic relations, Asians and Latinos would fall closer to blacks than to whites in their experiences in the United States, suggesting that the reporting of and experiences with multiracial identification should be similar for Asians, Latinos and blacks.…

A Tri-Racial Divide

While some social scientists propose that America's color line will reflect a binary structure (i.e., white-nonwhite or black-nonblack), others offer an alternative possibility—a tri-racial stratification system similar to that of many Latin American and Caribbean countries. In the United States, this would be viewed as consisting of whites, honorary whites and collective blacks (Bonilla-Silva 2004a, 2004b). Included in the "white" category would be whites, assimilated white Latinos, some multiracials, assimilated Native Americans and a few Asian-origin people. "Honorary whites" would include light-skinned Latinos, Japanese Americans, Korean Americans, Chinese Americans, Asian Indians, Middle Eastern Americans and most multiracials. Finally, the "collective black" category would include blacks, Filipinos, Vietnamese, Hmong, Laotians, dark-skinned Latinos, West Indian and African immigrants, and reservation-bound Native Americans.

Because many of today's new immigrants hail from Latin America and the Caribbean, Bonilla-Silva argues that a more complex tri-racial order may naturally emerge given the "darkening" of the United States. In his view, a tri-racial order would also serve to help maintain "white supremacy" by creating an intermediate racial group to buffer racial conflict (Bonilla-Silva 2004b:5). While a few new immigrants might fall into the honorary white strata and may even eventually become white, the majority would be consigned to the collective black strata, including most Latino immigrants whom he labels as "racial others" or persons whose experiences with race are similar to those of blacks. In this regard, the tri-racial model differs fundamentally from the black-nonblack divide because Bonilla-Silva posits that many Latinos are racialized in a manner similar to African Americans, and therefore fall on the black side of the divide....

Data and Methods

To assess the placement of the new color line, we examine both 2000 U.S. Census data and 46 in-depth interviews of multiracial adults carried out in southern and northern California.[1]...

Lasting up to two hours, the interviews were semi-structured, open-ended and tape-recorded. Respondents were asked questions about their racial/ethnic identities and cultural practices: how and why they choose to identify themselves the way they do; whether these identities had changed over time and/or in different contexts; what their identities meant to them; and their cultural and linguistic practices....

RESULTS

America's Multiracial Population

For the first time, the 2000 U.S. Census allowed Americans to mark "one or more" races to indicate racial identification. This was a landmark change in the way the census measures race; it acknowledged the reality of racial mixing and no longer required Americans to claim one race exclusive of all others—a momentous shift considering that the United States has been historically hostile to racial mixture as evidenced by the legal invocation of the "one-drop" rule of hypodescent constraining racial identity options for multiracial blacks (Dalmage 2004; Davis 1991; Farley 2002; Haney-Lopez 1996; Hirschman et al. 2000; Hollinger 2003; Nobles 2000; Waters 2000; Williams 2006). As a result of this change, about 6.8 million Americans, or 2.4 percent of the population, identified themselves or members of their households as multiracial. Although this may not appear large, demographers estimate that multiracials could soar to one in five persons by the year 2050. By then, a recent National Academy of Sciences study estimates that the multiracial population could increase to 21 percent when as many as 35 percent of Asians and 45 percent of Hispanics might have multiracial backgrounds (Smith and Edmonston 1997)....

Wide variations in rates of multiracial reporting also occur across groups. As shown in Table 1, 12 percent of Asians and 16 percent of "Other" Americans (i.e., Latinos) identified multiracially, yet only 4 percent of the black population did. The black rate of multiracial reporting is much lower compared to other groups, even after controlling for differences in age, education, nativity, gender and region of the United States (Tafoya et al. 2005). Moreover, while the Census Bureau estimates that at least three-quarters of blacks in the United States are ancestrally multiracial, just over 4 percent choose to identify as such, indicating that most black Americans do not depend strictly on their genealogy to identify themselves, but instead, rely on the social construction of racial boundaries. That the rate of multiracial reporting is three to four times higher among Asians and Latinos than among blacks suggests that the historical absence of the constraining "one-drop" rule for these groups may provide more leeway in

T A B L E   1   **Multiracial Identification by Census Racial Categories**

	Racial Identification[a]	Multiracial Identification[b]	% Multiracial
White	216.5	5.1	2.3
Black	36.2	1.5	4.2
Asian	11.7	1.4	12.4
Other	18.4	3.0	16.4
American Indian and Alaska Native	3.9	1.4	36.4
Native Hawaiian or Other Pacific Islander	.7	.3	44.8

NOTE: All racial and multiracial identification figures in millions.

[a]Racial/Ethnic group totals do not sum to the total U.S. population because multiracial persons are counted here in more than one group.

[b]Multiracial persons are counted for each race category mentioned.

SOURCE: U.S. Census 2000.

exercising discretion in the selection of racial/ethnic identities (Harris and Sim 2002; Xie and Goyette 1997)....

Outsiders' Ascription and the Inclusivity/Exclusivity of Racial Categorization

Based on the interviews, we find that multiracial blacks are less likely to identify multiracially compared to their Asian and Latino counterparts, in large part, because of outsiders' ascription, which powerfully influences one's choice of identities. Sociologists have noted that racial/ethnic identity is a dialectical process—one that involves both internal and external opinions and processes (Nagel 1994; Rodríguez and Cordero-Guzman 1992; Waters 1990, 1999). Researchers have also shown that outsiders' ascription most powerfully constrains the racial/ethnic options for blacks. While blacks in the United States make distinctions based on ethnicity, class, nativity and skin tone, the power of race—and blackness in particular—often overrides these internal differences (Kasinitz 1992; Waters 1999).

Multiracial blacks are less likely to identify as such, in part, because others identify them as black. For example, when we asked a 33-year-old woman born to a white mother and black father why she chose to identify as black on the census form, she explained,

> "I feel if somebody is going to look at me, they're not going to think I'm white so I put black... I don't think I'd identify as white very often, but I guess if it's very specific then I'm going to indicate that I'm both black and white. I mean, I know that I'm mixed, but if it were to come up, and it were to be a choice, one or the other, I would say I'm black."

Other multiracial adults with one black and one white parent echoed similar sentiments. While they recognize the racial mixture in their backgrounds, they choose to identify as black because, as a 26-year-old male notes, "I think the main reason I identify as black is if someone looks at me, I don't really necessarily look white." Here, he maintains that if one looks black, one cannot be white. So powerful is the force of outsiders' ascription that he chooses to identify his son, whom he conceived with a white woman, as black rather than as multiracial or white....

By comparison, multiracial Asian-whites and Latino-whites feel that they have much more leeway to choose among different racial options, including multiracial and white identities. Some choose to identify as half-Asian or half-Latino and half-white, and more importantly, others do not challenge these identities, nor do they automatically ascribe a monoracial Asian or Latino identity to them. Instead, their multiracial identities are more readily accepted than the multiracial identities for blacks. Moreover, unlike black-white multiracials, Latino-white and Asian-white multiracials are often identified as white, which, in turn, affects the way they see themselves. For example, many of the multiracial Latino-whites feel that they look "white" without a hint of Latino ethnicity. Their perception that they look white is reinforced by others who are shocked to learn they have a Latino parent, as a 23-year-old Mexican-white multiracial woman explains,

> "I feel like I'm white with a hint of Mexican. That's not usually what I identify with, and that's not how people identify me either. I feel mostly Caucasian, but I do have a Mexican background and family and heritage, but I identify with being white more just because that's the way I look. I mean, people are always surprised to find that my Mom is Mexican. They say, 'Oh my god, I never would have known. You look like a total white girl!'"

Similarly, another young female, born to a white father and Mexican mother explains that others (including other Mexicans) often assume that she is non-Mexican because of her white phenotype, as she relays,

> "I don't look Mexican, but I feel it within me. I cook Mexican food, and I listen to my Spanish stations. But a lot people identify me as white because my appearance. I always shock them when I say, 'Oh yeah, I'm Mexican.' I go to a restaurant or food place that's Mexican and always order in Spanish and they go, 'Oh, habla Espanol?', and I go, 'Si!' and then we start speaking."

This type of response was typical of many of the Latino-white multiracials who we interviewed.... The surprised reaction that Latino multiracials receive stems, in part, from the fact that many non-Latinos have a very narrow vision of what a Latino *should* look like. While non-Latinos may believe that there is a stereotypical Latino or Mexican look, Latinos recognize that many do not fit this stereotypical image, as a multiracial Mexican-white woman elaborates,

> "In Mexico, there's no one look. It's not all dark skin or dark eyes or looking Indian. One of my sisters has green eyes, and my other sisters have fair skin. I think that here, people think all Mexicans have dark skin and brown eyes. In Mexico you see a lot of kids that look like Anthony [her son] or are lighter. There's this concept—stereotype—that you look Mexican."...

SYMBOLIC IDENTITIES

For the Asian-white and Latino-white multiracial respondents, claiming a white racial identity does not preclude them from also claiming an Asian or Latino ethnicity; they can be white, yet also be Asian Indian, Japanese, Hispanic or Mexican, signifying that Asian and Latino ethnicities are adopting the symbolic character of European ethnicity for white Americans. By contrast, the black multiracials we interviewed have not been able to do the same; they have not been able to claim a white or nonblack racial identity and have those identities accepted by others, signaling that black remains a relatively fixed racialized category. The experiences of Asian-white and Latino-white multiracials thus differ starkly from those of black multiracials. Not only are Latinos and Asians more likely to report multiracial identifications, but these multiracials are more likely to describe their Asian and Latino identities as being voluntary and optional rather than ascribed and instrumental, suggesting that the Asian and Latino identities are adopting the symbolic character of white ethnicity.

CONCLUSIONS AND DISCUSSION

What do the patterns of multiracial identification suggest about the future of America's color line? The findings indicate that group boundaries appear to be fading more rapidly for Latinos and Asians than for blacks, signaling that today's new nonwhites are *not* strongly assimilating as racialized minorities who see their experiences with race as akin to those of blacks, as would be predicted by the white-nonwhite divide model. Moreover, a tri-racial hierarchy model that would place Latinos and most new immigrants into the "collective black" category and label them as "racial others" does not seem to accurately characterize the racialization process of America's nonwhite newcomers. Instead, experiences with multiraciality among Latinos and Asians are closer to those of whites than to blacks. Furthermore, that racial and ethnic affiliations and identities are much less matters of choice for multiracial blacks indicates that black remains a significant racial category. The lower rate of black multiracial reporting and the racial constraints that many multiracial blacks experience suggest that blackness continues to constitute a fundamental racial construction in American society. Hence, it is not simply that race matters, but more specifically that *black* race matters, consistent with the African American exceptionalism thesis.

The findings thus suggest that a black-nonblack divide is taking shape, in which Asians and Latinos are not only closer to whites than blacks are to whites, but also closer to whites than to blacks at this point in time (Gans 1999, 2005; Glazer 1997; Lee and Bean 2007; Quillian and Campbell 2003; Sears 2003; Sears et al. 2003; Waters 1999; Yancey 2003). Hence, America's color line may have moved toward a new demarcation that places many blacks in a position of disadvantage similar to that resulting from the traditional black-white divide. In essence, rather than erasing racial boundaries, the country may simply be reinventing a color line that continues to separate blacks from other racial/ethnic groups.

While a black-nonblack divide may depict the color line at the moment, it is also possible that a black-white divide might re-emerge. Whiteness as a category has expanded over time to incorporate new immigrant groups in the past, and it appears to be stretching yet again (Gallagher 2004; Gerstle 1999; Warren and Twine 1997). Based on patterns of multiracial identification, Asians and Latinos may be the next in line to be white, with multiracial Asian-whites and Latino-whites at the head of the queue. If this is the case, a black-white line may re-emerge, and Du Bois' century-old forecast may become relevant once again. However, regardless of whether the divide falls along black-nonblack or black-white lines, the position of blacks remains the same.

This is ominous because a color line that more strongly separates whites from blacks than one that divides whites from other groups invites misinterpretation about progress of black-white relations in the United States. Because boundaries are loosening for *some* nonwhite groups, this could lead to the erroneous conclusion that race is declining in significance for *all* groups or that relations are improving at the same pace for *all* racial/ethnic minorities. However, the results of the present research suggest that the social construction of race is more rigid for blacks than for Asians and Latinos. Not accounting for this difference could easily lead to the endorsement of the flawed logic that if race does not significantly impede the process of incorporation for Asians and Latinos, then it must not matter much for blacks either. Not only is this line of reasoning incorrect, it also risks creating specious support for so-called "color-blind" policies that fail to recognize that race and the color line have different consequences for different minority groups (Bobo 1997; Brown et al. 2003; Guinier and Torres 2002; Loury 2002).

Moreover, a logic of presumed "color blindness" also risks overlooking the fact that boundary maintenance and change are two-sided processes that involve both choice, and perhaps more importantly, constraint (Alba 1999; Bobo 1997; Lamont 2000). This means that not only must members of racial/ethnic minority groups pursue entry and incorporation into social contexts occupied by the majority group, but also that members of the majority group must be willing to accept their admission. Based on patterns of multiracial reporting, it appears that Asians and Latinos are more actively pursuing entry into the majority group, and that whites are more willing to accept their entry compared to blacks. At this time, the boundaries for Asians and Latinos appear more elastic than they seem for blacks, consequently reinforcing the racial stigma attached to blackness (Loury 2002). The fact that boundary dissolution is neither uniform nor unconditional indicates that we cannot be complacent about the degree to which opportunities are improving for all racial/ethnic groups in the United States, particularly when a deep and persistent divide continues to separate blacks from all other groups.

NOTE

1. For purposes of discussion and analysis, we employ the often used terms Asian, Latino and black even though we recognize that these categories are socially constructed and a great deal of ethnic heterogeneity exists within them.

REFERENCES

Alba, Richard. 1999. "Immigration and the American Realities of Assimilation and Multiculturalism." *Sociological Forum* 14(1):3–25.

Alba, Richard, and Victor Nee. 2003. *Remaking the American Mainstream.* Harvard University Press.

Bean, Frank D., and Stephanie Bell-Rose. Editors. 1999. *Immigration and Opportunity*. Russell Sage Foundation.

Bean, Frank D., and Gillian Stevens. 2003. *America's Newcomers and the Dynamics of Diversity*. Russell Sage Foundation.

Bobo, Lawrence D. 1997. "The Color Line, the Dilemma, and the Dream." Pp. 31–55. *Civil Rights and Social Wrongs*. John Higham, editor. Pennsylvania State University Press.

Bonilla-Silva, Eduardo. 2004a. "From Bi-racial to Tri-racial." *Ethnic and Racial Studies* 27(6):931–50.

——. 2004b. "We are all Americans." *Race & Society* 5(1):3–16.

Brown, Michael K., Martin Carnoy, Elliot Currie, Troy Duster, David B. Oppenheimer, Marjorie M. Shultz and David Wellman. 2003. *Whitewashing Race*. University of California Press.

Dalmage, Heather M. Editor. *The Politics of Multiracialism*. State University of New York Press.

Davis, F. James. 1991. *Who is Black?* Pennsylvania State University Press.

Drake, St. Clair, and Horace R. Cayton. [1945] 1993. *Black Metropolis*. University of Chicago Press.

Du Bois, W.E.B. [1903] 1997. *The Souls of Black Folk*. David W. Blight and Robert Gooding-Williams, editors. Bedford Books.

Farley, Reynolds. 2002. "Racial Identities in 2000." Pp. 33–61. *The New Race Question*. Joel Perlmann and Mary C. Waters, editors. Russell Sage Foundation.

Gallagher, Charles A. 2004. "Racial Redistricting: Expanding the Boundaries of Whiteness." Pp. 59–76. *The Politics of Multiracialism*. Heather M. Dalmage, editor. State University of New York Press.

Gans, Herbert J. 2005. "Race as Class." *Contexts* 4(4):17–21.

——. 1999. "The Possibility of a New Racial Hierarchy in the Twenty-first Century United States." Pp. 371–90. *The Cultural Territories of Race*. Michèle Lamont, editor. University of Chicago Press and Russell Sage Foundation.

Gerstle, Gary. 1999. "Liberty, Coercion, and the Making of Americans." Pp. 275–93. *The Handbook of International Migration*. Charles Hirschman, Philip Kasinitz and Josh DeWind, editors. Russell Sage Foundation.

Glazer, Nathan. 1997. *We Are All Multiculturalists Now*. Harvard University Press.

Guinier, Lani, and Gerald Torres. 2002. *The Miner's Canary*. Harvard University Press.

Haney-López, Ian F. 1996. *White by Law*. New York University Press.

Harris, David R., and Jeremiah Joseph Sim. 2002. "Who Is Multiracial?" *American Sociological Review* 67(4):614–27.

Hirschman, Charles, Richard Alba and Reynolds Farley. 2000. "The Meaning and Measurement of Race in the U.S. Census." *Demography* 37(3):381–93.

Hollinger, David A. 2005. "The One-Drop Rule and the One-Hate Rule." *Daedalus* 134(1):18–28.

——. 2003. "Amalgamation and Hypodescent." *American Historical Review* 108(5): 1363–90.

Jacoby, Tamar. 2001. "An End to Counting Race?" *Commentary* 111(6):37–40.

Kasinitz, Philip. 1992. *Caribbean New York*. Cornell University Press.

Lamont, Michèle. 2000. *The Dignity of Working Men*. Harvard University Press and Russell Sage Foundation.

Lee, Jennifer, and Frank D. Bean. 2007. "Redrawing the Color Line?" *City & Community* 6(1):49–62.

———. 2004. "America's Changing Color Lines." *Annual Review of Sociology* 30:221–42.

Loury, Glenn C. 2002. *The Anatomy of Racial Inequality*. Harvard University Press.

Myrdal, Gunnar. 1944. *An American Dilemma*. Harper.

Nagel, Joane. 1994. "Constructing Ethnicity." *Social Problems* 41(1): 152–76.

Nobles, Melissa. 2000. *Shades of Citizenship*. Stanford University Press.

Quillian, Lincoln, and Mary Campbell. 2003. "Beyond Black and White." *American Sociological Review* 68(4):540–66.

Rodríguez, Clara E., and Hector Cordero-Guzman. 1992. "Placing Race in Context." *Ethnic and Racial Studies* 15(4):523–42.

Sears, David O. 2003. "Black-White Conflict." Pp. 367–89. *New York & Los Angeles*. David Halle, editor. University of Chicago Press.

Sears, David O., Mingying Fu, P.J. Henry and Kerra Bui. 2003. "The Origins and Persistence of Ethnic Identity among the 'New Immigrant' Groups." *Social Psychology Quarterly* 66(4):419–37.

Skrentny, John D. 2002. *The Minority Rights Revolution*. Harvard University Press.

Smith, James P., and Barry Edmonston. 1997. *The New Americans*. National Academy Press.

Tafoya, Sonya M., Hans Johnson and Laura E. Hill. 2005. "Who Chooses to Choose Two?" Pp. 332–51. *The American People: Census 2000*. Reynolds Farley and John Haaga, editors. Russell Sage Foundation.

U.S. Bureau of the Census. 2001. *United States Census 2000*. Washington, D.C.: U.S. Government Printing Office.

U.S. Immigration and Naturalization Service. 2002. *2000 INS Statistical Yearbook*. Washington, D.C.: U.S. Government Printing Office.

Waldinger, Roger, and Jennifer Lee. 2001. "New Immigrants in Urban America." Pp. 30–79. *Strangers at the Gates*. Roger Waldinger, editor. University of California Press.

Warren, Jonathan W., and France Winddance Twine. 1997. "White Americans, the New Minority?" *Journal of Black Studies* 28(2):200–218.

Waters, Mary C. 2000. "Immigration, Intermarriage, and the Challenges of Measuring Racial/Ethnic Identities." *American Journal of Public Health* 90(11):1735–37.

———. 1999. *Black Identities*. Harvard University Press.

———. 1990. *Ethnic Options*. University of California Press.

Williams, Kim. 2006. *Mark One or More*. University of Michigan Press.

Xie, Yu, and Goyette, Kimberly. 1997. "The Racial Identification of Biracial Children with One Asian Parent." *Social Forces* 76(2):547–70.

Yancey, George. 2003. *Who Is White?* Lynne Rienner.

DISCUSSION QUESTIONS

1. What changes in society are transforming the historic color line in the
    United States? Will the color line be maintained or fade away? What would
    make either possibility happen?

2. What does it take to become "whitened," as Lee and Bean suggest? What
    does this tell you about the social construction of race in this nation?

37

Ten Ways to Fight Hate

A Community Response Guide

SOUTHERN POVERTY LAW CENTER

Many social problems are centuries old, even if they have changed in their scope. What can an individual do to make a difference? The Southern Poverty Law Center has long worked for social justice, and they offer ten suggestions to fight hate and make a difference in your own community.

1 ACT

Do something. In the face of hatred, apathy will be interpreted as acceptance—by the perpetrators, the public and, worse, the victims. Decent people must take action; if we don't, hate persists.

2 UNITE

Call a friend or co-worker. Organize allies from churches, schools, clubs and other civic groups. Create a diverse coalition. Include children, police and the media. Gather ideas from everyone, and get everyone involved.

3 SUPPORT THE VICTIMS

Hate-crime victims are especially vulnerable, fearful and alone. If you're a victim, report every incident—in detail—and ask for help. If you learn about a hate-crime victim in your community, show support. Let victims know you care. Surround them with comfort and protection.

SOURCE: Southern Poverty Law Center. 2010. *Ten Ways to Fight Hate: A Community Response Guide.* Montgomery, Alabama.

4 DO YOUR HOMEWORK

An informed campaign improves its effectiveness. Determine if a hate group is involved, and research its symbols and agenda. Understand the difference between a hate crime and a bias incident.

5 CREATE AN ALTERNATIVE

Do not attend a hate rally. Find another outlet for anger and frustration and for people's desire to do something. Hold a unity rally or parade to draw media attention away from hate.

6 SPEAK UP

Hate must be exposed and denounced. Help news organizations achieve balance and depth. Do not debate hate-group members in conflict-driven forums. Instead, speak up in ways that draw attention away from hate, toward unity.

7 LOBBY LEADERS

Elected officials and other community leaders can be important allies in the fight against hate. But some must overcome reluctance—and others, their own biases—before they're able to take a stand.

8 LOOK LONG RANGE

Promote tolerance and address bias before another hate crime can occur. Expand your community's comfort zones so you can learn and live together.

9 TEACH TOLERANCE

Bias is learned early, usually at home. Schools can offer lessons of tolerance and acceptance. Sponsor an "I Have a Dream" contest. Reach out to young people who may be susceptible to hate-group propaganda and prejudice.

10 DIG DEEPER

Look inside yourself for prejudices and stereotypes. Build your own cultural competency, then keep working to expose discrimination wherever it happens—in housing, employment, education and more.

DISCUSSION QUESTIONS

1. Can you think about a time when you took action in the face of an injustice? How did you feel about the action that you took? Was your work effective? Why or why not?

2. Why do you think the Southern Poverty Law Center suggests that people organize rather than acting alone?

38

Poisoning the Planet

The Struggle for Environmental Justice

DAVID NAGUIB PELLOW AND ROBERT J. BRULLE

Pellow and Brulle identify how the neighborhoods of people of color are often the dumping grounds for toxic waste and other pollutants. These are hazards to people's health and well-being. The authors discuss how people are fighting back as part of an environmental justice movement. They place the activism of people of color within a context of national and international racial politics.

One morning in 1987 several African-American activists on Chicago's southeast side gathered to oppose a waste incinerator in their community and, in just a few hours, stopped 57 trucks from entering the area. Eventually arrested, they made a public statement about the problem of pollution in poor communities of color in the United States—a problem known as environmental racism. Hazel Johnson, executive director of the environmental justice group People for Community Recovery (PCR), told this story on several occasions, proud that she and her organization had led the demonstration. Indeed, this was a remarkable mobilization and an impressive act of resistance from a small, economically depressed, and chemically inundated community. This community of 10,000 people, mostly African-American, is surrounded by more than 50 polluting facilities, including landfills, oil refineries, waste lagoons, a sewage treatment plant, cement plants, steel mills, and waste incinerators. Hazel's daughter, Cheryl, who has worked with the organization since its founding, often says, "We call this area the 'Toxic Doughnut' because everywhere you look, 360 degrees around us, we're completely surrounded by toxics on all sides."

THE ENVIRONMENTAL JUSTICE MOVEMENT

People for Community Recovery was at the vanguard of a number of local citizens' groups that formed the movement for environmental justice (EJ). This movement, rooted in community-based politics, has emerged as a significant player at the local, state, national, and, increasingly, global levels. The movement's origins

SOURCE: Pellow, David Naguib, and Robert J. Brulle. 2007. "Poisoning the Planet: The Struggle for Environmental Justice." *Contexts* 6 (February): 37–41.

lie in local activism during the late 1970s and early 1980s aimed at combating environmental racism and environmental inequality—the unequal distribution of pollution across the social landscape that unfairly burdens poor neighborhoods and communities of color.

The original aim of the EJ movement was to challenge the disproportionate location of toxic facilities (such as landfills, incinerators, polluting factories, and mines) in or near the borders of economically or politically marginalized communities. Groups like PCR have expanded the movement and, in the process, extended its goals beyond removing existing hazards to include preventing new environmental risks and promoting safe, sustainable, and equitable forms of development. In most cases, these groups contest governmental or industrial practices that threaten human health. The EJ movement has developed a vision for social change centered around the following points:

- All people have the right to protection from environmental harm.

- Environmental threats should be eliminated before there are adverse human health consequences.

- The burden of proof should be shifted from communities, which now need to prove adverse impacts, to corporations, which should prove that a given industrial procedure is safe to humans and the environment.

- Grassroots organizations should challenge environmental inequality through political action.

The movement, which now includes African-American, European-American, Latino, Asian-American/Pacific-Islander, and Native-American communities, is more culturally diverse than both the civil rights and the traditional environmental movements, and combines insights from both causes.

ENVIRONMENTAL INEQUALITIES

Researchers have documented environmental inequalities in the United States since the 1970s, originally emphasizing the connection between income and air pollution. Research in the 1980s extended these early findings, revealing that communities of color were especially likely to be near hazardous waste sites. In 1987, the United Church of Christ Commission on Racial Justice released a groundbreaking national study entitled *Toxic Waste and Race in the United States*, which revealed the intensely unequal distribution of toxic waste sites across the United States. The study boldly concluded that race was the strongest predictor of where such sites were found.

In 1990, sociologist Robert Bullard published *Dumping in Dixie*, the first major study of environmental racism that linked the siting of hazardous facilities to the decades-old practices of spatial segregation in the South. Bullard found that African-American communities were being deliberately selected as sites for the disposal of municipal and hazardous chemical wastes. This was also one of the first studies to examine the social and psychological impacts of environmental pollution in a community of color. For example: across five communities in Alabama, Louisiana, Texas, and West Virginia, Bullard found that the majority of people felt that their community had been singled out for the location of a toxic facility (55 percent); experienced anger at hosting this facility in their community (74 percent); and yet accepted the idea that the facility would remain in the community (77 percent).

Since 1990, social scientists have documented that exposure to environmental risks is strongly associated with race and socioeconomic status. Like Bullard's *Dumping in Dixie*, many studies have concluded that the link between polluting facilities and communities of color results from the deliberate placement of such facilities in these communities rather than from population-migration patterns. Such communities are systematically targeted for the location of polluting industries and other locally unwanted land uses (LULUs), but residents are fighting back to secure a safe, healthy, and sustainable quality of life. What have they accomplished?

LOCAL STRUGGLES

The EJ movement began in 1982, when hundreds of activists and residents came together to oppose the expansion of a chemical landfill in Warren County, North Carolina. Even though that action failed, it spawned a movement that effectively mobilized people in neighborhoods and small towns facing other LULUs. The EJ movement has had its most profound impact at the local level. Its successes include shutting down large waste incinerators and landfills in Los Angeles and Chicago; preventing polluting operations from being built or expanded, like the chemical plant proposed by the Shintech Corporation near a poor African-American community in Louisiana; securing relocations and home buyouts for residents in polluted communities like Love Canal, New York; Times Beach, Missouri; and Norco, Louisiana; and successfully demanding environmental cleanups of LULUs such as the North River Sewage Treatment plant in Harlem.

The EJ movement helped stop plans to construct more than 300 garbage incinerators in the United States between 1985 and 1998. The steady expansion of municipal waste incinerators was abruptly reversed after 1990. While the cost of building and maintaining incinerators was certainly on the rise, the political price of incineration was the main factor that reversed this tide. The decline of medical-waste incinerators is even more dramatic.

Sociologist Andrew Szasz has documented the influence of the EJ movement in several hundred communities throughout the United States, showing that organizations such as Hazel Johnson's People for Community Recovery were instrumental in highlighting the dangers associated with chemical waste incinerators in their neighborhoods. EJ organizations, working in local coalitions, have had a number of successes, including shutting down an incinerator that was once the largest municipal waste burner in the Western Hemisphere. The movement has made it extremely difficult for firms to locate incinerators, landfills, and related LULUs anywhere in the nation, and almost any effort to expand existing polluting facilities now faces controversy.

BUILDING INSTITUTIONS

The EJ movement has built up local organizations and regional networks and forged partnerships with existing institutions such as churches, schools, and neighborhood groups. Given the close association between many EJ activists and environmental sociologists, it is not surprising that the movement has notably influenced the university. Research and training centers run by sociologists at several universities and colleges focus on EJ studies, and numerous institutions of higher education offer EJ courses. Bunyan Bryant and Elaine Hockman, searching the World Wide Web in 2002, got 281,000 hits for the phrase "environmental justice course," and they found such courses at more than 60 of the nation's colleges and universities.

EJ activists have built lasting partnerships with university scholars, especially sociologists. For example, Hazel Johnson's organization has worked with scholars at Northwestern University, the University of Wisconsin, and Clark Atlanta University to conduct health surveys of local residents, study local environmental conditions, serve on policy task forces, and testify at public hearings. Working with activists has provided valuable experience and training to future social and physical scientists.

The EJ movement's greatest challenge is to balance its expertise at mobilizing to oppose hazardous technologies and unsustainable development with a coherent vision and policy program that will move communities toward sustainability and better health. Several EJ groups have taken steps in this direction. Some now own and manage housing units, agricultural firms, job-training facilities, farmers' markets, urban gardens, and restaurants. On Chicago's southeast side, PCR partnered with a local university to win a federal grant, with which they taught lead-abatement techniques to community residents who then found employment in environmental industries. These successes should be acknowledged and praised, although they are limited in their socio-ecological impacts and longevity. Even so, EJ activists, scholars, and practitioners would do well to document these projects' trajectories and seek to replicate and adapt their best practices in other locales.

LEGAL GAINS AND LOSSES

The movement has a mixed record in litigation. Early on, EJ activists and attorneys decided to apply civil rights law (Title VI of the 1964 Civil Rights Act) to the environmental arena. Title VI prohibits all government and industry programs and activities that receive federal funds from discriminating against persons based on race, color, or national origin. Unfortunately, the courts have uniformly refused to prohibit government actions on the basis of Title VI without direct evidence of discriminatory intent. The Environmental Protection Agency (EPA) has been of little assistance. Since 1994, when the EPA began accepting Title VI claims, more than 135 have been filed, but none has been formally resolved. Only one federal agency has cited environmental justice concerns to protect a community in a significant legal case: In May 2001, the Nuclear Regulatory Commission denied a permit for a uranium enrichment plant in Louisiana because environmental justice concerns had not been taken into account.

With regard to legal strategies, EJ activist Hazel Johnson learned early on that, while she could trust committed EJ attorneys like Keith Harley of the Chicago Legal Clinic, the courts were often hostile and unforgiving places to make the case for environmental justice. Like other EJ activists disappointed by the legal system, Johnson and PCR have diversified their tactics. For example, they worked with a coalition of activists, scholars, and scientists to present evidence of toxicity in their community to elected officials and policy makers, while also engaging in disruptive protest that targeted government agencies and corporations.

NATIONAL ENVIRONMENTAL POLICY

The EJ movement has been more successful at lobbying high-level elected officials. Most prominently, in February 1994, President Clinton signed Executive Order 12898 requiring all federal agencies to ensure environmental justice in their practices. Appropriately, Hazel Johnson was at Clinton's side as he signed the order. And the Congressional Black Caucus, among its other accomplishments, has maintained one of the strongest environmental voting records of any group in the U.S. Congress.

But under President Bush, the EPA and the White House did not demonstrate a commitment to environmental justice. Even Clinton's much-vaunted Executive Order on Environmental Justice has had a limited effect. In March 2004 and September 2006, the inspector general of the EPA concluded that the agency was not doing an effective job of enforcing environmental justice policy. Specifically, he noted that the agency had no plans, benchmarks, or instruments to evaluate progress toward achieving the goals of Clinton's Order. While President Clinton deserves some of the blame for this, it should be no surprise that things have not improved under the Bush administration. In response, many activists, including those at PCR, have shifted their focus from the national level back to the neighborhood, where their work has a more tangible influence and where polluters are more easily monitored. But in an era of increasing economic and political globalization, this strategy may be limited.

GLOBALIZATION

As economic globalization—defined as the reduction of economic borders to allow the free passage of goods and money anywhere in the world—proceeds largely unchecked by governments, as the United States and other industrialized nations produce larger volumes of hazardous waste, and as the degree of global social inequality also rises, the frequency and intensity of EJ conflicts can only increase. Nations of the global north continue to export toxic waste to both domestic and global "pollution havens" where the price of doing business is much lower, where environmental laws are comparatively lax, and where citizens hold little formal political power.

Movement leaders are well aware of the effects of economic globalization and the international movement of pollution and wastes along the path of least resistance (namely, southward). Collaboration, resource exchange, networking, and joint action have already emerged between EJ groups in the global north and south. In the last decade EJ activists and delegates have traveled to meet and build alliances with colleagues in places like Beijing, Budapest, Cairo, Durban, The Hague, Istanbul, Johannesburg, Mumbai, and Rio de Janeiro. Activist colleagues outside the United States are often doing battle with the same transnational corporations that U.S. activists may be fighting at home. However, it is unclear if these efforts are well financed or if they are leading to enduring action programs across borders. What is certain is that if the EJ movement fails inside the United States, it is likely to fail against transnational firms on foreign territory in the global south.

Although EJ movements exist in other nations, the U.S. movement has been slow to link up with them. If the U.S. EJ movement is to survive, it must go global. The origins and drivers of environmental inequality are global in their reach and effects. Residents and activists in the global north feel a moral obligation to the nations and peoples of the south, as consumers, firms, state agencies, and military actions within northern nations produce social and ecological havoc in Latin America, the Caribbean, Africa, Central and Eastern Europe, and Asia. Going global does not necessarily require activists to leave the United States and travel abroad, because many of the major sources of global economic decision-making power are located in the north (corporate headquarters, the International Monetary Fund, the World Bank, and the White House). The movement must focus on these critical (and nearby) institutions. And while the movement has much more to do in order to build coalitions across various social and geographic boundaries, there are tactics, strategies, and campaigns that have succeeded in doing just that for many years. From transnational activist campaigns to solidarity networks and letter-writing, the profile of environmental justice is becoming more global each year.

After Hazel Johnson's visit to the Earth Summit in Rio de Janeiro in 1992, PCR became part of a global network of activists and scholars researching and combating environmental inequality in North America, South America, Africa, Europe, and Asia. Today, PCR confronts a daunting task. The area of Chicago in which the organization works still suffers from the highest density of landfills per square mile of any place in the nation, and from the industrial chemicals believed to be partly responsible for the elevated rates of asthma and other respiratory ailments in the surrounding neighborhoods. PCR has managed to train local residents in lead-abatement techniques; it has begun negotiations with one of the Big Three auto makers to make its nearby manufacturing plant more ecologically sustainable and amenable to hiring locals, and it is setting up an environmental science laboratory and education facility in the community through a partnership with a major research university.

What can we conclude about the state of the movement for environmental justice? Our diagnosis gives us both hope and concern. While the movement has accomplished a great deal, the political and social realities facing activists (and all of us, for that matter) are brutal. Industrial production of hazardous wastes continues to increase exponentially; the rate of cancers, reproductive illnesses, and respiratory disorders is increasing in communities of color and poor communities; environmental inequalities in urban and rural areas in the United States have remained steady or increased during the 1990s and 2000s; the income gap between the upper classes and the working classes is greater than it has been in decades; the traditional, middle-class, and mainly white environmental movement has grown weaker; and the union-led labor movement is embroiled in internecine battles as it loses membership and influence over politics, making it likely that ordinary citizens will be more concerned about declining wages than environmental protection. How well EJ leaders analyze and respond to these adverse trends will determine the future health of this movement. Indeed, as denizens of this fragile planet, we all need to be concerned with how the EJ movement fares against the institutions that routinely poison the earth and its people.

DISCUSSION QUESTIONS

1. What is the vision of the environmental justice movement, and how is an understanding of racism central to this vision?

2. What do you learn from the environmental justice movement about the activism of communities of color?

3. Why would environmental justice activists in the United States benefit from connecting to activists in other nations?

39

Race, Reform, and Retrenchment

Transformation and Legitimation
in Antidiscrimination Law

KIMBERLÉ WILLIAMS CRENSHAW

Kimberlé Williams Crenshaw is a notable legal scholar and activist. In this article, written early in her career, she presents two views of antidiscrimination law. Both have the goal of eradicating racial oppression, but she sees limitations to the more restrictive of the two. You might think about which view of law currently prevails in the federal government, including the Congress, the Senate, the presidency, and the Supreme Court.

[A] ... basic conflict has given rise to two distinct rhetorical visions in the body of antidiscrimination law—one of which I have termed the expansive view, the other the restrictive view. The expansive view stresses equality as a result, and looks to real consequences for African-Americans. It interprets the objective of antidiscrimination law as the eradication of the substantive conditions of Black subordination and attempts to enlist the institutional power of the courts to further the national goal of eradicating the effects of racial oppression....

The restrictive vision, which exists side by side with this expansive view, treats equality as a process, downplaying the significance of actual outcomes. The primary objective of antidiscrimination law, according to this vision, is to prevent future wrongdoing rather than to redress present manifestations of past injustice. "Wrongdoing," moreover, is seen primarily as isolated actions against individuals rather than as a societal policy against an entire group. Nor does the restrictive view contemplate the courts playing a role in redressing harms from America's racist past, ... as opposed to merely policing society to eliminate a narrow set of proscribed discriminatory practices. Moreover, even when injustice is found, efforts to redress it must be balanced against, and limited by, competing interests of white workers—even when those interests were actually created by the subordination of Blacks. The innocence of whites weighs more heavily than do the past wrongs committed upon Blacks and the benefits that whites derived from those wrongs.... In sum, the restrictive view seeks to proscribe only certain kinds of subordinating acts, and then only when other interests are not overly burdened....

SOURCE: Crenshaw, Kimberlé Williams. 1988. "Race, Reform, and Retrenchment: Transformation and Legitimation in Antidiscrimination Law." *Harvard Law Review* 101 (May): 1331–1387.

As the expansive and restrictive views of antidiscrimination law reveal, there simply is no self-evident interpretation of civil rights inherent in the terms themselves. Instead, specific interpretations proceed largely from the world view of the interpreter.... For example, to believe ..., that color-blind policies represent the only legitimate and effective means of ensuring a racially equitable society, one would have to assume not only that there is only one "proper role" for law, but also that such a racially equitable society already exists. In this world, once law had performed its "proper" function of assuring equality of process, differences in outcomes between groups would not reflect past discrimination but rather real differences between groups competing for societal rewards.... Unimpeded by irrational prejudices against identifiable groups and unfettered by government-imposed preferences, competition would ensure that any group stratification would reflect only the cumulative effects of employers' rational decisions to hire the best workers for the least cost.... The deprivations and oppression of the past would somehow be expunged from the present. Only in such a society, where all other societal functions operate in a nondiscriminatory way, would equality of process constitute equality of opportunity.

This belief in color-blindness and equal process, however, would make no sense at all in a society in which identifiable groups had actually been treated differently historically and in which the effects of this difference in treatment continued into the present. If employers were thought to have been influenced by factors other than the actual performance of each job applicant, it would be absurd to rely on their decisions as evidence of true market valuations. Arguments that differences in economic status cannot be redressed, or are legitimate because they reflect cultural rather than racial inferiority, would have to be rejected; cultural disadvantages themselves would be seen as the consequence of historical discrimination.... One could not look at outcomes as a fair measure of merit since one would recognize that everyone had not been given an equal start. Because it would be apparent that institutions had embraced discriminatory policies in order to produce disparate results, it would be necessary to rely on results to indicate whether these discriminatory policies have been successfully dismantled....

... The passage of civil rights legislation nurtured the impression that the United States had moved decisively to end the oppression of Blacks. The fanfare surrounding the passage of these Acts, ... however, created an expectation that the legislation would not and could not fulfill. The law accommodated and obscured contradictions that led to conflict, countervision, and the current vacuousness of antidiscrimination law.

Because antidiscrimination law contains both the expansive and the restrictive view, equality of opportunity can refer to either. This uncertainty means that the societal adoption of racial equality rhetoric does not itself entail a commitment to end racial inequality. Indeed, to the extent that antidiscrimination law is believed to embrace color-blindness, equal opportunity rhetoric constitutes a formidable obstacle to efforts to alleviate conditions of white supremacy. As Alfred Blumrosen observes, "it [is] clear that a 'color-blind' society built upon the subordination of persons of one color [is] a society which [cannot] correct that subordination because it [can] never recognize it."[1] In sum, the very terms used to proclaim victory contain within them the seeds of defeat. To demand "equality of opportunity" is to demand nothing specific because "equality of opportunity" has assimilated both the demand and the object against which the demand is made; it is to participate in an abstracted discourse which carries the moral force of the movement as well as the stability of the institutions and interests which the movement opposed.

Society's adoption of the ambivalent rhetoric of equal opportunity law has made it that much more difficult for Black people to name their reality. There is no longer a perpetrator, a clearly identifiable discriminator. Company X can be an equal opportunity employer even though Company X has no Blacks or any other minorities in its employ. Practically speaking, all companies can now be equal opportunity employers by proclamation alone. Society has embraced the rhetoric of equal opportunity without fulfilling its promise; creating a break with the past has formed the basis for the neoconservative claim that present inequities cannot be the result of discriminatory practices because this society no longer discriminates against Blacks.

Equal opportunity law may have also undermined the fragile consensus against white supremacy.... To the extent that the objective of racial equality was seen as lifting formal barriers imposed against participation by Blacks, the reforms appear to have succeeded. Today, the claim that equal opportunity does not yet exist for Black America may fall upon deaf ears—ears deafened by repeated declarations that equal opportunity exists....

... [M]any things have changed under the political, legal, and moral force of the civil rights movement. Formal barriers have constituted a major aspect of the historic subordination of African-Americans and ... [T]he elimination of those barriers was meaningful. Indeed, equal opportunity rhetoric gains its power from the fact that people can point to real changes that accompanied its advent. ... [H]owever, what at first appears an unambiguous commitment to antidiscrimination conceals within it many conflicting and contradictory interests. In antidiscrimination law, the conflicting interests actually reinforce existing social arrangements, moderated to the extent necessary to balance the civil rights challenge with the many interests still privileged over it.

The recognition on the part of civil rights advocates that deeper institutional changes are required has come just as the formal changes have begun to convince people that enough has been done....

The flagging commitment of the courts and of many whites to fighting discrimination may not be the only deleterious effect of the civil rights reforms. The lasting harm must be measured by the extent to which limited gains hamper efforts of African-Americans to name their reality and to remain capable of engaging in collective action in the future. The danger of adopting equal opportunity rhetoric on its face is that the constituency incorporates legal and philosophical concepts that have an uneven history and an unpredictable trajectory. If the civil rights constituency allows its own political consciousness to be completely replaced by the ambiguous discourse of antidiscrimination law, it will be difficult for it to defend its genuine interests against those whose interests are supported by opposing visions that also lie within the same discourse. The struggle, it seems, is to maintain a contextualized, specified world view that reflects the experience of Blacks.... The question remains whether engaging in legal reform precludes this possibility....

NOTE

1. A. Blumrosen, Twenty Years of Title VII Law: An Overview 26 (April 18, 1985) (unpublished manuscript on file in the Harvard Law Library).

DISCUSSION QUESTIONS

1. What does Crenshaw mean by the expansive and the restrictive view of antidiscrimination law? What difference does it make in how you think about the role of law in pursuing racial justice?

2. The law can be both a protector of the status quo and an instrument of social change. Historically and now, which do you think has been most true?

40

Citizenship in Name Only

The Coloring of Democracy While Redefining Rights, Liberties, and Self Determination for the 21st Century

E. EARL PARSON AND MONIQUE MCLAUGHLIN

Voting rights are an essential part of national citizenship, yet they have historically been and are still being curtailed by various practices that especially affect the votes of people of color. This article identifies contemporary and historical voter suppression tactics, but also argues for the importance of a fully enfranchised citizenry if we are to realize a racial democracy.

[V]oting [is] a mechanism to advance a citizen's place in society. Although ... in many cases, voting is used by the majority to prevent the minority from exercising the rights of full citizenship. Along with jury participation, voting is limited only to United States citizens. Since Reconstruction, there has been a concerted and pervasive effort to block and to prevent minority citizenry from exercising their rights of citizenship in the voting booth. Recent efforts to prevent citizens from voting must be looked at through the prism of the long history of the United States in its attempts to preclude and disqualify minorities, particularly African Americans from voting, and thus, to dilute the African American's electorate and, in effect, to make African Americans second tier citizens in their own country. Without the vote, African Americans cannot participate in exercising their full citizenry or effect change to ensure their own prosperity....

Historically, state officials invented race neutral voting standards with pretextual questions to eliminate African Americans from registering to vote. African Americans were subject to poll tax payments in advance and aforementioned absurd literacy tests as mechanisms for discrimination.... The insidious discrimination was perpetuated by registrars and state officials determined not to allow African Americans the right to participate in the voting process.... In 1882, the mailbox system was adopted in South Carolina, requesting voters to use different boxes for different ballots.... A voting reform method of secret ballots continued to frustrate uneducated individuals and systematically disenfranchised the African American vote.... Voting requirements of poll taxes, grandfather clauses, and

SOURCE: Parson, E. Earl, and Monique McLaughlin. 2013. "Citizenship in Name Only: The Coloring of Democracy While Redefining Rights, Liberties, and Self Determination for the 21st Century." *Columbia Journal of Race and Law* 3: 103–117.

literacy tests, all under an umbrella of color-blind voting regulations, were in actuality officials' selective procedures to discriminate....

Grandfather Clause laws allowed illiterate whites to continue to vote, while African Americans were foreclosed from voting.... In *Guinn v. United States*, the court held that "grandfather clauses" violated the Fifteenth Amendment.... The Supreme Court deemed this a corrective means for disenfranchised African Americans, ... whose ancestors were slaves until 1863.

Under the 1965 Voting Rights Act, African Americans were given the full opportunity to exercise their full citizenship rights. This hallmark Act is considered the most successful Civil Rights legislation in history for African Americans. Over the past century, the United States Congress, through the Voting Rights Act, expanded the right to vote and knocked down innumerable barriers to full electoral participation. Since 2000, with tricks and intimidation tactics and, since 2010, with the first voter ID laws, that momentum to allow all citizens the unencumbered right to vote abruptly shifted. African Americans, after many years of exercising their right to vote, are now facing challenges in the voting process that are affecting their participation in the electorate system. Challenges to the African American vote include intimidation, trickery, and voter identification laws.

Recent attacks on African American voting rights began during the 2000 presidential election. The U.S. Commission on Civil Rights issued a report on discriminatory practices against African Americans on Election Day in Florida and found that thousands were denied the right to vote.... The report pointed out that African Americans were ten times as likely to have their ballots rejected than non-African Americans.... The report stated that about 14.4 percent of black voters' ballots were rejected compared to 1.6 percent of non-black voters.... The report's data further stated, that African Americans made up 11 percent of all Florida voters, but 54 percent of the spoiled ballots, some 187,000, were of black voters.... The data further indicated that eighty-three out of 100 precincts with spoiled ballots were from majority black precincts. These spoiled ballots were rejected and the votes were not counted....

In addition to claiming that African American ballots were spoiled, detractors have also used intimidation and tricks to rule the voting process. In Milwaukee, in November 2004, fliers were distributed in black neighborhoods under the false name of "Milwaukee Black Voters League."... The fliers inaccurately pronounced that "anyone convicted of any offense, however minor, is ineligible to vote"; "if any family member has any conviction, it also disqualifies other family members from voting"; and that "it's too late for unregistered voters to vote."... The flier further stated, "if these rules are violated, you are facing a conviction of ten years in prison and your children will be taken away from you."... Also, in Milwaukee, in October 2010, a billboard showed people in jail and behind bars referencing "We voted Illegally."...

A candidate for office in Orange Country, California sent fliers in Spanish to the Latino community prior to the November 2006 election stating; "You are advised that if your residence in this country is illegal or you are an immigrant, voting in a federal election is a crime that could result in jail time."... In response to the flier, California's Secretary of State, Bruce McPherson, mailed letters to the Latino community informing them that all U.S. citizens have a right to vote and stated that "voter intimidation in any form is completely unacceptable."... A federal judge in East Texas issued an order during the November 2006 elections stopping the Attorney General and Texas election officials from prosecuting people of color for helping the elderly, the disabled and other minorities cast their vote.... The Lone Star Project called this behavior "a voter suppression scheme" designed to impart fear and intimidation. During the November 2006 elections, Latinos in Virginia and Colorado were told that their ancestry would make them ineligible to vote.... In 1988, Republican Assembly candidate Curt Pringle settled a civil rights suit for voter intimidations. Pringle posted "security guards" at predominately Latino voting stations in Orange County, California to prevent non-citizens from casting ballots. In 2006, Republicans in New York challenged late registered voters by using a check challenge of sending police officers out to check listed addresses....

Because of these intimidation tactics, voter registration is down in many states....

Over twenty-one million citizens do not have a government-issued ID in the United States.[1] These restrictive laws will disproportionately affect the poor, minority, elderly and Blacks. For example, the Brennan study indicated that 25 percent of African Americans do not have the required government-issued ID.... At the same time, in Texas, a voter ID law allows a person to produce a concealed hand gun license as proof of identify but will not allow a state university ID for proof of identity.... Moreover, 92 percent of all concealed handgun owners are non-African American.... Twelve states in 2011 have introduced proof of citizenship laws and bills to eliminate early voting, same-day voter registration, voting registration drives and Sunday voting.... These laws inhibit the efforts of many African American churches that organize "souls to the polls" drives, which allow their members to collectively go to the polls after Sunday services to vote....

... Government-issued ID requirements, the reductions in early voting and the imposition of new restrictions on voter registration drives, threatens citizens' exercise of their full citizenship rights to participate in the political process by these new laws. In addition, the birther movement challenging President Obama's citizenship status, as well as bills and laws addressing birthright citizenship for illegal immigrant children born in the United States, indicate a clear movement toward second-tier citizenship....

One must be a United States citizen in order to vote in America. Normally, one must also be at least eighteen years old and swear by affidavit that he or she is a United States citizen and meets all voting requirements. However, states are now requiring citizens to produce documents proving citizenship status.... These laws are a direct outgrowth of the Arizona legislation's attack on Mexican immigrants and the false perception that immigrants were voting illegally.... The Arizona bill, which was called Proposition 200 and went into effect in 2006, authorized officials to reject a voter's registration that was not accompanied by an application with citizenship documentation....

These proof of citizenship laws have arisen based on the growing immigrant population in states like Alabama, Georgia and South Carolina.... Again, the same recycled argument is used, that it will prevent non-citizens from registering to vote and will combat voter fraud. However, these laws will exclude a large portion of eligible voters who do not have ready access to citizenship documents....

With the resurrection of these old challenges to citizenship rights for African Americans, there is a growing concern of voter disillusionment. A 2006 Pew Research Center Report found African Americans were twice as likely to have no confidence in the voting process as from previous elections.[2] Another study in 2004 found that African Americans felt less confident than white voters that their votes were accurately counted.[3]

With voter ID laws and other voting suppression tactics, the opportunities given by the Voting Rights Act may be in danger for the first time since its passage. The right to vote will always be an important conduit for economic property for all people of color....

NOTES

1. Wendy R. Weiser & Lawrence Norden, *Voting Law Changes in 2012*, BRENNAN CTR. FOR JUSTICE 2 (2011), *available at* http://brennan.3cdn.net/ 92635ddafbc09e8d88_i3m6bjdeh.pdf.

2. *Who Votes, Who Doesn't and Why: Regular Voters, Intermittent Voters, and Those Who Don't*, PEW RESEARCH CTR. (Oct. 18, 2006), http://people-press.org/reports/ display.php3?ReportID=292.

3. R. Michael Alvarez, Thad E. Hall, & Morgan Llewellyn, *American Confidence in Electronic Voting and Ballot Counting: A Pre-Election Update*, CALTECH/MIT VOTING TECHNOLOGY PROJECT (Nov. 3, 2006), at 4, *available at* http://www.vote.caltech .edu/sites/default/files/american_confidence_ev_bc.pdf.

DISCUSSION QUESTIONS

1. What specific means have been used to restrict voting rights for people of color? What can be done to prevent this?

2. Look up information on the voter turnout rates for different racial-ethnic groups. What factors might prevent and/or encourage people to vote?

Student Exercises

1. There is a long history of people working for social change. Investigate your campus and/or your community for an organization involved in working for social change. The group can be involved in any number of causes: the environmental justice movement, fair housing, housing the homeless, voting rights, political reform, or educational reform. Find out some details about the mission and objective of the organization. Arrange to talk with members of the organization about their activities. What do you see as the primary contributions of this organization? Are there limitations to what they can achieve?

2. Look up the data on voting rates for your state for the most recent national election, which you can find on the website for the United States Election Project (www.electproject.org). What percent of the population voted in your state (the voter turnout rate), and how does this compare to the voter turnout rate for the nation as a whole? Other states? Have there been movements in your state or local area to expand and/or contract voting? Are there organizations working to ensure access to the vote?

Index